# Degraded Forests in Eastern Africa

## Management and Restoration

*Edited by*
*Frans Bongers and Timm Tennigkeit*

earthscan
from Routledge

First published 2010 by Earthscan

2 Park Square, Milton Park, Abingdon, Oxfordshire OX14 4RN
52 Vanderbilt Avenue, New York, NY 10017

*Routledge in an imprint of the Taylor & Francis Group, an informa business*

First issued in paperback 2018

ISBN 978-1-84407-767-0 (hbk)
ISBN 978-0-415-85304-0 (pbk)

Typeset by MapSet Ltd, Gateshead, UK
Cover design by Susanne Harris

A catalogue record for this book is available from the British Library

Library of Congress Cataloging-in-Publication Data

Degraded forests in Eastern Africa : management and restoration / edited by Frans
Bongers and Timm Tennigkeit.
    p. cm.
Includes bibliographical references and index.
ISBN 978-1-84407-767-0 (hardback)
 1. Forest conservation—Africa, Eastern. 2. Forest degradation—Africa, Eastern. 3.
Sustainable forestry—Africa Eastern. I. Bongers, Frans. II. Tennigkeit, Timm.
SD414.A354D44 2010
634.9'209676—dc22

2010005564

# Contents

# Preface

Forested landscapes in Africa show great diversity. Of course the vast lowland areas dominated by forests in the Congo basin in central Africa are well known. And the savanna-dominated landscapes in eastern and southern Africa also traditionally receive much attention, mainly because of the attractiveness of the large animal species inhabiting them. But although maybe less well known, east African countries also show a large variety of forest types and forested landscapes. These vary from coastal mangrove forests and dry woodlands in the lowlands to rainforests at low and intermediate altitudes and Afroalpine forests on the roof of Africa, for instance in the Ethiopian highlands.

Many of these forests are under strong pressure from people: nearly everywhere the forested landscapes show clear evidence of human impact and most, if not all, forests are human-affected. Thus the structure and composition of the remaining forest is the product of both environmental and human factors. Human beings have interacted with the forests through collection of forest products, shifting cultivation, permanent or semi-permanent agriculture, and all kinds of agroforestry systems. This interaction has led to a mosaic of patches of different vegetations in the landscape.

During the last half-century or so the pressure from people on forests has intensified, not least as a result of a strongly increasing human population. Forest loss and forest degradation have now reached near crisis proportions in some areas and measures to counteract this development are being taken slowly and in many cases unsatisfactorily. Although conservation is high on national and international political agendas, at the local level, forest management, and especially sustainable forest management, remains a challenge.

This challenge is the topic of this book. We address several key ecological aspects underlying major forest ecosystems in the region and important social factors driving their management, and thereby aim to provide a background for the development of management tools for the rehabilitation and sustainable management of these ecosystems. We focus on crucial examples embedded in a general context, adopting a pragmatic and problem-solving approach. The focus of much of the book is on various plant species. We hope that the information we provide stimulates improved forest management and restoration approaches.

*Frans Bongers and Timm Tennigkeit*

# Contributors

## EDITORS

**Frans Bongers**, Forest Ecology and Forest Management Group, Wageningen University, PO Box 47, 6700 AA Wageningen, The Netherlands

**Timm Tennigkeit**, Institute of Silviculture, Albert-Ludwigs-University, Tennenbacher Strasse 4, 79106 Freiburg, Germany, and UNIQUE forestry consultants, Schnewlinstrasse 10, D-79098 Freiburg, Germany

## CHAPTER AUTHORS

**Abrham Abiyu**, Amhara Region Agricultural Research Institute (ARARI), PO Box 527, Bahir Dar, Ethiopia, and Institut für Waldökologie, University of Natural Resources and Applied Life Sciences, Peter Jordan strasse 82, A-1190, Vienna, Austria

**Jacob G. Agea**, Department of Community Forestry and Extension, Faculty of Forestry and Nature Conservation, Makerere University, PO Box 7062, Kampala, Uganda

**Girma Amente**, Oromia Forest and Wildlife Enterprise, PO Box 6182, Addis Ababa, Ethiopia

**Tesfaye Bekele**, Ethiopian Institute of Agricultural Research (EIAR), PO Box 31709, Addis Ababa, Ethiopia

**Thomas Buchholz**, Institute of Silviculture, Albert-Ludwigs-University, Tennenbacher Strasse 4, 79106 Freiburg, Germany. Present address: Department of Forest and Natural Resources Management, State University of New York, College of Environmental Sciences and Forestry (SUNY-ESF), One Forestry Drive, Illick 340 13210, Syracuse, New York, USA

**Shabani A. O. Chamshama**, Department of Forest Biology, Faculty of Forestry and Nature Conservation, Sokoine University of Agriculture, PO Box 3010, Chuo Kikuu, Morogoro, Tanzania

**Camille Couralet**, Forest Ecology and Forest Management Group, Wageningen University, PO Box 47, 6700 AA Wageningen, The Netherlands. Present address: Laboratory for Wood Biology and Xylarium, Royal Museum for Central Africa, Leuvensesteenweg 13, 3080 Tervuren, Belgium, and Laboratory for Wood Technology, Gent University, Coupure Links 653, 9000 Gent, Belgium

**Abeje Eshete**, Ethiopian Institute for Agricultural Research, Forestry Research Centre, PO Box 30708, Addis Ababa, Ethiopia, and Forest Ecology and Forest Management Group, Wageningen University, PO Box 47, 6700 AA Wageningen, The Netherlands

**Sisay Feleke**, Ethiopian Institute of Agricultural Research, Forestry Research Center, Forest Products Utilization Research coordination office, PO Box 2322, Addis Ababa, Ethiopia

**Kindeya Gebrehiwot**, Mekelle University, PO Box 231, Mekelle, Tigray, Ethiopia

**Alfred de Gier**, International Institute for Geo-information Science and Earth Observation, PO Box 6, 7500 AA, Enschede, The Netherlands

**John B. Hall**, School of Environment, Natural Resources and Geography, University of Wales, Bangor, Gwynedd LL57 2UW, United Kingdom

**John R. Healey**, School of Environment, Natural Resources and Geography, University of Wales, Bangor, Gwynedd LL57 2UW, United Kingdom

**Tesfaye Hunde**, International Network for Bamboo and Rattan, East-Africa Regional Office, PO Box 1463, Addis Ababa, Ethiopia

**Jürgen Huss**, Albert-Ludwig-Universität Freiburg, Institut für Waldbau, Tennenbacherstrasse 4, 79085 Freiburg, Germany

**Mengistie Kindu**, Ethiopian Institute of Agricultural Research, Forestry Research Center, PO Box 30708, Addis Ababa, Ethiopia

**Mulugeta Lemenih**, Debub University, Wondo Genet College of Forestry, PO Box 128, Shashamane, Ethiopia, and Forest Ecology and Forest Management Group, Wageningen University, PO Box 47, 6700 AA Wageningen, The Netherlands

**Lars Markesteijn**, Forest Ecology and Forest Management Group, Wageningen University, PO Box 47, 6700 AA Wageningen, The Netherlands

**Yitebitu Moges**, Ethiopian Institute for Agricultural Research, Forestry Research Centre, PO Box 30708, Addis Ababa, Ethiopia

**Grace Nangendo**, Wildlife Conservation Society, Plot 802 Kiwafu Road, Kansanga, PO Box 28144, Kampala, Uganda

**Demeke Nigussie**, Ethiopian Institute of Agricultural Research, PO Box 2003, Addis Ababa, Ethiopia

**John A. F. Obiri**, School of Environment, Natural Resources and Geography, University of Wales, Bangor, Gwynedd LL57 2UW, United Kingdom, and Centre for Disaster Management and Humanitarian Assistance, Masinde Muliro University of Science and Technology, PO Box 190 Kakamega-Webuye Road, Kakamega, Western Province, Kenya 50100

**Joseph Obua**, Makerere University, PO Box 7062, Kampala, Uganda, Currently Regional Coordinator, Lake Victoria Research Initiative (VicRes), The Inter-University Council for East Africa, PO Box 7110 Kampala, Uganda

**Woldeselassie Ogbazghi**, Agricultural College, University of Asmara, PO Box 1220, Asmara, Eritrea

**Yishak Sahle**, Ethiopian Institute of Agricultural Research, Holeta Research Centre, PO Box 2003, Addis Ababa, Ethiopia

**Ute Sass-Klaassen**, Forest Ecology and Forest Management Group, Wageningen University, PO Box 47, 6700 AA Wageningen, The Netherlands

**Julia Schmitt**, University of Hohenheim, Department of Agricultural Communication and Extension (430a), 70593 Stuttgart, Germany

**Hans ter Steege**, Ecology and Biodiversity, Institute of Environmental Biology, Utrecht University, Padualaan 8, 3584 CH Utrecht, The Netherlands

**Frank J. Sterck**, Forest Ecology and Forest Management Group, Wageningen University, PO Box 47, 6700 AA Wageningen, The Netherlands

**Wubalem Tadesse**, Ethiopian Institute of Agricultural Research, PO Box 2003, Addis Ababa, Ethiopia

**Demel Teketay**, University of Botswana, Harry Oppenheimer Okavango Research Centre (HOORC), Shorobe Road, Sexaxa, Private Bag 285, Maun, Botswana

**Vincent G. Vyamana**, Department of Forest Biology, Faculty of Forestry and Nature Conservation, Sokoine University of Agriculture, PO Box 3010 Chuo Kikuu, Morogoro, Tanzania

**Alemayehu Wassie**, Organization for Rehabilitation and Development in Amhara (ORDA) PO Box 132, Bahir Dar, Ethiopia

**Axel Weinreich**, UNIQUE forestry consultants, Schnewlinstrasse 10, D-79098 Freiburg, Germany

**K. Freerk Wiersum**, Forest and Nature Policy Group, Wageningen University, PO Box 47, 6700 AA Wageningen, The Netherlands

**Yonas Yemshaw**, The AFORNET secretariat, African Academy of Sciences, PO Box 24916-00502, Nairobi, Kenya

# Acknowledgements

During the course of preparing this book we had interaction with many colleagues and collaborators. We are very grateful to the large number of people involved in writing chapters, reviewing chapters, and coordinating and organizing work related to them. Thank you all for contributing your knowledge and your precious time and energy:

Abrham Abiyu, Jimmy Acidri, Girma Amente, A. Awio, Fred Babweteera, Wilson Bahemuka, Hank Bartelink, Jürgen Bauhus, John Ayongyera Begumana, Tesfaye Bekele, Tesgeye Bekele, Chris Bekker, Tom Blomley, Jean-Marc Boffa, Thomas Buchholz, Neil Burgess, Sebastian Büttner, S. A. O. Chamshama, Ariani Charles, Kizza Charles, Camille Couralet, Ronnie Cox, Arthur Mloka Dallu, Peter Dirninger, Klaus-Dieter Düformantel, Samuel Ebert, Franz Eichinger, Francis Osoto Esegu, Abeje Eshete, Bill Farmer, Clemens Fehr, Sisay Feleke, Christoph Fink, Kunihira Florence, Roland Freyer, Kindeya Gebrehiwot, Alfred de Gier, Nabanoga Nsubuga Gorettie, Sylvana Grabitzki, Markus Grulke, Deribe Gurmu, Sam Gwali, Maarit A. Haavisto, John Hall, K. F. S. Hamza, Tim Hardwick, John Healey, Oliver Heintz, Tesfaye Hunde, Jürgen Huss, Corinne Ingels, Paul Jacovelli, Anna Jaschok, Walaita Sebastian Javan, Acobo Jimton, John R. S. Kabogozza, Shizu Kaga, George C. Kajembe, Peter Karani, Severin K. Karonga, Margaret Kasekende, Wilson Kasolo, Israel Kikangi, Mengistie Kindu, Gaster Kawuubye Kiyingi, Robert Kundgjui, Edward Kyobe, Fred Lali, Mulugeta Lemenih, Rolf Link, Lars Markesteijn, Ryno Martyn, Aza S. C. Mbaga, Michael S. Mbogga, Alastair McNeilage, Bric Milligan, Yitebitu Moges, Frits Mohren, Ruth Mubiru, Ancelm G. Mugasha, Xavier Mugumya, Denis Mununuzi, Nelson Wajja Musukwe, Grace Nangendo, Peter Ndemera, P. Ngobi, Demeke Nigussie, J. Nkosi, Steve Nsita, John Obiri, Joseph Obua, Geoffrey G. O. Odokonyero, Woldeselassie Ogbazghi, G. Okethwengu, Gift O. Okojia, John Okorio, Peter Ongima, Andy Plumptre, Diana Pretzell, Warren Rance, Eckart von Reitzenstein, Nele Rogiers, Yishak Sahle, Ute Sass-Klaassen, Andrea Schäfer, Julia Schmitt, Henning Schrader, Matthias Seebauer, Douglas Sheil, Dieter Speidel, Heinrich Spiecker, William Gombya-Ssembajjwe, Hans ter Steege, Frank J. Sterck, Tsegaye Tadesse, Wubalem Tadesse, Demel Teketay, R. Tolith, Joy M. B. Tukahirwa, Levand Turyomurugyendo, Vincent G. Vyamana, Charles Walaga, Alemayehu Wassie, Axel Weinrich, Gabriela Hafke-Wessel, Freerk Wiersum, Ojuri Wilson, Kai Windhorst, Yonas Yemshaw and Kumelachew Yeshitela.

We are most grateful to the European Commission for the financial support provided towards the fieldwork underlying the preparation of a number of chapters in this book. The International Scientific Cooperation Project on 'Tools for Restoration and Sustainable Management of Forests in East Africa' was funded by the European Commission within its Fifth Research Framework Programme (project number ICA4-CT-2001-10097). We specifically thank Nicole Riveill, Sergio Micheli and Roberto Santoriello from the European Commission for their continuous support. This project was based at Freiburg University and was a close collaboration between researchers from Uganda, Ethiopia and Tanzania in eastern Africa and from Germany, the UK and The Netherlands in Europe. A special thanks to Jürgen Huss from Freiburg University, who was the project coordinator of this EU project.

Several other chapters were strongly supported by the Dutch–Ethiopian research programme 'FRAnkincense, Myrrh and Gum Arabic: Sustainable Use of Dry Woodland Resources in Ethiopia (FRAME)', largely funded by the Dutch National Science Foundation (NWO-WOTRO Integrated Programme project number W01.65.220.00), in collaboration with Dutch and Ethiopian Universities and research institutes.

Publication of this book was financially supported by the International Union for the Conservation of Nature (IUCN), specifically their Dutch branch IUCN-NL (grant number 600216), and the Universities of Bangor, Freiburg and Wageningen.

Ron Eijkman (de Vormgeverij) prepared a number of location maps in this book.

*We thank all contributors and hope that this book fulfils their expectations.*

# List of Abbreviations

| | |
|---|---|
| AAC | annual allowable cut |
| ANOVA | analysis of variance |
| asl | above sea level |
| CBD | United Nations Convention on Biological Diversity |
| CBFM | community-based forest management |
| CCD | United Nations Convention to Combat Desertification |
| CFR | central forest reserve |
| CIFOR | Center for International Forestry Research |
| DBH | diameter at breast height |
| DCA | detrended correspondence analysis |
| EFAP | Ethiopian Forestry Action Plan |
| EOTC | Ethiopian Orthodox Tewahido Church |
| ETB | Ethiopian Birr (the national currency of Ethiopia) |
| FAO | Food and Agriculture Organization of the United Nations |
| FBD | Forestry and Beekeeping Division (Tanzania) |
| FES | Forest Escarpment Company |
| FNCMP | Forest Nature Conservation Master Plan (Uganda) |
| FWS | forest, woodland and savannah |
| GDP | gross domestic product |
| GTZ | German Technical Cooperation |
| HASHI | Hifadhi Ardhi Shinyanga (Kiswahili) (Soil Conservation Project in Shinyanga, Tanzania) |
| IFMP | Integrated Forest Management Project |
| ITTO | International Tropical Timber Organization |
| IUCN | International Union for the Conservation of Nature |
| JFM | joint forest management |
| JICA | Japan International Cooperation Agency |
| KVTC | Kilombero Valley Teak Company |
| MAI | mean annual increment |
| MFP | minor forest produce |
| MNRT | Ministry of Natural Resources and Tourism |
| MSAVI | modified soil adjusted vegetation index |
| NEMA | National Environment Management Authority (Uganda) |
| NFA | National Forest Authority (Uganda) |

| | |
|---|---|
| NFPA | National Forest Priority Area |
| NGO | non-governmental organization |
| NTFP | non-timber forest product |
| NWFP | non-wood forest product |
| OFWE | Oromia Forest and Wildlife Enterprise (Ethiopia) |
| PAR | photosynthetically active radiation |
| PCT | potential crop trees |
| PFM | participatory forest management |
| PRA | participatory rural appraisal |
| SNNPRS | Southern Nations, Nationalities and Peoples Regional State (Ethiopia) |
| SOM | soil organic matter |
| TANWATT | Tanganyika Wattle Company |
| TAS | traditional agroforestry system |
| TCBFM | traditional community-based forest management |
| TLU | tropical livestock unit |
| UNFCCC | United Nations Framework Convention on Climate Change |
| USh | Uganda shilling |
| UWA | Uganda Wildlife Authority |
| VLFR | village land forest reserve (Tanzania) |
| WAJIB | forest-dwellers' association (Ethiopia) |
| WBISPP | Woody Biomass Inventory and Strategic Planning Project |

# 1

# Degraded Forests in Eastern Africa: Introduction

*Frans Bongers and Timm Tennigkeit*

## FORESTS IN THE REGION

Forests in eastern Africa show an amazing variety of structure and composition. This is to be expected, of course, as it is the result of the interplay between a large variety in environmental conditions and all kinds of human interactions. The altitudinal variation in the area is large, ranging from over one hundred metres below sea level to five thousand above. The area hosts the major mountainous area of the African continent and has a large number of peaks above 4000 metres above sea level. This gives rise to a large gradient in temperature. Rainfall conditions also vary tremendously, from very dry deserts to extremely wet rainforest and cloud forest systems in most countries. And soils vary from highly weathered old and poor soils to nutrient-rich volcanic soils that may support lush vegetation and are well suited for agriculture.

East Africa is perceived as the cradle of humanity and for millennia people have been living in the area, continuously changing and affecting the natural environment. In fact, currently most, if not all, lands have been seriously affected by mankind. This interaction, be it through permanent or semi-permanent agriculture, shifting cultivation, all kinds of agroforestry systems, or just collecting products from forests and other natural vegetation, has led to a mosaic of patches of different vegetations in the landscape. At most locations the present landscape is thus definitely the product of the natural variation in vegetation combined with human-induced variation. Consciously and uncon-

1

sciously human beings have changed the forests in their neighbourhoods. A good example of this is the human-modified natural coffee-containing forests in southwest Ethiopia (see Wiersum, Chapter 15) that are heavily modified to increase coffee production while maintaining forest structure as much as possible. An example of reasonably well-preserved forests is the eastern Arc forests in Tanzania, and in some areas the well-known miombo woodlands (see Obiri et al, Chapter 5).

The pressure of people on forests has intensified during the last century, not least as a result of a strongly increasing human population. In several areas in the region the loss and degradation of forests have now reached near crisis proportions. Measures to counteract this development are being taken slowly and in many cases unsatisfactorily. Conservation is high on the national and international political agendas, but at the local level forest management, and especially sustainable forest management, remains a challenge. In recent decades institutional reforms have been developed in many countries, partly as a result of large national and international pressure. Reforms may include development of new laws for forest and natural resource management, decentralization of forest and natural resources management responsibilities, increased integration or separation of forest and agriculture, combination or separation of production and conservation, and large changes in ideas about environmental protection. International organizations like the United Nations play a key role and actively stimulate international agreements (like the Convention on Biological Diversity, CBD).

# FOREST PRODUCTS

Forests harbour many plant products of direct use to people and harvesting and consumption of such plant products from natural forests is known to account for a large proportion of the livelihood of people living close. Worldwide, wood is arguably the most important forest product, and large areas of forests have been cut, and are being cut, for their wood. In the world an average of around one-third of forest area is primarily designated for production purposes. In Africa the figure is 30 per cent and has been declining since 1990 (FAO, 2007). In eastern Africa, the focal area of this book, the forest area designated for production declined at a constant rate of about 300,000 hectares per year during 1990–2005 (FAO, 2007). But the total wood removed from forests increased from ca. 185 to ca. 320 million $m^3$ per year between 1990 and 2005, despite a limited supply from forest plantations. 90 per cent of this wood is used for fuel (Figure 1.1).

The majority of plant products that are used from and collected in forests are non-timber forest products (NTFPs), meaning all biological materials used for purposes other than commercial timber. Wild plants are an important source of edible fruits, leafy vegetables and herbs, and are particularly important in ensuring food security and maintaining the nutritional balance in people's diets. During famine, wild plants become essential to human survival, and at other

**Figure 1.1** *Forests provide large amounts of fuelwood for local people*

*Source:* Photographs by R. Bäcker

times they both prevent the need for cash expenditure and provide a source of income to cash-poor households. It is estimated that between 4000 and 6000 non-timber plant species are of commercial importance worldwide (Iqbal, 1993; SCBD, 2001). However, inadequate information on the ecological productivity, growth forms, life history and conservation of the various species involved complicates management scenarios, the setting of conservation priorities and defining sustainable harvest levels (Ticktin, 2004; SCBD, 2001).

Although it has been suggested that intense harvesting of NTFPs is feasible because of the supposedly low associated ecological impacts, excessive extraction of forest products is likely to have a negative impact on population dynamics of the plants being exploited. Such impacts may lead to changes in community structure. Exploitation impact depends on the parts of plants that are harvested for NTFPs. Harvesting of some NTFPs, for instance leaves and fruits, may have a negligible effect on the plant population being exploited, depending on the intensity of the harvest. Harvesting of bark (Guedje et al, 2003; Botha et al, 2004), roots or bulbs, on the other hand, usually kills or fatally weakens the exploited plant species. For other products, such as palm heart, trees have to be cut to be able to harvest the product.

An important group of NTFP is east Africa is formed by the gums and resins of *Acacia*, *Boswellia* and *Commiphora* species (ABC species) that produce gum arabic, frankincense and myrrh, respectively. Sustainable production of these products depends on maintenance of healthy populations of the species involved. In Chapter 7 Abiyu and colleagues focus on frankincense as an example of such a species and address population status and factors that threaten these populations.

# THREATS

Many forests, wet and dry, lowland and montane, are severely under threat. Conversion of forest into agricultural land leads to fragmentation and isolation; legal and illegal logging for timber and extensive collection of wood for firewood and NTFPs lead to forest degradation and product exhaustion.

In general terms threats can be divided into threats for habitats and threats for species. Threats to habitats include habitat loss, habitat fragmentation, habitat disturbance, uncontrolled logging, increasing rates of fire, overharvesting of fuelwood, overgrazing of sensitive habitats, increasing populations, continuing poverty and debilitating diseases. Additionally climate change is having an effect on habitats (MEA, 2005).

Habitat loss is high all over Africa. The Food and Agriculture Organization of the United Nations (FAO, 2007) estimated annual forest decrease rates for east Africa of 0.94 and 0.97 per cent for the periods 1990–2000 and 2000–2005 respectively. This is higher than the average for Africa (0.64 and 0.62 per cent) and far higher than the world average (0.22 and 0.18 per cent). East Africa has lost between 770 and 800 thousand hectares of forest per year since 1990. This habitat loss is despite the fact that quite a lot of forest area is protected, although this is decreasing: in 2005 in east Africa 3,500,000 ha of forest were designated for protection, down from 3,750,000 in 1990 (FAO, 2007). Tropical rainforest loss levels in west Africa were among the highest in the world (Achard et al, 2002), and also in east Africa loss of forest area is high (see Chapters 2, 3 and 4 for data on the three focal countries of this book).

The alarming levels of forest loss in African countries has been doubted by some researchers (Fairhead and Leach, 1998; Nyssen et al, 2009), who focus on the bad quality of forest extension data before the era of remote sensing (roughly more than half a century ago). Notwithstanding these shortcomings, which indeed may overestimate the forest loss, it remains clear that over the last 50 years or so the area of forests has been reduced drastically in most countries, including east African ones.

Threats to species include increasing extinction risks (Thuiller et al, 2006), loss of species range (McClean et al, 2005), bushmeat hunting (Milner-Gulland et al, 2003) and competition by invasive species (Gurevitch and Padilla, 2004). Additionally climate change is having an effect on species (Thomas et al, 2004). Many African plant species have been shown to be reduced in range already and climate change predictions account for many more (Brooks et al, 2001; McClean et al, 2005), but this is not particular to Africa (Rodrigues et al, 2004). Bushmeat hunting is severe in many African countries and is also associated with the logging industry, which provides facilities for hunters and a transport system to the towns (Fa et al, 2003; Milner-Gulland et al, 2003), possibly also with long-term effects on plant species composition and diversity of tropical forests (Muller-Landau, 2007). Invasive species get increased attention as they become more and more dominant. The mean percentage of alien plants in a given flora is 12 per cent and on islands even close to 28 per cent (Vitousek et al, 1997). Examples for east Africa include forestry trees (*Pinus*, *Acacia* and *Casuarina* species) and of a database of 2000 species used in agroforestry, some 135 were reported to be invasive (for example *Acacia*, *Prosopis* and *Sesbania* spp.) (Richardson, 1998).

# FOREST MANAGEMENT, RESTORATION AND CONSERVATION

Forest management is concerned with the technical aspects of forests (silviculture, forest protection, forest conservation and forest regulation) and with all administrative, economic, legal and social aspects of forests. Management thus should include a wide array of aspects such as wood and non-wood products, wildlife, recreation, forest genetic resources, forest resource values, and water provision and protection. In general, forest management can be based on economics or on conservation of resources, or a mixture of the two. Internationally, concerns regarding forest management have shifted from a focus on extraction of timber to a focus on other forest resources, including biodiversity conservation, climate change, watershed management, wildlife management and recreation. This shift, however, is not uniform across and within countries. At the local level in many countries, including in east Africa, economic and livelihood aspects are still the dominant forces determining actual use and management of forests. That this local management is not always in line with national and international rules and priorities is evident.

Sustainable forest management (SFM) is a special type of forest management and is based on principles of sustainable development ('meeting the needs of the present without compromising the ability of future generations to meet their own needs' – Brundtland Report, UN, 1987). It takes broad social, economic and environmental goals into account, in a continuous evaluation and adaptation to the changing social, economic and natural environments.

Ecological restoration may be defined as the intentional activity that initiates or accelerates the recovery of an ecosystem with respect to its health, integrity and sustainability (SER, 2004; Andel and Aronson, 2006). Forest restoration may include reforestation, erosion control, revegetation of disturbed areas, enrichment planting of degraded areas with valuable species, habitat improvement for targeted species, removal of weeds and exotic (non-native) species, and reintroduction of native species.

In this book various aspects of forest management and forest restoration are addressed, but overall the focus is on applied forest ecology, rather than on issues such as forest policy, forest laws, land tenure and forest economics. In many chapters, however, some of these aspects are touched upon, as far as needed for understanding the specific issues being addressed.

Forest management directed at conservation is a challenge for countries in east Africa (and elsewhere). Although management of forests is as old as people trying to control their environment, institutionalized forest management has developed largely during the last century, in most countries established by colonial powers. Most colonial governments established their own system for management of natural resources, in most cases organized in four branches: wildlife, forestry, fisheries and agriculture. While wildlife conservation was strongly focused on large animals (mostly savanna animals), forestry departments aimed at supply of forest products. In this period the first forest reserves were established, mainly focused on resource exploitation and water catchment conservation. Biological diversity was not an issue. Much forest and savanna area was also transformed into agricultural land.

After the colonial period (roughly since the 1960s), the new powers continued this tradition and have developed their departments for better management of their countries' resources. Only in the last 25 or so years have international organizations, in collaboration with country governments, set up agreements for management of natural resources across countries according to international rules and agreed-upon conventions like the United Nations Conventions on Biological Diversity (CBD), to Combat Desertification (CCD) and on Climate Change (UNFCCC). These have led to strategy development and action plans, for instance the Tropical Forest Action Plan. These conventions and action plans have been adopted by most African countries as well, and currently in most of these countries a wide array of conservation strategies, action plans and implementation ideas are active. The reality on the ground, however, is far from ideal, and most countries struggle with effective implementation. This indeed looks likely to remain a challenge for decades to come.

To avoid every country developing its own ideas and strategies, and country borders inhibiting effective conservation of resources that in many cases extend beyond them, at the international level the focus has shifted towards integrative management ideas regarding ecoregions. Ecoregions are large units of land and water that contain a distinct assemblage of species, habitats and processes, and whose boundaries attempt to depict the original extent of natural communities before major land-use change (Olsen et al, 2001; Burgess et al, 2004). Conservation strategies increasingly consider that biogeographic units at the scale of ecoregions are ideal for protecting a full range of representative areas, conserving special elements, and ensuring the persistence of populations and ecological processes, particularly those that require the largest areas or are most sensitive to anthropogenic alterations (Olson et al, 2001).

A good example of this is the vast Serengeti ecosystem, crossing the border between Kenya and Tanzania, where each year 1.3 million wildebeest, 200,000 plains zebras and 400,000 Thomson's gazelle seasonally migrate. The Serengeti is in fact an example of a transboundary protected area (TBPA), which over the last few years have been established in several places. Other examples are the Great Limpopo Transfrontier Park (linking South Africa, Zimbabwe and Mozambique), the Transfrontier Protected Area Network of the central Albertine Rift (DR Congo, Rwanda and Uganda; signed October 2005), the Gola Forest Transboundary Peace Park (Liberia and Sierra Leone; signed 2009) and the Selous Niassa Wildlife Corridor (Kenya and Tanzania). Currently plans are being developed across the borders between Ethiopia and Kenya and Ethiopia and Sudan, among others. Recent studies estimate that worldwide there are now 188 transboundary conservation areas in 112 countries, making up about 17 per cent of the designated protected areas around the world (Mittermeier et al, 2005).

The basis for the ecoregions of the African tropics is formed by White's (1983) phytogeographic regions. Of the 867 ecoregions of the world (Olson et al, 2001), 119 are found in Africa and its islands (Burgess et al, 2004), many of which are forest, woodland or woody thickets. Many groups of ecoregions cover forest types: tropical and subtropical moist broadleaf forests (30 types), tropical and subtropical dry broadleaf forests (3), temperate coniferous (1) and mangroves (5). Several other groups include woodlands and woodland–savanna mosaics.

East Africa harbours a large number of these, ranging from high mountain afroalpine coniferous forests (Figure 1.2) in, for instance, Ethiopia, Kenya and Tanzania, through wet and dry lowland forests, to mangrove forests along a large part of the east African coast. World famous are the wet montane forests of the Eastern Arc forests in southern Kenya and Tanzania (Lovett and Wasser, 1993). These forests harbour large numbers of endemic species and are currently under severe threat. Many of these forests are islands in a sea of arid and agricultural lands.

**Figure 1.2** *Afromontane forests in Simien national park, Ethiopia*

*Source:* F. Bongers

Despite large-scale national and international efforts to protect and conserve forests in east Africa, the reality on the ground is challenging. Almost everywhere there is a continuous tradeoff between long-term conservation goals and short- and long-term livelihoods of people. The 'tragedy of the commons' continues, as the number of people inhabiting most countries continues to increase. Less and less area is available for more and more mouths to feed. The fact that food is largely cooked on wood in east African countries aggravates the consequences for forests and the people that depend on them.

Forest landscapes in east Africa and many other parts of the tropics are under large pressures, and mismanagement, overexploitation and degradation is not uncommon. Since forests are traditionally an important element of the livelihood strategies of rural dwellers and the source of income to many, degradation today seriously affects rural economies and national income from forest revenue collection. This forest degradation is accompanied by loss of biodiversity, loss of carbon storage, watershed degradation, soil loss, fertility decline and climatic change well beyond the boundaries of the forested areas.

A promising forest development in eastern Africa is the increased attention on tree planting. After decades of forest decline, community and commercial tree planting has increased substantially over the last five years in the target countries of this book. In Uganda, for instance, where only 2000 hectares of planted timber forests remained a few years ago, but the equivalent of 45,000 to 50,000 hectares is required to meet the domestic demand for timber (UNIQUE forestry consultants, 2005), about 20,000 hectares of new forests have been established within the last eight years. Additionally, the carbon finance providing incentives for Reduced Emissions from Deforestation and Forest Degradation (REDD) may release some pressure from precious natural forests, may enable forest restoration, and may provide new rural income generation and livelihood opportunities. National park development, and maybe especially the transboundary parks, with their large international attention and support, may be instrumental as well for stimulating positive and future directed forest landscape management.

Historically, forestry research in east Africa was focused on commercial utilization of natural forests, mostly high forests under state control and to a lesser degree the widely prevailing miombo woodlands (Campbell, 1996; Chidumayo, 1997). Later this was followed by research in plantation forestry, mostly with exotic species and by a wave of research on social issues surrounding forestry, particularly communal involvement in forest management. In recent years the focus has shifted to issues of biodiversity and climate change. Very little research is geared towards management of marginal forests and restoration of degraded forests, despite the fact that they now constitute the majority of remaining forests in east Africa.

# THIS BOOK

This book emphasizes several key ecological aspects underlying major forest ecosystems in the region and important social factors driving their management. It provides a background for the development of management tools for the rehabilitation and sustainable management of these ecosystems. Of course it would be over-ambitious to assume that one book could comprehensively deal with all forest degradation-related issues in the whole region. We have therefore opted for a strategy of focusing on crucial examples embedded in a general context. Apart from general country overviews for the three focal countries – Ethiopia, Uganda and Tanzania – we highlight particular forest ecosystems with their related ecology, management and restoration possibilities. Our focus is a pragmatic and problem-solving approach, based on the examples from the featured case studies. This is intended to have direct leverage effects with regards to the adoption of improved forest management and restoration approaches, and the book aims to direct new research initiatives to improve the scientific basis and education required.

In this book we focus on three countries in east Africa: Ethiopia, Tanzania and Uganda (Figure 1.3; Table 1.1), mainly because the basis of this book was a large international cooperation research project that focused on these countries. Of course other countries in the region, and their forests, are of equally high importance. As many forests and forest types are shared between countries, and as many of the topics and problems addressed in this book are common, we

**Figure 1.3** *Map of eastern Africa with the focal countries of this book:*
*Ethiopia, Uganda and Tanzania*

believe that the analyses and messages in this book will be of high importance
to these other countries as well. Several chapters also cover studies from other
areas.

As not much general information is available in the international literature
on forest and forestry issues in these countries, we here provide a comprehensive overview of these aspects in the three countries (Chapters 2–4). These
overviews detail the current situation of the forests and the major challenges
for forest resource management in Ethiopia, Uganda and Tanzania respectively.
All three chapters give an overview of the most dominant forest types in the
country.

**Table 1.1** *Forests in Ethiopia, Uganda and Tanzania*

| Parameter | Ethiopia | Uganda | Tanzania |
|---|---|---|---|
| Land area in 1000 ha | 100,000 | 19,710 | 88,350 |
| Population in millions (2009) | 79.2 | 32.4 | 43.7 |
| Forest area in 1000 ha | 13,000 | 3,600 | 35,000 |
| Annual change in extent of forest, 2000–2005 (%) | –1.1 | –2.2 | –1.1 |

*Source:* FAO (2007)

Chapter 2 (Teketay et al) additionally widely reviews available information on forest cover losses, forest products and production, industries and enterprise, management and conservation efforts, socio-economic roles, and policy and legislation related to forest resources in Ethiopia. The chapter also highlights the challenges facing its forestry sector and provides various recommendations that can help to address these challenges. The authors highlight the fact that for sustainable management and conservation of forest resources in Ethiopia, a stable institutional set-up is required and that political recognition of its socio-economic and ecological significances is warranted. They find that in the absence thereof, small-scale farm household-based tree-planting practices will continue to play a dominant role in the forest development direction of Ethiopia.

Chapter 3 (Obua and Agea) presents the status, structure and distribution of forests in Uganda, outlines the historical profile of forest resource management and the institutional set-up for forest management, and highlights the evolution and landmarks of forest policy in Uganda. The chapter further analyses the contribution of forests to socio-economic development, the forest revenue systems, the main causes of forest degradation, the importance of invasive species and their impact on forest management, and finally the attempts for restoration of degraded forests in Uganda. The chapter stresses that the Ugandan government needs to link forest landscape restoration to economic activities and has to involve the corporate bodies that cause a large part of the current forest degradation and loss of tree cover in Uganda. Finally, the authors solicit more research in forestry and related fields, notably on the impacts of sectoral and macroeconomic policies and legislations on forest degradation in Uganda, on socio-economic evaluation of successfully rehabilitated/restored forest areas, on nursery and field trials of single and mixed tree/shrub species for degraded land planting, on harmonization of demands on land resources, notably agriculture, on forest production, especially in conflict-prone areas, and on integrated and holistic approaches, including industrial and other off-farm livelihood opportunities to reduce pressure on forest resources. Sustainable agroforestry production systems urgently need to be designed that are affordable to the resource-poor.

In Chapter 4 Chamshama and Vyamana give an account of Tanzania's forest resources and their importance, and then outline the legal status of forest in the country, highlighting current provisions for forest management as relevant context. They conclude with brief reports on two encouraging initiatives taken to restore the forest resource role of degraded dryland, the biome where the need for restoration has become most urgent. They stress that sustainable forest management can only be realized if the forestry sector aims to optimize the dual objective of improving forest condition and conserving the environment, while at the same time improving livelihoods of the people. This is particularly important for the poor, who largely depend on forest resources for their livelihoods. Success stories of forest landscape restoration in Tanzania are strongly associated with situations where communities were actively involved, and where their interests, local knowledge and practices were taken into account.

This notion nowadays is part of most current policies and legislation, providing a good formal basis for restoration of degraded lands.

John Obiri and co-workers (Chapter 5) address the situation of miombo woodland in Tanzania, focusing on regeneration problems (but see also Chapter 12 for other aspects). The miombo forests in Tanzania are suffering heavy exploitation and degradation reflected in negative effects on species composition and ecosystem services. In much of the miombo of the Morogoro Region, pioneer tree and shrub species are increasing in abundance at the expense of the large canopy trees typical of the undisturbed community. The authors compare areas under three different forest management responsibilities, and associated levels of disturbance, and show that their species composition differs, and that tree regeneration varies along a disturbance gradient from the forest edge towards the forest core, with invasive species dominating the more disturbed areas.

To reduce the current rate of disturbance, active interventions are required. The authors argue that these are only likely to be successful if local stakeholders are empowered with more secure tenure over forest resources as a component of more sustainable livelihoods. Awareness needs to be improved, and needs to be based on enhanced monitoring of forests. Highlighting which components of their biodiversity have been declining as a result of current exploitation and disturbance is of crucial importance. The authors suggest that more forest needs to be under participatory forest management.

Church forests are the last remaining forests in many areas in northern Ethiopia and are under threat of fragmentation and degradation. These forests need more care and an improved management. Wassie and co-workers (Chapter 6) suggest that legal protection of church forest is badly needed, and that church forest areas have to be gazetted after being clearly demarcated and marked in the field. They promote training on silvicultural and forest management techniques and species-based tree propagation, and argue that interventions on church forests should be designed and applied in line with community traditions. Species-specific silvicultural interventions are needed to facilitate species regeneration, for instance opening of trampled soil, opening of dense thickets of shrubs and lianas, addition of seeds and seedlings from outside, liberation of specific tree seedlings and saplings in dense shrub or liana vegetation, and protection against grazing. The authors strongly believe that church forests should be expanded in area, because their current small patch size might introduce extinctions of tree species in the longer run. They conclude that the ecological and social status of church forests provides a strong opportunity to conserve natural forests and to restore degraded areas into more productive and diverse natural forests.

Also dryland forests in the region are under pressure, and an important example is the incense tree-dominated forest found in northern Ethiopia and neighbouring Eritrea and Sudan, and also in southern and southeastern Ethiopia and adjoining areas in Somalia and Kenya. In Chapter 7 Abiyu and collaborators analyse the regeneration problems and restoration possibilities of such forests, using *Boswellia papyrifera*-dominated forests as an example. They show regeneration levels in these forests are very low, and that this is the case almost all over the growing niche of B. *papyrifera*. The low regeneration rate is particularly worrying because of the failure to successfully grow and raise seedlings in the nursery and get them established in the field. Long-term monitoring studies and experiments are needed to pinpoint the effects of possible underlying factors. The lack of regeneration can have several reasons. Regeneration under natural conditions may be episodic, dependent, for example, on the availability of good climatic conditions. Climatic variability may play a role in this respect. However, human-controlled factors (for example grazing and fire) may also be inhibiting *Boswellia* regeneration. Together these factors may jeopardize the long-term provision of its important non-timber product, frankincense. The international concerns for the future of the frankincense tree put forward by different parties (for example TRAFFIC) are justified.

In large parts of eastern Africa the landscape is dominated by a mosaic of forest, woodland and savanna (FWS) and such mosaics need special, landscape-level management. As an example of such systems Nangendo and collaborators (Chapter 8) analyse a forest–woodland–savanna mosaic in Uganda, with a focus on vegetation cover and woody plant composition and changes therein over time. Fire had a strong driving role in these changes, as their 46-year long-term effect study (Nangendo et al, 2005) has shown. The chapter provides important strategies that could guide the management of such FWS landscapes. For biodiversity conservation, for instance, a part of each vegetation cover type needs to be conserved, as these types complement each other in terms of species composition. The chapter clearly indicates that fire is essential for conserving FWS mosaics and that a well-balanced fire management system is needed to control forest expansion while allowing the existence of varying fire disturbance regimes. Long-term fire regimes, however, cause species composition to converge towards a dominance of species better adapted to the existing fire regime.

Fragmentation and degradation of the forested landscape always lead to reduced forest services, among which the provision of wood is a well-known example. Enrichment planting and its extreme case, planting of complete forests, be they mono-dominant or mixed species forests, is able to alleviate part of this reduction. In east Africa still very few trees and hardly any indigenous trees are being raised (more are planted but do not survive), although the possibilities and the advantages of such planting may be large. Lemenih and Bongers (Chapter 9) synthesize the available information on plantation forestry in, particularly, eastern Africa, with the aim of providing evidence for the potential of plantation forests in re-establishing a sufficiently rich diversity of flora and fauna and in improving soil conditions. Additionally they show that plantation establishment and management techniques and networks that promote sufficient diversity need to be communicated to promote successes in practical applications. Although the chapter focuses on experiences from east Africa, it also covers a broad range of studies from elsewhere to capture the major findings of recent biodiversity and soil property studies in plantation forests, and thus provides a more comprehensive review. The chapter ends with a box

13

of guidelines indicating how plantation forests may be employed in biodiversity restoration. Among other aspects, they advocate that a simple start, with one to three species only, may be the most rewarding and most easily socially accepted system, whereafter complexity is introduced over time to achieve the desired ecosystem. Such a desired ecosystem should systematically be discussed and negotiated with stakeholders, especially people living in the area.

Agroforestry is a very important mixture of agriculture and forestry: in fact it is the combined use of both agricultural crops and trees on the same area of land. It involves a lot of tradeoffs, both ecological and social. Well-designed agroforestry practices are seen as important contributors to long-term sustainability of land use. Correct species selection is critical for successful agroforestry, but may be hampered by lack of information and by changing societal importance. Hall (Chapter 10) documents these aspects for *Maesopsis eminii*, a major agroforestry species, and shows that conflicts in interest may be guiding the discussion. He details the increase in research and importance of this species in east Africa, and at the same time documents the increase of critique, especially with respect to the invasiveness of the species. His account is an example of the kind of information that we need on a large number of potentially important agroforestry species, and he pleas for decisions to be made based on hard scientific data. He finds the conservation impact of *Maesopsis* as an aggressive invasive not becoming catastrophic. He concludes that *Maesopsis* is not a great agroforestry asset from a biodiversity perspective, but that the species definitely has potential, especially for timber production in agroforestry systems.

This timber aspect is further developed in Chapter 11, where Buchholz and co-workers focus on the importance of single-tree management models and use *Maesopsis* as an example as well. In their chapter the timber aspects of this species are highlighted and they show that single-tree management concepts, compared to stand- or basal area-focused systems, are better suited considering different production targets and local site conditions. Single-tree models can be used to predict timber production and standing crop under a large variety of conditions. *Maesopsis* indeed is one of the few indigenous tree species in eastern Africa that has substantial potential to produce high-quality timber in short rotation periods.

Silvicultural management is also the topic of Amente and collaborators (Chapter 12), but they address much broader issues and want management to be directed towards multiple uses, instead of focusing on timber only. They report on an integrative project in the Bale Mountains in southern Ethiopia, where they promote multiple-use forestry in close collaboration with local communities. This community-based approach has had a major impact on forest policy over the last years in that part of Ethiopia and is currently seen as one of the major developments in the country. Of course, participatory forest management needs much time, and dedication, especially in a country with a short history in this field. The authors suggest that forestry should be more focused on future productivity in terms of products and services. Silvicultural interventions are crucial for that future and should continuously be monitored to be

14

able to adjust management interventions when needed. Active and long-term participation and collaboration between forest organizations and local people is a must.

Such a future-directed attitude is also clear in Chapter 13, where Sterck and colleagues focus on the future development of *Juniperus*, African pencil-wood, a major Afromontane multipurpose tree species. This species is of very high importance as, among other uses, construction timber. Regeneration is scarce, however, and the lack of systematic establishment of plantations of this species, for various reasons, results in continued cutting of the species from the last remaining natural forests, from secondary forests and from church forests. These forests, however, remain important resources where restoration programmes can be started. Harvesting of *Juniperus*, especially illegal cutting, needs to be controlled. Silvicultural interventions are badly needed (for example thinning) and, particularly, grazing in the last *Juniperus* forests urgently needs to be reduced.

The last three chapters focus on social aspects of forest management, probably the most critical of all. Examples from three different systems are developed. In Chapter 14 Obiri and co-workers analyse the attitudes of local forest stakeholders towards degradation, plant species invasion, regulation of tree harvesting and alternative structures for management responsibility in the dry forests of east-central Tanzania. Maybe against expectations, invasive weedy plant species, whose increase in abundance is associated with forest degradation, are of major concern to local communities. Invasive species therefore merit greater attention in ecologically informed forest management plans. Of course such plants can be weeded out of the system, but that is expensive. It is better to adjust management in such a way that forests are less degraded and thus are less easy to invade for such weedy species. Intriguing is the fact that women see forest management different than men do, most probably as a result of their greater dependency on forest products for their livelihoods. Together with Vyamana (2009), the results of this chapter criticize the widely held idea that participatory forest management reduces poverty and social exclusion, and the authors plea for better and more systematic research integrated into participatory forest management project activities.

In Chapter 15 Wiersum describes the multifaceted nature of the biocultural process of domestication of coffee forests in southwest Ethiopia and assesses the tradeoffs between ecological degradation and resource enrichment by analysing the major dimensions of the domestication process. He stresses that the process of forest domestication in most cases results in a forested landscape mosaic including a combination of both natural, modified and transformed forests as well as cultivated fields. Such diversified forest landscape offers good options for the incorporation of forests in local livelihoods. He uses the forest coffee systems to show that different interpretations of forest domestication have their consequences for forest restoration. For effective forest restoration not only ecological practices for ecosystem rehabilitation should be considered, but also innovative practices for acculturalization of the new forest systems. Both provisioning, supporting, regulatory and cultural services that the restored forests should have for local livelihoods and the question of how local people can actually benefit from those services as a result of locally adjusted regulatory and marketing conditions need to be addressed. Rather than focusing on the restoration of forests in the form of specific ecosystems, restoration efforts should focus on creating forested landscape mosaics, as these offer good options for combining ecological conservation with human development.

Finally, in Chapter 16 Schmitt describes a recent development of communal management of degraded Afromontane forests. Groups of families are responsible for a given area of forest, where each family is endowed with sufficient fuel and construction wood and may use cash income from forest products, similarly to what an average farmer can achieve through agriculture. In reality, however, the forest-dwellers combine forest production with subsistence agriculture and livestock production and have to prevent over-utilization by outsiders to maintain their exclusive forest user rights. In principle, a small amount of stand increment is available for utilization and timber production is profitable. Yet limited silvicultural knowledge and management capacity prevent forest-dwellers from using the full production potential. Especially in times of resource shortages, 'scientific' forest management knowledge is needed to produce sustainable outcomes, in other words balancing increment and utilization. Schmitt is positive about conditions for sustainable forest management by the community members but stresses that technical advice on silvicultural management issues and access to credits to implement improved timber-processing techniques are crucial. She further recommends improving the productivity of the forest and the utilization techniques to increase the timber recovery rates and supporting vertically integrated local timber manufacturing to increase profitability of forest production for forest-dwellers. Finally, she challenges the reader in asking whether the unequal distribution of forest resources between 'forest owners' and 'outsiders' and within user groups reflecting age, gender and ability can be accepted as a reasonable tradeoff between social and ecological needs. Can we indeed maintain the forest at limited cost?

This book presents examples of forest management and restoration possibilities in a few countries in east Africa, and definitely does not pretend to be comprehensive. Of course, many initiatives have been started and many more results wait to be described and analysed. We hope that the information provided in this book helps the reader to understand the challenges that are faced when the degraded forests in east Africa are managed. We also hope that it is clear that restoration of these degraded forests is badly needed, both for the sake of the forest and the forest landscapes and for the sake of the people living in these landscapes.

# REFERENCES

Achard, F., Eva, H. and Stibig, H. J. (2002) 'Determination of deforestation rates of the world's humid tropical forests', *Science*, vol 297, pp999–1002

Andel, J. van and Aronson, J. (eds) (2006) *Restoration Ecology*, Blackwell Science, Oxford, UK

Botha, J., Witkowski, E. T. F. and Shackleton C. M. (2004) 'The impact of commercial harvesting on *Warburgia salutaris* ("pepper-bark tree") in Mpumalanga, South Africa', *Biodiversity and Conservation*, vol 13, pp1675–1698

Brooks, T., Balmford, A., Burgess, N. Fjeldså, J., Hansen, L. A., Moore, J., Rahbek, C. and Williams, P. (2001) 'Toward a blueprint for conservation in Africa', *Bioscience*, vol 51, no 8, pp613–624

Burgess, N., D'Amico Hales, J., Underwood, E., Dinerstein, E., Olson, D., Itoua, I., Schipper, J., Ricketts, T. and Newman, K. (2004) *Terrestrial Ecoregions of Africa and Madagascar. A Conservation Assessment*, Island Press, Washington, DC

Campbell, B. (ed) (1996) *The Miombo in Transition: Woodlands and Welfare in Africa*, Center for International Forestry Research (CIFOR), Bogor, Indonesia

Chidumayo, E. N. (1997) *Miombo. Ecology and Management. An Introduction*, Intermediate Technology Publications, London, and Stockholm Environment Institute, Stockholm

Fa, J. E., Currie, D. and Meeuwig, J. (2003) 'Bushmeat and food security in the Congo Basin: Linkages between wildlife and people's future', *Environmental Conservation*, vol 30, pp71–78

Fairhead, J. and Leach, M. (1998) *Reframing Deforestation: Global Analysis and Local Realities: Studies in West Africa*, Routledge, London

FAO (2007) *State of the World's Forests*, Food and Agricultural Organization of the United Nations, Rome

Guedje, N. M., Lejoly, J., Nkongmeneck, B. A. and Jonkers, W. B. J. (2003) 'Population dynamics of *Garcinia lucida* (Clusiaceae) in Cameroonian Atlantic forests', *Forest Ecology and Management*, vol 177, pp231–241

Gurevitch, J. and Padilla, D. K. (2004) 'Are invasive species a major cause of extinctions?', *Trends in Ecology and Evolution*, vol 19, pp470–474

Iqbal, M. (1993) *International Trade in Non-Wood Forest Products. An Overview*, Food and Agriculture Organization, Rome

Lovett, J. C. and Wasser, S. K. (eds) (1993) *Biogeography and Ecology of the Rain Forests of Eastern Africa*, Cambridge University Press, Cambridge, UK

McClean, C. J., Lovett, J. C., Küper. W., Hannah, W. L., Sommer, J. H., Barthlott, W., Termansen, M., Smith, G. F., Tokumine, S. and Taplin, J. R. D. (2005) 'African plant diversity and climate change', *Annals of the Missouri Botanical Garden*, vol 92, pp139–152

MEA (2005) *Millennium Ecosystem Assessment. Ecosystems and Human Well-Being: Current State and Trends 2005*, Island Press, Washington, DC

Milner-Gulland, E. J., Bennett, E. L. and the SCB 2002 Annual Meeting Wild Meat Group (2003) 'Wild meat: The bigger picture', *Trends in Ecology and Evolution*, vol 18, no 7, pp351–357

Mittermeier, R. A., Kormos, C. F., Mittermeier, C. G., Robles Gil, P., Sandwith, T. and Besançon, C. (2005) *Transboundary Conservation: A New Vision for Protected Areas*, University of Chicago Press, Chicago, IL

Muller-Landau, H. C. (2007) 'Predicting the long-term effects of hunting on plant species composition and diversity in tropical forests', *Biotropica*, vol 39, pp372–384

Nangendo, G., Stein, A., Steege, H. ter and Bongers, F. (2005) 'Changes in woody plant composition of three vegetation types exposed to a similar fire regime for over 46 years', *Forest Ecology and Management*, vol 217, pp351–364

Nyssen, J, Haile, M., Naudts, J., Munro, N., Poesen, J., Moeyersons, J., Frankl, A., Deckers, J. and Pankhurst, R. (2009) 'Desertification? Northern Ethiopia re-photographed after 140 years', *Science of the Total Environment*, vol 470, pp2719–2755

Olson, D. M., Dinerstein, E., Wikramanayake, E. D., Burgess, N. D., Powell, G. V. N., Underwood, E. C., D'amico, J. A., Itoua, I., Strand, H. E., Morrison, J. C., Loucks, C. J., Allnutt, T. F., Ricketts, T. H., Kura, Y., Lamoreux, J. F., Wettengel, W. W., Hedao, P. and Kassem, K. R. (2001) 'Terrestrial ecoregions of the world: A new map of life on Earth', *BioScience*, vol 51, pp933–938

Richardson, D. M. (1998) 'Forestry trees as invasive aliens', *Conservation Biology*, vol 12, pp18–26

Rodrigues, A. S. L., Andelman, S. J., Bakarr, M. I., Boitani, L., Brooks, T. M., Cowling, R. M., Fishpool, L. D. C., da Fonseca, G. A. B., Gaston, K. J., Hoffmann, M., Long, J. S., Marquet, P. A., Pilgrim, J. D., Pressey, R. L., Schipper, J., Sechrest, W., Stuart, S. N., Underhill, L. G., Waller, R. W., Watts, M. E. J. and Xie Yan (2004) 'Effectiveness of the global protected area network in representing species diversity', *Nature*, vol 428, pp640–643

SCBD (2001) *Sustainable Management of Non-Timber Forest Resources*, CBD Technical Series 6, Secretariat of the Convention on Biological Diversity, Montreal, Canada

SER (2004) *The SER Primer on Ecological Restoration*, Version 2, Society for Ecological Restoration Science and Policy Working Group, Tucson, AZ

Thomas, C. D., Cameron, A., Green, R. E., Bakkenes, M., Beaumont, L. J., Collingham, Y. C., Erasmus, B. F. N., Ferreira de Siqueira, M., Grainger, A., Hannah, L., Hughes, L., Huntley, B., van Jaarsveld, A. S., Midgley, G. F., Miles, L., Ortega-Huerta, M. A., Peterson, A. T., Phillips, O. L. and Williams S. E. (2004) 'Extinction risk from climate change', *Nature*, vol 427, pp145–148

Thuiller, W., Broennimann, O., Hughes, G., Alkemade, J. R. M., Midgley, G. F. and Corsi, F. (2006) 'Vulnerability of African mammals to anthropogenic climate change under conservative land transformation assumptions', *Global Change Biology*, vol 12, pp424–440

Ticktin, T. (2004) 'The ecological implications of harvesting non-timber forest products', *Journal of Applied Ecology*, vol 41, pp11–21

UN (1987) 'Report of the World Commission on Environment and Development', General Assembly Resolution 42/187, 11 December 1987

UNIQUE forestry consultants (2005) 'Reducing the uncertainty for forest investors in Uganda: Value chain assessment for timber and timber products', Report to the Sawlog Production Grant Scheme

Vitousek, P. M., D'Antonio, C. M., Loope, L. L, Reymánek, M. and Westbrooks, R. (1997) 'Introduced species; a significant component of human-caused global change', *New Zealand Journal of Ecology*, vol 21, no 1, pp1–16

Vyamana V. G. (2009) 'Participatory forest management in the Eastern Arc Mountains of Tanzania: Who benefits?', *International Forestry Review*, vol 1, pp239–253

White, F. (1983) *The Vegetation of Africa: A Descriptive Memoir to Accompany the UNESCO/AETFAT/UNSO Vegetation Map of Africa*, United Nations Educational, Scientific and Cultural Organization, Paris

2

# Forest Resources and Challenges of Sustainable Forest Management and Conservation in Ethiopia

*Demel Teketay, Mulugeta Lemenih, Tesfaye Bekele, Yonas Yemshaw, Sisay Feleke, Wubalem Tadesse, Yitebetu Moges, Tesfaye Hunde and Demeke Nigussie*

## INTRODUCTION

Ethiopia occupies the interior of the Horn of Africa, stretching between 3° and 15°N and 33° and 48°E (Figure 2.1). It covers a total area of 1.13 million km$^2$ (CSA, 2000) that spans over a wide range of altitude, from 110m below sea level to over 4600m above sea level (asl). The wide altitudinal coverage enhances the diversity of climate, topography, soil, and thus vegetation resources. Ethiopia is the second most populous country in sub-Saharan Africa, with 74 million people, and 84 per cent of these live in rural areas (CSA, 2008), depending mainly on mixed crop/animal and forest/tree production for livelihoods.

The Ethiopian landscape contains a range of vegetation resources, from tropical rain and cloud forests in the southwest to the desert scrubs in the east and northeast and parkland agroforestry on the central plateau. The large terrestrial land surface with biologically productive climate and soil offers the country a huge potential for developing forest resources. Most of the existing vegetation resources in the country are natural, while limited plantation forests also exist.

In recent decades, private tree planting by smallholder farmers, a practice called farm forestry or forest farming, is increasingly covering the rural highlands.

The structure and composition of the natural vegetation are diverse, reflecting their distribution over wide physiognomic and climatic landscapes. Ethiopia is one of the top 25 biodiversity-rich countries in the world (WCMC, 1994), and hosts at least two of the world's biodiversity hotspots, namely the Eastern Afromontane and the Horn of Africa hotspots. The vegetation comprises over 7000 species, of which 1150 are recognized as endemic to the country. Endemism is particularly high in the high mountains (Afroalpine region) and the southeastern lowlands. The vegetation resources also harbour diverse fauna. There are 240 species of mammals and 845 species of birds, of which 22 species of mammals and 24 species of birds are endemic (Anonymous, 1997a).

The vegetation resources contribute to the production, protection and conservation functions in the country. They supply most of the wood products (industrial and non-industrial) consumed within the country and provide diverse non-timber forest products (NTFPs), important among these being wild coffee, gum resins, honey and beeswax, herbal medicines and bamboo. They also provide varied ecosystem services, some of which are global environmental goods. These services include watershed protection, biodiversity conservation and carbon sequestration. The vegetation ecosystems in the Ethiopian plateau and mountains are the sources of a number of great rivers, including the Nile, Omo and Wabi Shebelle. Some of these rivers are the only sources of permanent water for the surrounding arid and semi-arid environments in and outside Ethiopia. The high floristic endowment and ecological diversity of Ethiopia's vegetation provides a vital genetic reservoir for at least 197 species of crops, including grains, pulses, oil seeds, vegetables, tubers, fruits, spices, stimulants, fibres, dyes and medicinal plants. For instance, Ethiopia (and some neighbouring districts in Sudan) is the world's only reservoir of the wild gene pools of coffee (*Coffea arabica*),[1] 'teff', 'enset', sesame and 'noug' (*Guizotia abyssinica*). This is the reason for the recognition of Ethiopia as a centre of agro-biodiversity, designated as one of eight Vavilov Centers in the world.

The role of the forestry sector in the national and local economies in Ethiopia is significant. However, aggregate data on economic values and productions of forest products and services are fragmentary and inadequate. A recent assessment put the contribution of the forestry sector to gross domestic product (GDP) at 9.0 per cent, which is quite high compared to the 5.7 per cent estimate provided in the national accounting system (Nune et al, 2009).

Perhaps the most important service that vegetation ecosystems in Ethiopia have been offering is the provision of fertile croplands upon conversion. The Ethiopian economy still largely depends on subsistence agriculture. The traditional agricultural system is based on extensification (horizontal expansion) rather than intensification, and vegetated areas have been providing productive croplands for millennia (Lemenih et al, 2008; Lemenih and Bongers, in press). When unconverted to croplands, forest and vegetated lands serve as natural rangelands for one of the largest livestock population in sub-Saharan Africa (FAO, 2004). The middle elevation ranges (1500–2500m) of the vegetation

**Figure 2.1** *Map of Ethiopia showing major forest areas*

*Source:* Adapted from Reusing (1998)

ecosystems in Ethiopia are densely inhabited by sedentary subsistent agrarian populations, and the vegetation ecosystems here have been predominantly serving the provision of agricultural lands. The lowlands (<1500m asl) vegetation ecosystems are occupied by nomadic and agro-pastoralists that make up 15–20 per cent of the Ethiopian population. Here the vegetation ecosystems are extensively utilized as natural rangelands.

Second to agricultural services, the main use of the forests of Ethiopia is as a source of fuelwood (firewood and charcoal). Biomass is the major source of energy, accounting for 97 per cent of the total domestic energy consumption, out of which woody biomass covers 78 per cent (WBISPP, 2004). The volume of fuelwood demand at national level is nearly 20 times greater than the demand for other forest products combined. In the 1990s, the demand was 80 million m³/yr and a recent estimate puts the demand at 109 million m³/yr (FAO, 2005).

Ethiopia's forests have been subject to intense human use for millennia, and this pressure has intensified since the last century. The loss and degradation of forests has now reached near crisis proportions. A number of measures, though unsatisfactory, have been taken to conserve, manage and develop the vegetation resources of the country. In recent decades, more institutional reforms have also taken place that directly or indirectly affect the forestry sector. These reforms include decentralization of political administration and the devolution of forest and natural resources management responsibilities, the enacting of a new forestry policy, and a number of other policy documents related to rural lands, investment and environmental protection.

This chapter reviews available information on the types, cover losses, products and production, industries and enterprise, management and conservation efforts, socio-economic roles, policy, and legislation related to forest resources in Ethiopia as well as the challenges facing its forestry sector. We conclude with various recommendations that can help to address these challenges.

# FORESTS AND OTHER VEGETATION RESOURCES

## Types of vegetation resources

Ethiopia's vegetation resources comprise both natural and man-made components distributed in the highlands and lowlands. The natural vegetations are of different types, ranging from desert scrubs to tropical rainforests and cloud forests. Several attempts have been made to classify the vegetations of Ethiopia in general and that of its forest vegetations in particular. The classifications by different authors have followed different systems, with some employing climate and physiognomy as their basis, while others combined climate, physiognomy and species composition (Senbeta, 2006). The vegetation classification by Demissew (1996), for instance, grouped the vegetation resources of the country into nine broad categories, namely:

1  Afroalpine and Sub-Afroalpine vegetation;
2  Dry evergreen montane forest;
3  Moist evergreen montane forest;
4  Wetlands;
5  Evergreen scrub;
6  *Combretum-Terminalia* woodland;
7  *Acacia-Commiphora* woodland;
8  Lowland dry forest; and
9  Lowland semi-desert and desert scrubs.

Forests have been classified by a number of authors. Prominent among these are Logan (1946), Chaffey (1979), Friis (1986 and 1992), Friis and Tadesse (1990) and Tadesse (1993). Here we present the forest vegetation classification provided by Friis and Tadesse (1990) and Tadesse (1993) (Table 2.1).

**Table 2.1** *Forest types in Ethiopia with their altitudinal ranges, annual average temperatures and major species*

| Forest type | Category | Altitudinal range (m) | Annual rainfall (mm) | Rainfall period | Average temperature (°C) |
|---|---|---|---|---|---|
| Upland dry evergreen forest of Eastern Escarpment | | 1600–2400; 1500–2000 (in Sidamo) | 400–700 | October–March; April–May and September–October (in Sidamo) | 15–20 |
| Upland dry evergreen forest of the Ethiopian Plateau | | 1600–3300 | 500–1500 | March–April; July–September | 12–18 |
| | Plateau in Tigray | | | | |
| | Plateau in Gonder | | | | |
| | Plateau in Welo | | | | |
| | Plateau in Shewa | | | | |

**Table 2.1** *continued*

| Annual rainfall (mm) | Rainfall period | Average temperature (°C) | Major species |
|---|---|---|---|
| | | | *Juniperus procera, Podocarpus falcatus, Albizia schimperiana, Celtis africana, Cordia africana, Ficus vasta, Juniperus procera, Millettia ferruginea* and *Mimusops kummel* |
| 700–1100 | July–September | 14–20 | *Juniperus procera, Olea europaea* subsp. *cuspidata* and *Podocarpus falcatus* |
| 700–2000 | April–May; July–September | 15–20 | *Podocarpus falcatus* (replaced in Kefa and Ilubabor by *Pouteria adolfi-friederici*), *Hagenia abyssinica, Schefflera volkensii, Hypericum revolutum, Croton macrostachyus, Ilex mitis, Olea europaea* subsp. *cuspidata, Schefflera abyssinica, Albizia gummifera, A. schimperiana, A. grandibracteata, Blighia unijugata, Cassipourea malosana, Ekebergia capensis, Euphorbia ampliphylla, Ficus sur, F. vasta, F. thonningii, Hallea rubrostipulata, Ilex mitis, Macaranga capensis, Ocotea kenyensis, Olea welwitschii, Polyscias fulva, Prunus africana, Sapium ellipticum* and *Syzygium guineense* subsp. *Afromonatanum* as well as *Filicium dicipiens* and *Alangium chinense* (known only in Harenna forest in Ethiopia) |

| Transitional and lowland (semi)-evergreen forest of Southwest Ethiopia | | 450–1500 | | NA | |
| --- | --- | --- | --- | --- | --- |
| | Transitional forests | | 2000 (at Yeki) | NA | 20–25 |
| | Lowland forests | | 1300–1800 | NA | 18–20 |
| Riverine forest | | NA | NA | NA | NA |

*Source:* Modified from Friis and Tadesse (1990)

## Spatial coverage and regional distribution

In Ethiopia, probably like most other developing countries, detailed measures of vegetation spatial coverage, standing volume, changes over time (deforestation or regrowth), annual incremental yields, regeneration and recruitment status, and other essential information are not available or those available are inadequate. In the absence of national programmes or mandated institutions to undertake regular inventories and monitor the vegetation resources of the country, providing sufficient detail and precise statistics is hardly possible. Published data on the forest and vegetation resources of Ethiopia are confusing, with different institutions providing different statistics. The data from different institutions are not only inconsistent but also lack methodological clarity on how, by whom and when the data were collected. Vegetation statistics by different agents show a staggering high disparity (Table 2.2) – even data from a single organization during different times show considerable inconsistency. For instance, the 2001 report by the FAO estimated the high forests at about 5.75 million hectares, while its report in 2005 put the same forest types at about 13 million hectares, which is more than double the figure in the first report, despite the fact that the natural forests of the country were shown to decline over the same period at a rate of 141,000 hectares per year by the FAO itself. The discrepancies between the different data illustrate the difficulties of reaching strong conclusions on the forest resource base of the country.

Perhaps, the only credible national-scale vegetation assessment in Ethiopia is the work by the Woody Biomass Inventory and Strategic Planning Project (WBISPP, 2004). But data from remotely sensed images could be an alternative way to provide relatively accurate information on the forest/vegetation resource base. Reusing (1998), for instance, collated various fragmentary historical inventory records and combined these with remote sensing techniques to provide a reasonably good estimate of the high forest resources of the country. Both estimates give a better forest cover in the country than earlier estimates reported by various bodies such as the Ethiopian Forestry Action Plan (EFAP, 1994).

**Table 2.2** *Vegetation resources in Ethiopia by region*

| Region | High forest | | Woodlands | | Shrublands | |
|---|---|---|---|---|---|---|
| | Total (ha) | % of total | Total (ha) | % of total | Total (ha) | % of total |
| Oromia | 2,547,632 | 62.5 | 9,823,163 | 34 | 7,750,422 | 29 |
| SNNPRS | 775,393 | 19.0 | 1,387,759 | 5 | 2,434,779 | 9 |
| Gambella | 535,948 | 13.2 | 861,126 | 3 | 146,103 | 1 |
| Amhara | 92,744 | 2.3 | 1,040,064 | 4 | 4,352,672 | 16 |
| Tigray | 9332 | 0.2 | 294,455 | 1 | 1,841,182 | 7 |
| Beneshangul-Gumuz | 68,495 | 1.7 | 2,473,064 | 8 | 1,422,191 | 5 |
| Afar | 39,197 | 1.0 | 163,657 | 1 | 3,024,697 | 11 |
| Somali | 4257 | 0.1 | 13,199,662 | 45 | 5,384,022 | 20 |
| Others (Harari, Dire Dawa) | 216 | 0.0 | 0 | 0 | 44,132 | 0 |
| Total | 4,073,214 | 100 | 29,242,949 | 100 | 26,400,200 | 100 |

*Source:* WBISPP (2004)

**Table 2.3** *Forest resources statistics of Ethiopia as provided by different bodies*

| Forest resource | EFAP (1994) | | Reusing (1998) | WBISPP (2004) | | FAO (2001) | FAO (2005) | |
|---|---|---|---|---|---|---|---|---|
| | Area (M ha) | Growth stock (m³/ha) | Area (M ha) | Area (M ha) | Growth stock (m³/ha) | Area (M ha) | Area (M ha) | Stock (m³/ha) |
| Natural high forest (total) | 2.3 | | 4.506 | 4.072 | 130.5 | 5.755 | 13.000 | 22 |
| Natural high forest (slightly disturbed) | 0.7 | 90–120 | 0.235 | | | 1.680 | | |
| Natural High forest (highly disturbed) | 1.6 | 30–100 | 4.271 | | | 4.075 | 12.509 | 22 |
| Woodlands | 5.0 | 10–50 | | 29.24 | 21 | 31.554* | 44.650 | 2.3 |
| Bushlands | 20.0 | 5–30 | | 26.40 | 15 | | | |
| Plantations** | 0.2 | | | 0.96 | 178.8 | 0.216 | 0.419 | 22 |
| Farm forests | NA | NA | | NA | 15 | NA | NA | NA |
| Relative reliability ranking of sources | 3 | | 2 | 1 | | 4 | 4 | |

*Notes:* * Both woodlands and bushlands combined; ** the FAO statistics on plantation refers to productive plantation, while the others did not distinguish the type of plantation.

According to WBISPP (2004), a total of 59.7 million hectares of woody vegetation resources exist in Ethiopia, among which 6.8 per cent is forest, 49 per cent is woodland and 44.2 per cent is shrubland (Table 2.3). The high forests of the country are shown to be about 4.07 million hectares (3.6 per cent of the total landmass), which is very close to the estimate of 3.9 per cent by Reusing (1998). Woodlands and shrublands cover 29.24 and 26.4 million hectares or 25.5 and 23.1 per cent of the country's territorial surface respectively (WBISPP, 2004). Regarding regional distribution, Oromia (62.5 per cent), the Southern Nations, Nationalities and Peoples Regional State (SNNPRS) (19 per cent) and Gambella (9 per cent) are the three largest natural high forest owners, while Somali (33 per cent), Oromia (32 per cent) and Amhara (10 per cent) regions share the largest woodlands and shrublands (Table 2.2).

Ethiopia has long been engaged in plantation forest development. However, even the plantation forest statistics in Ethiopia are inconsistent. While the national inventory by Teklu (2003) gave an estimate of about 210,000 hectares, which, however, was not a complete inventory, the FAO (2005) report presented an estimate of 419,000 hectares and that of the WBISPP (2004) an estimate of 955,705 hectares.

## Species composition and productivity

Species richness in natural forests varies from eight in the lowland woodlands to about 420 in the rainforests such as the Yayu forest (Yebeyen, 2006; Senbeta et al, 2007; Woldemariam et al, 2008). The Yayu rainforest is perhaps the most species-diverse forest in Ethiopia, hosting about 420 species (Woldemariam, 2003; Schmitt, 2006; Senbeta, 2006; Woldemariam et al, 2008). An important fact about the species composition of the vegetations of Ethiopia is that they harbour gene pools of crop wild relatives for at least 197 species of crops. However, continuous forest degradation has resulted in the dominance of a few species such as C. *arabica* in the rainforests or pioneer species like *Croton macrostachyus* in the dry Afromontane forests.

In their current state, the productivity of the natural forests and woodlands of Ethiopia is very low, although significant variation occurs between forest types and stands based on species composition, geographical location, altitude and level of anthropogenic disturbance. Productivity is low due to poor stocking as a result of the continuous open-access harvest for timber, construction material and fuelwood. Illegal logging in the high forests is widespread and is very selective, in practice confined to very few timber species such as *Cordia africana, Hagenia abyssinica, Podocarpus falcatus, Pouteria adolfi-friederici* and a few others (NRMRD, 2001; Nigatu, 2004). For instance, half of the 47,590m$^3$ of timber consumed per annum by carpentry workshops in Addis Ababa originates from natural forests, and 30 per cent of this originates from protected species such as *Cordia africana* (22 per cent), *Pouteria adolfi-friederici* (17 per cent), *Hagenia abyssinica* (2 per cent) and *Podocarpus falcatus* (5 per cent) (Nigatu, 2004). The timbers from these species are supplied through private dealers who trade in illegal pit-sawn boards (Nigatu, 2004; Haile et al, 2009). Such creaming of valuable timber species from the natural forests has degraded the quality, species composition and thus productivity of the natural forests. Individuals in the standing stock are predominantly shrubs, pioneers or over-aged and deformed trees. For instance, in the Tiro natural forest, selective illegal logging has reduced the stocking of 12 commercial species such as *Pouteria adolfi-friederici, Cordia africana, Syzygium guineense* and *Ekebergia capensis*, which are supposed to predominate in the forest, to make up only 25 per cent of the total stand density (Mengesha, 1996). In the same forest, 75 per cent of the stocking is dominated by species that are commercially less attractive pioneers or shrubs common to disturbed forests, such as *Bersema abyssinica, Maytenus senegalensis* and *Teclea nobilis*. Similarly, in the Munessa-Shashamane forest, the density of *Croton macrostachyus*, a pioneer species common to disturbed sites, is more than twice the density of *Podocarpus falcatus*, a species that should have dominated the upper storey of the forest (Abate, 2004). Senbeta et al (2007) also showed that over 80 per cent of the plant density in the Sheko rainforest is contributed by 20 species, most of which are shrubs such as *Coffea arabica*. Dryland woodlands are also similarly affected, but from unregulated fuelwood (firewood and charcoal) harvest.

The predominance of low-quality stems and pioneer species in the forests and woodlands is resulting in low tree stem density, basal area and standing volume. Density of trees (>10cm diameter at breast height; DBH) range from about 268 stems/ha in the woodlands through 360 stems/ha in the dry

Afromontane high forests to about 500 stems/ha in the moist forests of the southwest (see, for example, Mengesha, 1996; Yebeyen, 2006; Senbeta, 2006; Tesfaye, 2008). Basal area also varies from less than $20m^2$/ha in the woodlands to about $44m^2$/ha in dry Afromontane and up to $54m^2$/ha in the moist forests. Stand volume per hectare in the rainforests can reach up to 350 $m^3$/ha. Mean annual volume increment for the dry Afromontane forests in Bale is estimated to be $1m^3$/ha/yr (Ameha, 2002); that of *Acacia-Commiphora* woodlands is about $0.0015m^3$/ha/yr, with the standing volume of about $6.5m^3$/ha, and for the moist high forests the increment figures are about 2 to $5m^3$/ha/yr, depending on level of disturbance and species composition. The desert and semi-desert scrubs and evergreen scrubs provide a much less incremental yield. Tesfaye (2009) investigated biomass productivity of the primary rainforests of Haranna (southeastern Ethiopia) and found it to range between 9.92 and 11.42 tons/ha/yr. The variation is mainly related to altitude. The same study also investigated the biomass productivity of forest regrowth after fire in the same area, and found it to be 6.93–14.56 tons/ha/yr in 3-year-old and 14.00–19.28 tons/ha/yr in 7-year-old stands.

## Regeneration and recruitment status

The forests and woodlands of Ethiopia are generally characterized by poor regeneration and recruitment. Heavy and continued anthropogenic disturbances such as forest grazing, fire, and unregulated harvest for timber and firewood have been negatively affecting their regeneration and recruitment (Mengesha, 1996; Tesfaye, 2008). However, regeneration patterns of plant communities and species also vary depending on environmental variables, such as altitude, slope, aspect, canopy light (for high forests), edaphic conditions, and the intrinsic and adaptation behaviour of species. Indeed, in the same plant community some species may show a healthy regeneration and normal population structure, while others do not. In the dry evergreen Afromontane forests of Munessa (central Ethiopia), species such as *Celtis africana, Prunus africana, Croton macrostachyus* and *Podocarpus falcatus* produce reasonably large seedling numbers (160–7180 individuals/ha, individuals with less than 150cm height or <10cm DBH), while species like *Polyscias fulva* are without seedlings (Tesfaye, 2008). Similarly, in the *Combretum-Terminalia* deciduous woodlands in Metema (northwest Ethiopia), species like *Boswellia papyrifera* and *Acacia polyacantha* (Asfaw, 2006), and in the *Acacia–Commiphora* deciduous woodlands in the central Rift Valley species such as *A. senegal* and *A. tortilis* produce high seedling populations (>500 individuals/ha), while species like *A. etbaica* are without seedlings (Yebeyen, 2006).

Some disturbances influence seedling survival more than seedling emergence, although the effect is also species-specific. In the dry evergreen Afromontane forests of Munessa, *Pouteria adolfi-friederici, P. africana, Celtis africana* and *Syzygium guineense* lost between 10 and 70 per cent of their seedlings due to herbivory, while *Croton macrostachyus* and *Podocarpus falca-*

*tus* experienced no seedling mortality under the same disturbance regime (Tesfaye, 2008). These last species possess tough and unpalatable leaves as an adaptive defence to herbivory that contributed to their low seedling mortality (Tesfaye, 2008). Yebeyen (2006) also found that regeneration of *Acacia senegal* in the *Acacia-Commiphora* deciduous woodland is favoured by a high disturbance regime where open grazing site and farmlands had 620 and 600 seedlings per hectare respectively, as compared to 170 seedlings per hectare on controlled grazing and only 62 seedlings per hectare on undisturbed (protected) sites. However, at community level, the undisturbed site had an inverse J-shaped population structure, where proportionally good numbers of individuals in all diameter classes are found, while the disturbed sites had extra large populations in the lower DBH classes of <8cm but very few individuals in the higher DBH classes (Yebeyen, 2006). On the other hand, a study in the *Combretum-Terminalia* deciduous woodland reported grazing to suppress seedling emergence, while fire was favouring it (Asfaw, 2006). Burnt and grazed woodland sites hosted 30 per cent lower seedling density than burnt but ungrazed woodlands, while unburnt but grazed woodlands had 50 per cent fewer seedlings. This study showed that fire at lesser intensity stimulates seedling emergence, while grazing has the opposite effect. However, species like *Boswellia papyrifera*, despite a good seedling emergence, show very high seedling mortality, approximating 100 per cent, both in burnt and grazed forests, and even when protected against fire, grazing and human disturbances (Ogbazghi et al, 2006; Negussie et al, 2008; Abiyu et al, Chapter 7 of this volume). The causes for such seedling mortality have not been identified yet. In some woodlands (for example in the Borana rangeland), interruption of traditional management such as a ban on fire use has favoured species like *Acacia drepanalobium*, a native species, to become invasive (Dalle, 2004; Angassu, 2007; Woldu and Namomissa, 2006; Homann et al, 2008; Terefe, 2009).

In most cases assessment at community level reveals patterns of vegetation population structure of an inverted J-shape. This shape apparently seems to reveal a healthy population pattern. However, the inverted J-shape population structure for the forest community is attributed to the prevalence of shrubs, pioneers and other woody species with naturally small diameter rather than a normal regeneration of climax species. When disaggregated at individual species level, a range of population patterns are revealed. In general, at species level, six different population structures are observed in most of the vegetation communities in Ethiopia (Senbeta, 2006; Worku, 2006; Yebeyen, 2006; Tesfaye, 2008):

1 structure displaying an inverted J-shape, representing those species with healthy regeneration;

2 a U-shaped structure, representing species exposed to periodic disturbances and thus interruption of regeneration; this is common to species that are harvested mostly at intermediate sizes such as *Olea welwitschii* (Senbeta et al, 2007);

3   J-shape structure, representing species suffering from absence of regenera-
    tion; these types of species are either not producing enough seeds or seeds
    are highly recalcitrant, suffering high predation, or experiencing new
    environmental and/or anthropogenic interfaces; one such species is
    *Syzygium guineense* (Senbeta et al, 2007);
4   bell-shaped structure, representing species with many individuals in the
    middle diameter classes and few in the lower and higher diameter classes
    (for example *Boswellia papyrifera*);
5   broken inverted J-shape structure; and
6   broken J-shape distribution patterns, which occur in conditions where
    individuals are selectively removed from one or two size classes but not
    other size classes.

# PRODUCTION AND SUPPLY OF FOREST PRODUCTS

## Wood products

The limited forest resources of the country are supplying most of the wood
products utilized therein. The demand–supply gaps for industrial woods are
filled with imports (Table 2.4). There has also been intermittent export of
*Eucalyptus* poles to the neighbouring countries of Sudan, Djibouti, Yemen and
Eritrea, although the amounts have not been registered. Industrial wood
products (for example round wood and sawn wood) are predominantly
produced from industrial plantations. The natural forests also supply quite
significant volumes of various industrial woods, but mostly through illegal
harvest. Farm forests are supplying the bulk of construction woods (poles and
posts) and some industrial round woods, while fuelwoods are supplied from
diverse sources such as natural high forests, woodlands, bushlands, industrial
and peri-urban plantations, and farm forests.

Except for fuel and construction wood, wood production, particularly of
industrial wood, is dwindling as the forest coverage declines. For instance,
lumber and plywood production per year have declined from 65,000m$^3$ and
1900m$^3$ respectively in the 1970s (Kvarnbäck and Natvig, 1979) to 16,000m$^3$
and 1600m$^3$ in the early 2000 (FAO, 2005). Consequently, importing began to
play an important role in the supply of industrial woods such as sawn wood,
wood-based panels, veneer and wood pulp. On the other hand, the bulk of wood
use in Ethiopia goes to fuelwood (Table 2.4).

## Non-timber forest products (NTFPs)

NTFPs are defined here as all products of biological origin other than timber
extracted from forests, woodlands and trees outside forests for human use.
Ethiopia's forest and other vegetation resources offer diverse NTFPs that
provide substantial inputs into the livelihoods of a very large number of people

**Table 2.4** *Average annual wood products production, import and consumption in Ethiopia*

| Product type | Unit | Import Qty. (1000) | Import $US (1000) | Export Qty. | Export $US (1000) | Production Qty. (1000) | Production $US (1000) | Consumption Qty. (1000) | Consumption $US (1000) |
|---|---|---|---|---|---|---|---|---|---|
| Sawn wood | m³ | 1.8 | 455 | 0 | 0 | 60 | 15,167 | 61.8 | 15,622 |
| Wood-based panels | m³ | 15.1 | 3913 | 0 | 0 | 10.1 | 2617 | 25.2 | 6530 |
| Veneer sheets | m³ | 3.1 | 1030 | 0 | 0 | 0 | 0 | 3.1 | 1030 |
| Industrial round Wood (logs) | m³ | 0 | 0 | 0 | 0 | 2459 | 38,251 | 2459 | 38,251 |
| Fuel wood | m³ | 0 | 0 | 0 | 0 | 84,135 | 420,673 | 84,135 | 420,673 |
| Round wood (Poles, posts, Construction wood) | m³ | 0 | 0 | 0* | 0* | 86,532 | 1,047,999 | 86,532 | 1,047,999 |
| Wood pulp | Mt | 12 | 9960 | | | 9 | 7470 | 21 | 17,430 |
| Other fibre [ulp | Mt | 0 | 0 | 0 | 0 | 9.4 | 2350 | 9.4 | 2350 |
| Recovered paper | Mt | 0 | 0 | 0 | 0 | 2.5 | 625 | 2.5 | 625 |
| Total | | | 15,358 | | 0 | | 1,535,152 | | 1,550,510 |

*Note:* * There is some intermittent export of round wood, for example to the Sudan, but quantities are not known (Haile et al, 2009).
*Source:* Lemenih (2008)

in the country. The most important NTFPs include wild coffee, gum resins, products of apiculture (honey and beeswax), herbal medicines, bamboo, spices, civet musk, and forest grazing (Table 2.5).

Ethiopia is one of the major coffee-producing countries and the only country with a wild coffee population. About 30–35 per cent of annual coffee production in the country originates from either wild or semi-managed coffee forests (Workaffess and Kassu, 2000), which in 2005, for instance, provided a

**Table 2.5** *Annual production of non-wood products and their gross financial values*

| Product type | Annual production (tons) | Estimated annual turnover in $US (1000)* |
|---|---|---|
| Wild coffee | 70,000–90,000 | 210,000 |
| Gum/incense (gum arabic, Olibanum, myrrh, etc.) | 5000 | 6800 |
| Honey | 30,000–50,000 | 86,500 |
| Beeswax | 4000 | 19,840 |
| Herbal medicine | 56,000 | 2,055,484 |
| Bamboo | ND | 10,555 |
| Forest grazing (Fodder) | ND | ND |
| Forest food (wild food) | ND | ND |
| Essential oils | ND | ND |
| Spices | 1208 | 2700 |
| Civet musk | 400 | 183 |
| Total | | 2,305,122 |

*Note:* * Includes sales on export and domestic markets; ND denotes no data available.
*Source:* Lemenih (2008)

total of 69,681 tons of coffee (Nune et al, 2009). Coffee yield in the forest coffee system is on average 225kg/ha, while in the semi-coffee forest system it is about 450kg/ha (Teketay et al, 1998; Teketay, 1999a).

About 30,000–50,000 tons of honey and 4000 tons of beeswax are annually produced in Ethiopia, most or all of which is forest/vegetation-based in terms of nectar provision, bee colony hosting and construction material supply. After China, Mexico and Turkey, Ethiopia is the fourth largest wax exporter to the world market (Lemessa, 2006).

Various kinds of gum resins, namely gum arabic, frankincense, myrrh and opoponax, are produced from a number of *Acacia*, *Boswellia* and *Commiphora* species. Ethiopia belongs to the gum belt of Africa and is one of the major gum-resin producing countries in the Horn. Production has been increasing since the early 1990s due to market reforms. Before 1991, the sub-sector was monopolized by one government agent called the Natural Gum Processing and Marketing Enterprise (NGPME). However, following changes in the political environment and subsequent market reforms, several private entrepreneurs have engaged in the production and marketing, with a consequent significant increment in both. Annual production is about 5000 tons, half of which is exported and the rest traded on the domestic market. Ethiopia's gum-incense exports amount to 28 per cent of the African total (ITC-P-Maps, 2006).

Nearly 84 per cent of the Ethiopian population that dwell in rural areas and a very significant number of those in urban areas and 90 per cent of livestock depend on traditional medicine for primary healthcare (WHO, 1998). So far, about 1000 species of indigenous plants, most of which are wild plants, have been recorded to have herbal medicinal applications. These plant species have been used in the traditional healthcare system to treat nearly 300 physical disorders, from childhood leukaemia to toothaches and mental disorders. About 56,000 tons of medicinal plants are harvested and used per annum in Ethiopia, most of which are harvested largely from wild plant stocks. Traditional healers who use traditional medicines are estimated to number 80,000 in the country, with about 9000 being officially registered.

Ethiopia owns the largest livestock resources in Africa. There are about 35 million tropical livestock units (TLU) or about 80 million head (ca. 30 million cattle, over 42 million sheep and goats, and 7 million equines) of livestock in Ethiopia (FAO, 2004). Forest grazing and browsing is the major source of feed for the vast population of livestock in Ethiopia. Some 175,000km$^2$ or nearly 35 per cent of Ethiopian rangelands are found under forest cover of mainly bush and shrub, and fodder deriving from forest lands provides 10 per cent and 60 per cent of livestock feed in the wet and dry seasons respectively. In pastoral areas, forest grazing and browsing constitute the sole land-use system.

The forests of Ethiopia also host several commercial spice plant species such as *Aframomum corrorima* and *Piper capense*. These wild spices have been harvested and traded in many areas of southern Ethiopia. The total supply of spices from the Shekicho-Keficho Zones to the regional and national markets in 1999 was about 1208 million tons (Vivero, 2002).

Bamboo is treated in most literature under NTFPs, although its culms can be put into use as timber or poles. Due to the global literature classification of it as an NTFP, we also treated it here in the same way. Two indigenous species of bamboo, namely the African alpine bamboo or *Arundinaria alpina* and the lowland bamboo or *Oxytenanthera abyssinica*, are recognized in Ethiopia. Ethiopia has one of the largest bamboo resources in the world, with an estimated area of over 1.1 million hectares (150,000 of highland and 959,000 of lowland) (Kelbessa et al, 2000; Nune et al, 2009), though settlements and cropland expansion could have reduced the resources significantly (Kebede, 2006; Andargatchew, 2008). Recently about 119,580 hectares of *Arundinaria alpina* forest has been mapped in the Bale Mountains forest alone (Andargatchew, 2008). If the above estimates are correct, the bamboo resources of Ethiopia are nearly 67 per cent of all African bamboo resources or about 7 per cent of the world total (Embaye, 2000 and 2003).

Bamboo grows with incredible speed and great density per square metre, and its culms mature and are ready for utilization every three to four years. One hectare of the highland bamboo forest is estimated to carry an average of 6000 culms (Embaye, 2003), while that of a lowland bamboo carries an average of 8124 living culms and 4185 dead culms (Kelbessa et al, 2000). Bamboo culms are versatile in application and have numerous industrial and local uses. The present commercial utilization of bamboo in Ethiopia is very low, although efforts to promote its commercial utilization are under way. Local uses are limited to fences, rafts, and vessels for carrying and storing water, water pipes, and splits for baskets, beehives, hats, mats, furniture, flutes, household utensils and agricultural tools.

## Forest-based industries and enterprises

The wood processing enterprises in Ethiopia can be divided into primary and secondary wood processing industries or enterprises. The primary wood industries consist of larger sawmills, predominantly operating today under the Oromia Forest and Wildlife Enterprise (OFWE). Prior to the establishment of OFWE, the state-based Ethiopian Sawmills and Joinery Enterprise (ESJE) was the dominant owner of the primary wood industries in the country. OFWE has recently begun to supersede ESJE, and today manages 28 sawmills, 22 of which are currently operational. However, the majority of the primary wood processing factories are equipped with old machinery and some can be evaluated as obsolete.

Secondary forest enterprises or wood processing industries operating in Ethiopia are diverse and predominantly are in the class of small and medium forest enterprises (SMFEs). They include more than 737 carpenters producing furniture and construction timber in Addis Ababa alone (Nigatu, 2004), one paper factory, and a few newer establishments such as particleboard factories at Awassa and Michew and two new bamboo-based manufacturing industries. Compared to the primary forest industries, secondary wood processing industries are mostly loosely regulated. The sub-sector is characterized by low capacity (less than 50m$^3$ lumber per annum) and is highly influenced by the semi-legal lumber supply. A recent survey of six major towns, excluding Addis Ababa, registered 7415 SMFEs, 30 per cent of them informal (unregistered for tax) (Haile et al, 2009). Large furniture industries are few and include the Finfinne Furniture Factory, Salvatore de Vita & Family and the Wanza Furniture Industry.

Bamboo processing and marketing in Addis Ababa and other major towns has also been proliferating in recent years. 42 privately owned semi-modern and 16 traditional bamboo entrepreneurs' workshops have been inventoried in Addis Ababa alone (Eastern Africa Bamboo Project, 2007). Moreover, two big new bamboo manufacturing companies have recently been established, namely the Land and Sea Development – Ethiopia PLC (LSDE) and Adal Industrial PLC, both near Addis Ababa (Eastern Africa Bamboo Project, 2007).

# SOCIO-ECONOMIC IMPORTANCE

## Contribution to national economy

The contribution of the forestry sector to the national economy through export, import substitution, employment generation and expansion of the gross domestic production is considerable. Forests, woodlands and trees outside forests are the major suppliers of energy and wood-based products for national consumption. The contribution of the forestry sector to the GDP is reported to amount 9 per cent (Nune et al, 2009).

Some of the NTFPs supplied by forests and other wooded vegetation of the country occupy key positions in the state's economy, particularly in foreign currency earning through export. The most renowned of these are wild coffee, gum resins, honey and beeswax, and ecotourism. Coffee is Ethiopia's identity in terms of culture and economy. It contributes about 60 per cent of the country's foreign currency earnings and 10 per cent of the gross domestic product, and it supports the livelihoods of around 20 million people in one way or another (Teketay et al, 1998; Teketay, 1999a). Coffee production in Ethiopia is predominantly the occupation of smallholder farmers, whose production is mostly forest-based. The contribution of forest-based coffee to the national economy is estimated at US$130 million per year (Lemenih, 2008).

Over the last decade average annual gum-resin production has increased in Ethiopia and so has earnings from the sub-sector. The amount of foreign currency earned from export sales is about US$4.1 million per year on average. At the national level fuelwood entrepreneurs receive about US$420 million per year, and honey and beeswax are worth US$86 million per year (Lemenih, 2008). The total value added to the economy from traditional medicine in 2005 was estimated at ETB$^2$2 billion (Mender et al, 2006). About 56,000 tons of medicinal plants are used per annum in Ethiopia, most of which are harvested largely from wild plant stocks. The forestry sector also provides quite a large employment opportunity at the national level, formally and informally. Official estimates put forest employment to be 0.29 per cent of total workforce but this ignores a huge number of self-employed citizens in the sector. For instance, over 35,000 women have been recorded in Addis Ababa alone to engage in fuelwood collection business, with 82 per cent of them fully dependent on the business (WBISSP, 2004).

## Contributions to household economy

Ethiopia is a country with a predominantly rural structure. Over 60 million people reside in rural areas, depending on crop and livestock farming and extraction of various products from forest, tree and other vegetation resources. Forests, woodlands and tree resources contribute to the cash income and subsistence needs of households. This income is the second largest source of non-agricultural income, only after petty trade, for rural households in the country (see, for example, Turnbull, 1999; Jagger and Ponder, 2000; Mekonnen et al, 2007).

More than 300 species of wild trees and shrubs have been recorded as important traditional food sources in Ethiopia, including forest plants (Zemede and Mesfin, 2001; Teketay et al, in press). The majority of these species (ca. 72 per cent) have edible fruits and/or seeds, while in the remaining their vegetative parts – leaves, stems and tubers/roots – are eaten. *Moringa stenopetula* and *M. oleifera*, for example, provide edible as well as nutrient- and vitamin-rich leaves and shoots, which also have medicinal values. *Moringa* is widely used as a source of food to households in the semi-arid regions of southern Ethiopia, particularly in Konso. Fruits of *Cordia africana, Balanites aegyptiaca, Dovyalis abyssinica, Ficus* spp., *Carissa edulis* and *Rosa abyssinica* are commonly consumed in rural Ethiopia. Fruits of *Opuntia ficus-indica* and *Borassus aethiopum* are consumed and traded in the market for cash generation in Tigray and Afar. Similarly, a large number of species of wild animals, including fish, mammals and birds, are utilized for food or as a trophy, and in many cases it provides direct income.

Most of the NTFPs provided by the vegetation resources of the country are harvested by rural communities both for cash income and subsistence. NTFP extraction generates considerable cash for rural households. In the Bale Mountains, various NTFPs provide about half of household annual income (Andargatchew, 2008; Aliyi, 2008). The study of Andargatchew (2008) shows that 47 per cent of annual cash income of households in the Shedem Peasant Association (PA) in Goba district is derived from sale of bamboo. Farmers of the PA provide about 17,000–23,000 bamboo culms each market day to Goba town to earn cash (Andargatchew, 2008). Aliyi (2008) in the same region reports that various NTFPs extracted from the vegetation of the region contribute on average 54 per cent to household total annual income. Similarly, for Goba town alone annual firewood turnover was worth US$887,790 (rate calculation at the time of reporting was US$ ≈ 9.5 Ethiopian Birr), with 70 per cent of the supply being collected by women. In Bench Maji, 52 per cent of annual cash income of households is obtained from NTFPs, while in Sheka it contributes to about 41 per cent of household income (Adilo, 2007). In Gore district 88 per cent of households collect NTFPs and thereby generate 23 per cent of their average annual income of ETB1895 (Debela, 2004). NTFPs also contribute a similar figure of 27.4 per cent to the average annual income of households around the Menagesha Forest (Fetene, 2006). The mean annual income from beekeeping among households in Walmara district was between ETB450 and 3300 (US$47–347) or between 11.6 and 81.9 per cent of total household income depending on the wealth of the households (Lemessa, 2006). Fuelwood, fodder, honey and construction material productions from Chilimo forests contribute significantly to the livelihoods of households in Dendi district, contributing an average of 39 per cent of the annual household income (Mamo et al, 2007).

Furthermore, cash income from tree farming is the main incentive driving tree planting throughout rural Ethiopia (Teklay, 1996; Asnake, 2001; Toru, 2002; Negash, 2002; Mekonnen et al, 2007). Private tree farming is also a growing source of household income in Ethiopia. In areas such as the northern, central and southern highlands, where natural forests have been impoverished, Eucalyptus farming is contributing up to 25 per cent of household cash income (see, for example, Teklay, 1996; Teshome, 2004; Mekonnen et al, 2007). The lucrative cash income from tree farming is driving conversion of farmlands in many of these areas (see, for example, Jenbere, 2009).

# DEFORESTATION AND FOREST DEGRADATION

Data on deforestation and forest degradation in Ethiopia show a staggeringly high rate. However, similar to the problem on vegetation resource statistics discussed above, obtaining reliable statistics on historical accounts of forest cover changes in the country is difficult. For many years the estimate of 40 per cent forest cover of the country around the beginning of the last century and a deforestation rate of 150,000–200,000 ha/yr were almost considered established facts of Ethiopian forestry. Only a few authors (for example Gebre-Egziabher, 1986; Bekele, 1992 and 2003; Pankhurst, 1995; Wøien, 1995; McCann, 1995 and 1999; Ritler, 1997; Nyssen et al, 2004 and 2009) critically questioned these myths. Pankhurst (1995) states that there have been no reliable records of the extent of forest cover of Ethiopia in the past. Some scholars have taken further steps to prove or disprove these claims by employing techniques such as pollen analyses and soil stratigraphy, both of which revealed a much longer history of deforestation and less vegetation cover in the country than often stated. The studies of McCann (1995), Pankhurst (1995) and Ritler (1997) indicate that open grasslands and wooded savanna predominated in the central, northern and southern highlands and that only western parts of the country were dominated by dense hardwood forests during historic times. Based on analysis of notes, reports and diaries of numerous European travellers in

Ethiopia for the period 1699–1865, Bekele (1992) and Ritler (1997) suggested that closed forests of significant size were rare, except in areas unfavourable for cultivation in the central and northern highlands of Ethiopia. Ritler's study also indicates that the use of dried dung mixed with earth and straw for fuelwood dates back prior to the 20th century. Butzer (1981) and McCann (1999) also argue that forest vegetation cover changes due to anthropogenic impacts in northern Ethiopia have a long history. These analyses suggest that the overall vegetation cover declines and land degradation in Ethiopia are rather gradual processes that have occurred over longer periods of time (Nyssen et al, 2004), and possibly pre-date the southward expansion of the Ethiopian empire (McCann, 1995; Hagos et al, 2002). However, accelerated forest cover declines since the middle of the last century, though not as dramatic as often stated, are real (Bekele, 1992 and 2003), as is revealed in several local studies using modern tools such as remote sensing.

Nonetheless, even with the inadequate database, it is very clear that Ethiopia has lost and is losing much of its vegetation covers at an alarming rate (Reusing, 1998; WBISPP, 2004; FAO, 2007). Reusing (1998) indicates a defor-estation rate of 163,600ha/yr between 1986 and 1990, and the report by the FAO (2007) indicates a deforestation rate of 141,000ha/yr or 0.93 per cent loss per year between 1990 and 2000, which has increased during the period between 2000 and 2005 to 1.04 per cent per year. According to this latter report, Ethiopia lost around 2,114,000ha of forest cover in the 15 years between 1990 and 2005.

Local-scale studies, which are much more realistic, confirm the same alarm-ing rate of deforestation. Tadesse (2007) investigated forest cover change in four districts in southwestern rainforest for the period between 1973 and 2005, and found 67 per cent (2.1 per cent per year) forest cover decline. A similar study by Dessie and Kleman (2007) in Awassa watershed revealed 80 per cent ($\approx$ 4.4 per cent per year) forest cover decline between 1972 and 2000, which is a comparable rate with the 4.3 per cent per year decline reported by Seifu (1998) for the Munessa-Shashamane forest. Getaneh (2008) reported a forest decline rate of 1.3 per cent annually for the Yayu forest in the southwest. In southern Wollo, Kebrom and Hedlund (2000) also reported 3 per cent and 14 per cent forest and shrubland cover decline between 1958 and 1986. Near Ambo, bush and woodlands decreased from 42 per cent to 33 per cent between 1957 and 1994 (Van Muysen et al, 1998), while Zeleke and Hurni (2001) reported a 27 per cent natural forest cover disappearance in parts of Gojam since 1957. The report by Reusing (1998) shows annual rates of agricultural clearings for three regions to be 1.16 per cent per year in Oromiya, 2.35 per cent in SNNPRS and 1.28 per cent in Gambella. These deforestation rate figures in Ethiopia are high when compared to the range of annual deforestation rates reported for humid forests in Africa, Latin America and Southeast Asia of between 0.4 and 0.9 per cent per year (Mayaux et al, 2005).

# VEGETATION MANAGEMENT AND CONSERVATION EFFORTS

Institutional instability has never allowed long-term, sustained and successful forest management practices in Ethiopia. Although intermittent, however, there have been a number of efforts to conserve, develop and manage Ethiopian vegetation resources. Some of the efforts are government- and/or non-government-sponsored, while others are local initiatives. Some of the management measures have been directed towards preventing further degradation and deforestation, while others are efforts to restore degraded forest lands. Here we summarize some of the prominent forest management measures that have been practised in the country over the years.

## Plantation forest developments as buffer to natural forests

One of the earliest forest management interventions in Ethiopia were reforestation and afforestation (RA) practices that began over a century ago. The purpose of early plantation establishments was to substitute or augment natural forests for fuelwood and construction material supply. Gradually the plantation developments have taken various forms, such as industrial plantations (established to supply sawn woods and lumber), peri-urban plantations (meant for the supply of poles and firewood to urban centres), catchment protection plantations and farm forests.

Compared to the level of natural forest degradation and deforestation, plantation forest development in Ethiopia has been very slow and very small. The total area planted up to the end of 1973 was 42,300ha, of which 40,000ha were private woodlots and only 2300ha belonged to government plantations (FAO, 1979). The area had increased to more than 85,000ha by 1980 (FAO, 1988) and to about 250,000ha by 1985 (Davidson, 1989; Eldridge et al, 1997; Teklu, 2003). Since then, large-scale plantation development has almost ceased. But after nearly 20 years of pause, planting seems to be gaining momentum through a programme called the Millennium Tree Planting Program. The survival and further development of these recent plantations are yet to be seen.

### Government-owned plantation estates

State-owned plantation forests are either for production or protection. Planted species in the state-owned plantations are diverse and include both exotic and native species. The exotics, however, predominate in terms of area coverage. *Eucalyptus* spp. and *Cupressus lusitanica* cover 59.3 per cent and 20.6 per cent respectively of the planted forest area. These are followed by the indigenous *Juniperus procera* species, which covers 5.7 per cent (Teklu, 2003).

State plantations are mostly established from potted seedlings at a stem density of 1500–2500 stems/ha. Productivity of planted exotics in plantation

39

forests is very high: up to 40–55m$^3$/ha/yr for *Eucalyptus* on a 5–10 years rotation (Örlander, 1986; Kebebew, 2002) and about 30m$^3$/ha/yr for older ages (Pohjonen and Pukkala, 1990). As expected, productivity varies considerably with site quality, even within species. The national-scale average mean annual wood increment (MAI) of 30m$^3$/ha/yr for *E. globulus*, one of the most success-ful species in the country, is considered fair when managed on a short rotation (see, for example, Pohjonen and Pukkala, 1990). For species like *Pinus patula* and *Cupressus lusitanica*, productivity varies between 18 and 25m$^3$/ha/yr depending on site factors, intensity of silvicultural treatments (for example thinning) applied and when managed on a 20–25 years rotation. The total clear-cut yield is usually between 350 and 560m$^3$/ha (Örlander, 1986). *Eucalyptus* is managed mainly by coppicing method, while other exotic and indigenous planta-tions are clear-felled and replanted using seedlings.

## Private (farm household) tree planting

Rural landscapes, particularly close to human settlements, throughout the rural highlands of Ethiopia reflect more green cover than their surroundings due to trees planted by farm households, a practice often termed farm forestry or forest farming. Such practices involve tree plantings in rows, in patches as woodlots, or scattered on farmlands, farm boundaries, pasturelands or other open areas nearby homes and farms. Farm forestry practice has two major objec-tives: the first is to satisfy subsistent wood demand of households and the second is to generate cash and augment household income (see, for example, Teklay, 1996; Seifu, 2002; Asnake, 2001; Teshome, 2004). In some cases, farmers are planting trees for ornamental purpose and/or as a symbol of status (Achalu, 1995; Toru, 2002; Jenbere, 2009).

Tree farming is expanding in Ethiopia and is amongst the largest sources of non-agricultural income for rural households in the country (see, for example, Turnbull, 1999; Jagger and Ponder, 2000; Mekonnen et al, 2007). Studies made on over 450 households all over the country show that farmers practise tree farming, and that the practice is expanding at a high rate (Figure 2.2). Although retaining some indigenous tree species is a common compo-nent of farmers' tree cultivation tradition, in terms of density and area coverage, fast growing exotic species, principally *Eucalyptus* predominate (Teklay, 1997; Tolera et al, 2008). Farmers' choice for *Eucalyptus* is dictated by its several merits such as rapid growth and wood production, coppicing ability, established wood market with good price, easy management and unpalatability to animals (Abebe, 2005; Jagger and Pender, 2000; Mekonnen et al, 2007). Even in Sidama and Gedio, where traditional agroforestry is an age-old and advanced practice, *Eucalyptus* predominates in terms of stand density and wood volume. For the Sidama traditional agroforestry systems, Abebe (2005) showed that *E. camaldulensis* alone accounts for up to 61 per cent of the tree population, while 116 other tree species altogether repre-sented only 39 per cent of the stem density. Another study reported that the share of *Eucalyptus* in tree stocks of a household can be as high as 98 per cent

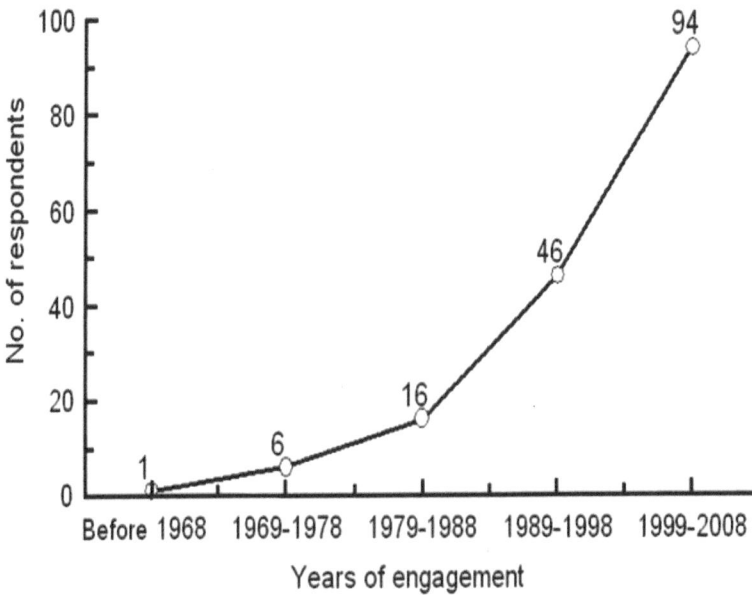

**Figure 2.2** *Expansion of eucalypt planting as farm forest in Arsi Negele district, Central Ethiopia*

*Source:* Jenbere (2009)

(Negash, 2002). When established as woodlot, farmers manage the species usually on a short rotation basis of 5 to 10 years, and plant it at an extraordinarily high density of 10,000 to 25,000 trees per hectare (Abebe, 2005; Kebebew, 2002; Mekonnen et al, 2007). This is four to seven times denser than the stem density used in state industrial plantations. In some case a density of up to 40,000 stems/ha can be found.

WBISPP (1995) estimated that there are 51 million on-farm trees in Ethiopia, predominantly *Eucalyptus* spp., with an estimated yield of 15m³/ha/yr. In some areas farm forests are probably the only forest estates found. For instance, in the Gurage highlands, an area characterized by widespread tree planting by farm households, small community woodlots and private farm forests represent the largest forest resources (Achalu, 1995). Financial evaluation of farm woodlots in various parts of the country show a highly lucrative return from eucalypt farming as compared to diverse agricultural crops including some of the cash crops such as coffee and chat, which is the most important incentive to farmers to engage in tree farming (Achalu, 2005; Negash, 2002).

Besides providing a significant contribution to household income, farm forests also play a considerable role in wood products supply in Ethiopia. In the Arsi highland, central Ethiopia, *Eucalyptus* wood grown by farmers supplies 86 per cent of firewood, 31 per cent of charcoal, 100 per cent of leaves and twigs for firewood, and 100 per cent of poles used by urban dwellers. Similarly, in rural areas the wood from eucalyptus contribute 92 per cent of poles, 74 per cent of timber, 85 per cent of firewood, 40 per cent of charcoal, 83 per cent of posts and 91 per cent of farm implements. On town market days on average 74 per cent of the firewood, 100 per cent of the poles, 100 per cent of the posts and 21 per cent of the charcoal is *Eucalyptus* grown on farm forests. In terms of cash income it contributes up to 5 per cent of household annual income for the relatively wealthy, 20 per cent for the medium and 72 per cent for the poor households, and on aggregate the income from sales of *Eucalyptus* products is second only after crop farming, contributing 28 per cent of the total annual household cash income (Mekonnen et al, 2007). Similar observations were reported for Tigray (Teklay, 1996).

## *Plantations as foster ecosystems*

Recent studies increasingly acknowledge plantation forests in Ethiopia for their positive ecological effects such as their catalysing and fostering role in rehabilitation of degraded forests and their ecosystems (see, for example, Woldu, 1999; Senbeta and Teketay, 2001; Senbeta et al, 2002; Yirdaw, 2002; Lemenih et al, 2004; Lemenih and Teketay, 2005). The study by Woldu (1999) in Entoto, near Addis Ababa, shows that regeneration of *Juniperus procera* under *Eucalyptus globulus* is 1750 stems/ha, 44 per cent of which is seedling, 13 per cent sapling and 43 per cent pole size. The overstorey *Eucalyptus* stand had a density of 2000 stumps/ha but with 7100 coppice stems/ha. Similar encouraging observations have been reported throughout Ethiopia (see Yirdaw, 2002; Alem and Woldemariam, 2009). This is an observation in line with a large body of studies from various geographical regions such as Africa, Asia, the Caribbean, Australia and Latin America (see overview in Lemenih and Bongers, Chapter 9 of this volume). This way, plantations are also contributing to the conservation of indigenous plant species against local extinction. Additionally, plantation development is offering high employment, both in plantation management and in product processing industries.

## Area exclosure for forest regeneration

Several area enclosure projects have already been launched in the various regional states of Ethiopia by both government organizations and NGOs. Area exclosure refers to the practice of land management that involves the exclusion of livestock and humans from open access to an area that is characterized by severe degradation. The purposes are to prevent further degradation of ecosystems and to advance restoration of the overall ecological conditions of the area. Although much of restoration/rehabilitation in area exclosure is natural regeneration, in some cases exclosure management in Ethiopia involves enrichment plantings of native and/or exotic species as well as soil and water conservation activities as supplementary rehabilitation efforts to foster the restoration processes. The technique is employed in wide forest ecosystems from the very dry woodlands to the sub-humid Afromontane forest ranges. In the Tigray region alone a total of 400,629 hectares or 8 per cent of the regional territory has been exclosed so far (Nyssen et al, 2004). Likewise, in Amhara and Oromia and southern regional states large areas of degraded lands have been set aside (Nedessa et al, 2005).

Studies show that area exclosure rehabilitates vegetation, restores wildlife populations and improves soil properties of degraded lands (see, for example, Mengistu et al, 2005; Yami, 2005; Hailu et al, 2006). From a replicated (across three sites) study, Hailu et al (2006) recorded 16 woody species per hectare in exclosure as compared to only 9 in open grazed land nearby. Mengistu et al (2005) also recorded woody plant density of 3705 stems/ha in exclosures compared to 3048 stems/ha in non-exclosed plots. Assessments in a chronosequence of closely located exclosures show increasing trends in species richness and density of plants and animals with age of exclosure. Area exclosures also improve soil chemical and physical properties, increase litter production and reduce soil erosion (see, for example, Mekuria et al, 2007; Descheemaekers et al, 2006), thus leading to ecosystem restoration. However, mismanagement, such as lack of a responsible managing body or loosely defined ownership over exclosed areas, has also led several area exclosure initiatives to fail, particularly in the central and southern parts of the country (Nedessa et al, 2005).

## Protected areas and forest priority areas for conservation

Another effort used for forest conservation in Ethiopia is the protected area approach, where most of the remnant natural forests of the country were designated as National Forest Priority Areas (NFPAs, Figure 2.1). A number of forest patches and blocks totalling an area of 2.8 million hectares were demarcated as NFPAs, with a principal objective of protection and conservation of their biodiversity (Table 2.6). The NFPAs comprise most of the major remnant forest patches and some plantations. The NFPAs covered an area of 4.8 million hectares and comprised natural forests, plantation forests and non-forested lands, which was planned to be used for expanding the forests. At that time, 13 of the NFPAs were accessible for commercial exploitation, while the remaining forests, covering 60 per cent of the total area, were envisaged for rehabilitation, protection and genetic conservation. Of the 58 NFPAs, 48 had been demarcated, 5 had been inventoried and 4 had management plans. Since none of them were gazetted, they have been exposed to uncontrolled and unsustainable exploitation (Teketay, 1999b). Since the designation of NFPAs, 24 more high forests have been identified, making a total of 82 important forest areas (Reusing, 1998). Following the regionalization of the country and the consequent devolution of power, the responsibility of NFPAs, which also became known as Regional Forest Priority Areas (RFPAs), was also transferred to the National Regional States. All of these forest priority areas have been under extreme pressure from settlement, land-use conversion to farming and grazing, excessive extraction, and neglect in terms of forest management and protection.

**Table 2.6** *National Forest Priority Areas with their sizes by National Regional States/Administrative City Councils*

| National Forest Priority Area* | Number | Area (ha) | | |
| | | Slightly disturbed | Heavily disturbed | Total |
| --- | --- | --- | --- | --- |
| Amhara National Regional State (ANRS) | 4 | - | 39,100 | 39,100 |
| Gambella National Regional State (GNRS) | 2 | 442,350 | 45,000 | 487,350 |
| Oromiya National Regional State (ONRS) | 31 | 288,200 | 720,200 | 1,008,400 |
| SNNP** Regional State (SNNPRS) | 5 | 132,000 | 243,000 | 375,000 |
| Tigray National Regional State (TNRS) | 1 | - | 11,500 | 11,500 |
| Dire Dawa City Council and ONRS | 1 | 1500 | - | 1500 |
| GNRS & SNNPRS | 1 | 80,000 | 35,000 | 115,000 |
| ONRS & SNNPRS | 4 | 407,900 | 165,000 | 572,900 |
| GNRS, ONRS & SNNPRS | 1 | 40,000 | 100,000 | 140,000 |
| Total | 50 | 1,391,950 | 1,358,800 | 2,750,750 |

*Note:* *Only NFPAs that contain natural forest areas were considered; **SNNP = Southern Nations, Nationalities and Peoples.
*Source:* EFAP (1994) and Reusing (1998)

Some lowland forests, woodlands and biodiversity-rich Afroalpine ecosystems have also been demarcated in the forms of national parks and wildlife sanctuaries (for example Awash, Nechi Sar, Mago, Abijata–Shalla, Yabello, Sinkile and Babile), which also forms part of the protected area management network in Ethiopia. The Wildlife Conservation Areas (WLCAs) include nine national parks, four sanctuaries, eight wildlife reserves and 18 controlled hunting areas (Anonymous, 2005). WLCAs cover a total of about 188,710km$^2$ (16.7 per cent of the country) (Hillman, 1993). Unfortunately, the current status of these protected areas, including the parks, is discouraging. They are suffering intense human pressure including encroachment and settlement within parks (Anonymous, 2005).

## Participatory forest management (PFM)

Participatory forest managements (PFM) was introduced to Ethiopia nearly two decades ago as a solution to the problem of forest resources open access and to promote sustainable forest management through community participation. PFM was introduced by NGOs, notably by FARM Africa, SOS Sahel, GTZ (German Technical Cooperation) and JICA (Japan International Cooperation Agency). These non-state actors attempted to respond to the prevailing forest management problems of Ethiopia through the introduction, adaptation and establishment of PFM projects, with the ultimate aim of mainstreaming it as one model of a forest management system in the country.

PFM pilot activities that started in Ethiopia include projects at Chilimo, Bonga, Borana and Bale Eco-Region run jointly by FARM Africa and SOS Sahel, the one at Adaba-Dodolla run by GTZ, and the project at Belete Gera run by JICA (Temesgen et al, 2007; Tolossa, 2002). Currently about 640,857 hectares

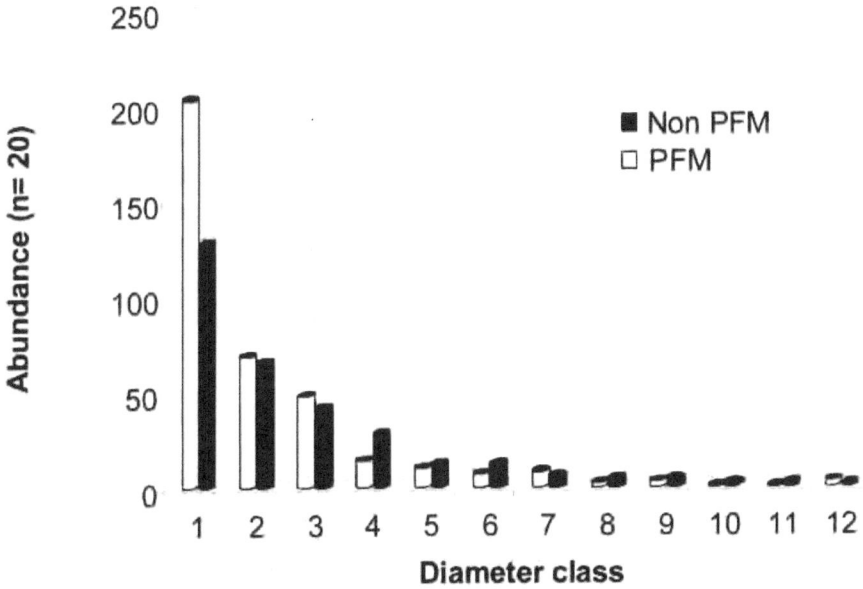

**Figure 2.3** *Diameter class distributions for naturally regenerating woody species in forests under PFM and non-PFM in Bonga, southwest Ethiopia*

Diameter class: 1= seedling, 2 = sapling, 3 = 10–19.5cm, 4 = 20–29.9cm, 5 = 30–39.5cm, 6 = 40–49.9cm, 7 = 50–59 9cm, 8 = 60–69.9cm, 9 = 70–79.9cm, 10 = 80–89.9cm, 11 = 90–99.9cm and 12 = >100cm.
*Source:* Gobeze et al (2009)

of forest lands in the country is managed under PFM schemes. These pilot PFM projects attempted to introduce:

- devolution of certain bundles of property rights from the state to the community;
- allowing local people to manage the forest resources sustainably; and
- partial utilization of the forest resources for livelihood support.

Experiences gathered from the existing PFM programmes demonstrate good successes in enhancing forest regeneration, improving forest protection, regulating access, and thus good forest conservation outcomes, at least during the project period (Gobeze, 2008). Gobeze et al (2009) compared forests under PFM with an adjacent non-PFM forest site in Bonga and reported an increment of woody species diversity and seedling and sapling densities in the PFM forests. Similarly, the population structure of the forests exhibited better structures and healthier populations compared with the non-PFM block (Figure 2.3). This seems to have been achieved because of the regulated access and the development work communities did in the forests. These findings on the forest resources following introduction of PFM conforms to several similar studies in other countries. Another study from Adaba-Dodola PFM project site also recorded more healthy seedlings that had been neither trampled nor browsed, and more regeneration of indigenous plant species in the PFM forests (Bekele et al, 2004).

These experiences have influenced recent policymaking processes in the forestry sector of Ethiopia. The fact that the new forestry policy (*Negarit Gazeta*, 2007) and the proclamation that followed it embraced community participation is assumed to reflect the influence of these PFM projects. New forestry institutions established in Ethiopia, such as the OFWE, have also mainstreamed participatory approaches as their strategy in forest management, which is attributed to the PFM projects outlined above. However, an important aspect of the PFM scheme in Ethiopia is the need to monitor for its post-project sustainability. Most of the information available about PFM today is that collected during project lifetime, which definitely will give a biased impression of its success. Furthermore, the diversity of PFM implementation modalities by the different actors, sometimes within the same region, means it is difficult to reach consensus on the most effective approaches in implementation. It is therefore worth assessing the forest and socio-economic impacts of the approaches in a comparative manner.

## Traditional forest management and conservation practices

Substantial forest and woodland resources are being managed effectively by local communities that exercise diverse traditional management practices either communally or privately. Moreover, over the last two decades the involvement of civil societies in forest management has been increasing, with commendable multidimensional successes such as policy lobbying and introducing and testing new community-based forest management schemes.

### Communal forest management

In many places, local people have developed a relatively advanced indigenous knowledge and well-organized indigenous institutions to manage their forest resources. Some of these traditional community-based forest management (TCBFM) systems involve communal efforts such as the *Acacia-Commiphora* woodlands managed as rangelands by the Borana people with the Gada institution (Watson, 2003) and the management of Afromontane forests in the southwest for non-timber forest product extraction by the Kobo system (Tadesse and Woldemariam, 2007). Others are smaller and private efforts involving various forms of traditional agroforestry. The Kobo system is a forest (tree) tenure institution that grants first claimers an exclusive use right to a block of forest, usually for collection of forest coffee, hanging beehives and other NTFPs. Once claimed, the forest block is de facto individual property, respected by fellow citizens of the area, and the owner has the right to exclude others. This way, the system has resolved what could have been an open-access

**Figure 2.4** *Eucalyptus in the Sidama traditional agroforestry system as one of the major tree components*

*Source:* M. Lemenih

system (see, for example, Tadesse and Woldemariam, 2007). The Borana Gada system imbeds different hierarchical rangeland management institutions within it. The most important part of the rangeland management institution is the obligation for animal movement according to the patterns outlined by elders based on range availability, rangeland condition and seasonal carrying capacity of the Borana plateau to avoid degradation (Watson, 2003). In this way the institution has managed the rangeland for generations.

## Traditional agroforestry systems

The other common form of TCBFM in Ethiopia is the traditional agroforestry systems (TAS). Diverse TAS are practised with extensive area coverage (Hoekstra et al, 1990). A typical example is the home garden agroforestry system in the drylands of south and southwest Ethiopia. In this region home gardens and other forms of TAS are estimated to cover 576,000 hectares (Abebe, 2005). Most of the home gardens evolved from forests, where farmers maintain the upper storey trees and clear the understorey vegetation to open up space for planting enset, coffee, and other food and cash crops. Abebe (2005), for instance, found 120 tree and shrub species in 144 home gardens, with 83 per cent indigenous species, an average number of tree species per farm of 20.7, an average wood standing volume of 50.4m$^3$ per farm (24m$^3$/ha), and an average density of 855 trees/farm (see Figure 2.4 for the example of Sidama TAS). Stem density in the TAS varies from area to area and according to the size of land holding. For instance in the TAS of the Sidama, stem density ranged from 13 to 64 per hectare, with species richness from 3 to 35 per hectare. Moreover, parkland agroforestry is almost the rule throughout the country (see, for example, Tolera et al, 2008). Households incorporate woody plants in these agroforestry systems mostly by preserving selected tree species in the processes of transforming forest lands into agriculture. Farmers also maintain diversity and density of woody plants in their TAS through enrichment planting, using indigenous as well as exotic species, with *Eucalyptus* predominating (Figure 2.4). Some of the TAS even host higher diversity of woody species than their nearby natural woodlands or forest lands, thus providing safe havens for conservation of diverse native plant species (Tolera et al, 2008).

## Church and sacred forests

Significant forest patches are conserved and managed as sacred groves in and around churches, monasteries, graveyards, mosque compounds and other sacred sites in several parts of Ethiopia. Particularly, the northern highlands are almost devoid of forests in other areas as these have been converted into farms and grazing lands, leaving few patchy remnants confined mainly around churches (see, for example, Wassie et al, 2005 and 2009; Aerts et al, 2006, Bongers et al, 2006; Wassie, 2007, also Chapter 6 of this volume). For instance, Wassie et al (2009), in their study of 28 Orthodox churches in northern Ethiopia found a total of 500.8 hectares of remnant forests around them (average of 17.9 ha/church) and recorded 160 indigenous and 8 exotic woody species (100 tree species, 51 shrubs and 17 lianas). The total number of species per church ranges from 15 to 78. The species composition of these church forests are old growth type predominated by *Juniperus procera*, *Olea europaea* subsp. *cuspidata* and *Celtis africana*. According to Wassie (2007) there are 35,000 similar churches throughout Ethiopia that are likely to contribute to the conservation of considerable areas of remnant dry forests in the country. These forests are not only remnants of old-growth vegetations but also provide diverse forest products and services, and may act as sources of genetic materials for restoration of degraded dry Afromontane forests. Linked through appropriate vegetation corridors they may form a unique landscape matrix for large-scale landscape restoration. Recent studies on management interventions (for example seed sowing, seedling planting, soil scarification and exclosing) in and around these forests (Wassie et al, 2009) show promising results in this respect.

# FOREST POLICY AND LEGISLATION

Ethiopia is yet to have a consolidated and formal forest policy. Current and past policies on forest resources can only be inferred from scattered instruments, such as institutional mandates, five-year or ten-year plans, national action programmes, and bodies of legislation, all of which constitute elements of the policy system. For instance, legislation that provides for development, conservation and sustainable utilization of forest resources in Ethiopia has been slowly and ineffectively developed. Formal forest legislation in Ethiopia dates back to the 1960s. In 1965 a proclamation was promulgated for each of private, state and protection forests. Following the 1974 revolution and the change of government, some basic policy changes necessitated change in forest law, and Proclamation 192/1980 was enacted. Since instalment in 1994, Proclamation

No. 94/1994 has been serving as the forest policy statement of the country. The principles and objectives enshrined in the 1994 law include forest development and protection for its service and economic role. It has also introduced the principle of benefit-sharing with the local people in forest management and public participation. Moreover, emphasis was put on sectoral coordination with the Ministry of Agriculture and other related sectors. This law recognized three types of forest ownership, namely federal forest, regional state forest and private forest. The last, however, needs to be plantation forest as under the constitution all natural resources including natural forest are public property.

The new legislation, Proclamation 542/2007, by and large maintains the broad policy direction of its predecessor. It goes a step further, however, in its people-centred approach by requiring communities to have a say in state forest management and by legalizing community access to forests for non-timber forest products. While it outlaws new settlements in state forests, the proclamation requires that community welfare be given priority in issues of evictions. Principles of scientific management and multiple use have been retained. An interesting departure, though, is the new provision for private (as individual or organization) ownership within existing natural forests and the strengthening of forest tenure. To its credit, the legislation emphasizes strong extension support, provision of germplasm, value adding and market support.

Like all past forest legislation, the current proclamation leaves a huge institutional void and suggests the status quo of an uneasy marriage between the forest and agriculture sectors continues for forest management, extension and regulatory functions. It makes no mention whatsoever of institutions for service functions (in other words research and education). The roles of federal bodies vis-à-vis individual states continue to be fuzzy. The forest industry sector and issues of trade and export have been omitted.

The new proclamation does appear to recognize the need for better forest governance and law enforcement. The two regulatory instruments employed in the legislation are business licensing and forest product movement permits. Penalties for forest offences have been raised to a maximum penalty of ETB30,000 (about US$300) and/or five years' imprisonment, depending on the type and degree of offence. Uniformed forest guards and forest products movement inspectors are to be the primary enforcers of the legislation, though custom officers and the regular police can enforce in their absence. A reward system is also suggested for informers against offenders. However, the bottlenecks in enforcement in the past were corruption, smallness of penalties, congestion of the legal system and lack of political support. It is not clear whether the new legislation has put in place measures to address these challenges.

The state of affairs in the forest sector in Ethiopia as a whole is precarious. The forest sector is considered as residual land use and given less priority, resulting in inadequate allocation of organizational, financial, human, facility and other resources. The existing production capacity of the remaining forests is small, while the demand for forest output is large with the gap growing by the day. Moreover, the public as a whole are not involved, and the mechanisms to widen the domain of popular participation are not yet on the horizon. As a result, there is a general antagonism between the ecological or sustainability concerns and economic concerns, as conservation is perceived to restrict free access to forest resources.

In-migration coupled with the population growth of original forest settlers has resulted in further land-use conversion from forest to settlement and farming. While this built up over the years, the public forest administration was progressively reduced from department to team of experts and now even less within the Ministry of Agriculture and Rural Development, where it remains paralysed from lack of financial and human resources in addition to inadequate policy, legislation and institutional backing.

Other than the forest Proclamation 542/2007, a number of policies, proclamations and strategies that directly relate to the forest sector have also been issued during the last 18 years in Ethiopia. These include:

- Ethiopian Forestry Action Plan (EFAP, 1994);
- Conservation Strategy of Ethiopia (Anonymous, 1997b);
- Environmental Policy of Ethiopia (Anonymous, 1997a);
- Forest Management, Development and Utilization Policy (Anonymous, 2007);
- Rural Land Administration and Land Use Proclamation No. 456/2005 (RLALUP, 2005);
- Environmental Impact Assessment Proclamation (Anonymous, 2002); and
- Plan for Accelerated and Sustainable Development to End Poverty (PASDEP, 2005).

The Environmental Policy of Ethiopia issued in 1997 has a strong element of encouraging people participation in forest management. Under the forest, woodland and tree resources sector, the policy addresses the complementary roles of communities, private entrepreneurs and the state in forestry development, integration of forestry development with land, water resources, energy resources, and ecosystem and genetic resources development in addition to crop and livestock production. Selection of suitable species for afforestation/reforestation, with particular emphasis on indigenous tree species, is one important statement included in this policy. The policy emphasizes that utilization of forests should be based on the regenerative capacity of the forest. Hence forest management that accounts for a sustainable supply without affecting environmental and social amenities derived from the forests is needed. The policy states that such sustainability is attained by formulating and implementing socially suitable, environmentally sound and economically acceptable management plans. Since free grazing affects natural regeneration of valuable indigenous trees, the policy restricts free grazing in protected forest areas.

Rural development policies, strategies and methods and the Rural Land Administration and Land Use Proclamation issued in 2002 clearly stated the need for proper land use in order to maximize the economic return of the land.

In line with this policy, Proclamation No.456/2005, 'Federal Democratic Republic of Ethiopia Rural Land Administration and Land Use', was issued. In this proclamation, rural lands with a slope of over 60 per cent shall not be used for farming and free grazing; they shall be used for development of trees, perennial plants and forage production. In PASDEP (2005), it was also planned to increase the forest cover of the country to 9 per cent between 2005 and 2010.

However, like the forest policy and its proclamation, most of the aforementioned policies suffer from weak institutional/organizational set-up and for most of them, except the written documents, there is no clearly defined implementing body. This has now forced scholars to state that the problem of the forestry sector in Ethiopia is not the lack of policies but the institutional arrangement.

A new major development in the forestry organizational arrangement since 1991 has been the decentralization of political administration and the devolution of forestry and natural resources administration. Forestry administration and management responsibility has been handed over to the Regional States, which have also produced their own regional forestry policies, strategies and organizational set-ups. Regional States appear to have been showing renewed interest in forestry sector development, although with high inter-regional variations. For instance, in Tigray Region area exclosure has been practised widely to rehabilitate degraded forest lands. In Oromia a radically new forestry institution has been established, called the Oromia Forestry and Wildlife Enterprise (OFWE), with a mandate to contribute to the conservation, management and production of the forest and wildlife resources of the region. Large-scale illegal forest extraction, political instability, lack of clearly defined property right (tree and land reforestation and afforestation programmes) are also ongoing in Amhara Regional State as part of the reform. The expanding forest product market, improving land/tree tenure arrangements, and other socio-economic and political changes are more or less driving the rapid expansion of farm forestry by private rural households at least. Today, farm forestry is supplying the bulk of wood products, principally poles and posts, consumed in Ethiopia.

## Challenges of the forest sector

A number of challenges beset the forestry sector of Ethiopia. Extensive forest degradation and deforestation is one. The driving forces for the extensive forest degradation are many and can be categorized into direct and proximate causes. The direct causes are forest clearance for crop and settlement, overgrazing, fire, and unsustainable harvest for timber and non-timber uses. The underlying causes are mostly population growth beyond the carrying capacity of the national economy, widespread poverty and absence or instability of forestry sector organization. In addition, the occurrence of intermittent but serious drought periods affects natural and plantation forests. Moreover, high malaria and tsetse infestations in the lowlands lead to burning of forests and woodlands and constrains tree-planting practices. Civil war and unrest in the country have led to widespread destruction of forest resources, creating insecurity for planting trees. Especially the 1974 and 1991 changes of government have resulted in vandalizing forests on a large scale (Teketay, 2004).

## Suggested key policy issues for improved forest management

Improving the state of forest resources and their sustained provision of goods and services demands a lot of work. The nature of interventions needed range from technical actions in the field to political actions such as institution reforms and law enforcement. The technical options for sustainable forest product and services flow may include establishment of alternative forms of forest management including plantation development, whether of 'forest' trees or of fruit or other tropical commercial species (coffee, *Rhamnus prinoides* L'Hér., citrus fruits and so on), agroforestry and trees on farms (with or without livestock), and grazing land improvement.

Ethiopia needs to place some priority on broadening the product base of forest resources beyond traditional timber products. The nature of the products may vary from locality to locality. Much needs to be learnt about the supply and the demand for such products, and their importance for family nutrition, food security, livestock, energy, commercial sale or processing. Issues on protection of rights for products or processes also arise. The role of the forest sector in rural employment and income also involves more general consideration of the potential of trees in land uses, including plantation crops, agroforestry systems and homestead lots. In some cases, the value of such products exceeds that from the natural forests. However, there are also areas of uncertainty, including the nature and scope of market demand and potential for commercialization of such products, which cover a wide range of wood, food, fuel, medicinal, fodder and other end uses, including service benefits such as shelter, windbreaks, nitrogen fixation and soil conservation, which are difficult to assess and value.

A clear distinction must be made between encroachment on forest resources by migrants from other areas and traditional cultivation and movements of people within forested areas. Migration to forested areas is generated by the external pressures of rapid population growth, poverty and lack of employment opportunities in other areas and is, therefore, an inter-sectoral issue. The movements of existing people within the forests are related to unsustainable shifting cultivation and, sometimes, pressures to move exerted by professionals in the forest sector or settlement policies. Related to this are issues of ownership, tenure, access and usufruct, especially concerning common property resources. In some regions such as Tigray, measures to transfer greater responsibility for forest management and afforestation from government agencies to local authorities, communities and individual stakeholders are already in hand, although much remains to be learnt about the risks, benefits and successful mechanisms for implementing them. A further extension of such measures is distribution of forest land to households, or alternatively rights to plant trees on land leased from the state, which is being exercised with respect to plantations in some regions. While there are elements of doubt concerning the risks of misuse and degradation of land once it is out of the direct control of the Bureaus of Agriculture, some foresters see the need to move towards community and private forestry as possible means of alleviating the free-for-all problems of common access, and also to relieve some of the pressures on the limited resources of these bureaus. The latter emerges as a constraint to efficient professional forest management in Ethiopia, because of a shortage of trained staff and advisory roles to play in oversight, research, supervision of nursery stock production and hygiene standards. Some traditional service tasks might be decentralized and some services privatized, such as nursery management.

The frequent restructuring and reshuffling of public forest administration is seen to be an important constraint to the development of focused, independent, more flexible and outward-looking forestry practices. Further, the fact that the forest sector is sheltered in agriculture has resulted in the latter assimilating the former. Isolation of senior forestry professionals from national policymaking is also a matter of concern, leading to low priority for the forest sector in national plans and financial allocations. Furthermore, the apportioning of responsibilities of the forest sector among different institutions, including the Ministry of Agriculture and Rural Development, the Environmental Protection Authority and the Ministry of Education is multiplying the problems besetting the sector.

# CONCLUSIONS

Owing to its unique physical conditions and variations in altitude, which have resulted in a great diversity of climate and soils, Ethiopia is endowed with diverse vegetation types, including different types of forest resources. The forest resources provide timber and wood, which are used for construction, manufacturing various wood products and fuel, as well as numerous non-timber forest products, for example frankincense, myrrh, gum arabic, honey and beeswax, spices and condiments, bamboos, civet musk, and medicine. Unfortunately, knowledge on growth rates, rotation age, biology and ecological requirements of most of the indigenous species has been deficient and hence most of the forests do not have sound management plans to guide the actual operations on the ground. Therefore the harvesting procedure employed has been dominantly selective cutting of quality and large trees of economic importance, leading to not only depletion of the populations but possibly also genetic erosion of the timber species. Moreover, systematic care to minimize damages to residual stock has been hardly considered and standard silvicultural procedures for forest renewal or regeneration following harvesting operations have been either inadequate or lacking. These gaps also provide interesting research opportunities for national and international researchers.

Furthermore, forests and other vegetation resources have been either shrinking in size or heavily degraded at an alarming rate over centuries as a result of not only the lack of or inadequacy of silvicultural and management systems but also several socio-economic and policy-related, anthropogenic and natural factors. The consequences of deforestation and forest degradation have been a chain of undesirable events involving land degradation (soil erosion and loss of soil fertility), decline, degradation and/or loss of water bodies and biodiversity, as well as a contribution to enhanced global warming. These, in turn, affect other sectors, for example agriculture, and exert serious impacts on the welfare of people, animals and micro-organisms.

The sustainable management and conservation of forest resources in Ethiopia require a stable institutional set-up and political recognition of the socio-economic and ecological significances. In the absence of these, small-scale farm household-based tree-planting practices will continue to play a dominant role in the forest development direction of the country.

# NOTES

1   Plant nomenclature in this article follows those of Friis (1992), Hedberg and Edwards (1989 and 1995), Edwards et al (1995, 1997 and 2000) and Hedberg et al (2003, 2004 and 2006).
2   ETB denotes Ethiopian Birr, the national currency of Ethiopia.

# REFERENCES

Abate, A. (2004) *Biomass and Nutrient Studies of Selected Tree Species of Natural and Plantation Forests: Implications for a Sustainable Management of the Munessa-Shashemene Forest, Ethiopia*, PhD thesis, Universidad Bayreuth, Bayreuth, Germany

Abebe, T. (2005) 'Diversity in homegarden agroforestry systems of southern Ethiopia', Tropical Resource Management Papers, no 59, Wageningen University, The Netherlands

Achalu, N. (1995) *A Monographic Review on* Juniperus Excelsa, Alemaya University of Agriculture, Alemaya, Ethiopia

Adilo, M. (2007) *The Contribution of Non-Timber Forest Products to Rural Livelihoods in Southwest Ethiopia*, MSc thesis, Wageningen University and Research Center, The Netherlands.

Aerts, R., Van Overtveld, K., Haile, M., Hermy, M., Deckers, J. and Muys, B. (2006) 'Species composition and diversity of small Afromontane forest fragments in northern Ethiopia', *Plant Ecology*, vol 187, pp127–142

Alem, S. and Woldemariam, T. (2009) 'A comparative assessment on regeneration status of indigenous woody plants in *Eucalyptus grandis* plantation and adjacent natural forests', *Journal of Forestry Research*, vol 20, pp31–36

Aliyi, N. (2008) *An Analysis of Socio-Economic Importance of Non-Timber Forest Products for Rural Households: A Case Study of Bale Mountain National Park*, MSc thesis, Copenhagen University, Copenhagen

Ameha, A. (2002) 'Sustainable supply of wood resources from Adaba-Dodola Forest Priority Area', paper presented at the Alumni Seminar, Addis Ababa University, 14–15 December, Addis Ababa

Andargatchew, A. (2008) *Value Chain Analysis for Bamboo Originating from Shedem Kebele, Bale Zone*, MBA thesis, Addis Ababa University, Addis Ababa

Angassu, A. (2007) *The Dynamics of Savanna Ecosystems and Management in Borana, Southern Ethiopia*, PhD dissertation, Addis Ababa University, Addis Ababa, Ethiopia

Anonymous (1997a) *Environmental Policy of Ethiopia*, Environmental Protection Authority, Addis Ababa

Anonymous (1997b) *The Conservation Strategy of Ethiopia, The Resources Base: Its Utilisation and Planning for Sustainability, Vol I*, Environmental Protection Authority in collaboration with Ministry of Economic Development and Cooperation, Addis Ababa

Anonymous (2002) *Environmental Impact Assessment Proclamation*, Environmental Protection Authority, Addis Ababa

Anonymous (2005) *National Biodiversity Strategy and Action Plan*, Institute of Biodiversity Conservation, Addis Ababa

Anonymous (2007) *Forest Management, Development and Utilization Policy*, MoARD, Addis Ababa

Asfaw, S. (2006) *Effects of Fire and Livestock Grazing on Woody Species Composition, Structure, Soil Seed Banks and Soil Carbon in Woodlands of North Western Ethiopia*, MSc thesis, University of Natural Resources and Applied Life Sciences, Vienna

Asnake, A. (2001) *Yield and Economics of Growing* Eucalyptus camaldulensis *by Smallholder Farmers of Amhara Region: The Case of Gondar Zuria District, North Gondar, Ethiopia*, MSc thesis, ISSN 1402-201X (2002:57), Swedish University of Agricultural Sciences, Uppsala, Sweden

Bekele, M. (1992) *Forest History of Ethiopia from Early Times to 1974*, MSc thesis, School of Agricultural and Forest Sciences, University College of North Wales, Bangor, UK

Bekele, M. (2003) *Forest Property Rights, the Role of the State, and Institutional Exigency: The Ethiopian Experience*, Doctoral dissertation, Swedish University of Agricultural Sciences, Uppsala, Sweden

Bekele, T., Senbeta, F. and Amaha, A. (2004) 'Impact of participatory forest management practice in Adaba-Dodola Forest Priority Area of Oromia, Ethiopia', *Ethiopian Journal of Natural Resources*, vol 6, pp89–109

Bongers, F., Wassie, A., Sterck, F. J., Bekele, T. and Teketay, D. (2006) 'Ecological restoration and church forests in northern Ethiopia', *Journal of the Drylands*, vol 1, pp35–44

Butzer, K. (1981) 'Pleistocene history of the Nile Valley in Egypt and Lower Nubia', in M. Williams and H. Faure (eds) *The Sahara and the Nile*, Balkema, Rotterdam, The Netherlands, pp238–252

Chaffey, D. (1979) *South-west Ethiopia Forest Inventory Project: A Reconnaissance Inventory of Forest in South-west Ethiopia*, Land Resources Development Centre, London

CSA (2000) *Statistical Abstracts*, Central Statistical Agency, Addis Ababa

CSA (2008) *Statistical Abstracts*, Central Statistical Agency, Addis Ababa

Dalle, G. (2004) *Vegetation Ecology, Rangeland Condition and Forage Resources Evaluation in the Borana Lowlands, Southern Oromia, Ethiopia*, dissertation, Cuvillier Verlag, Göttingen, Germany

Davidson, J. (1989) 'The *Eucalyptus* dilemma. Arguments for and against *Eucalyptus* planting in Ethiopia', The Forestry Research Centre Seminar Note Series No. 1, Forestry Research Center, Addis Ababa

Debela, B. (2004) *Contribution of Non-Timber Forest Products to the Rural Household Economy: Gore District, Southwestern Ethiopia*, MSc thesis, Swedish University of Agricultural Sciences, Skinnskatteberg, Sweden

Demissew, S. (1996) 'Ethiopia's natural resource base', in S. Tilahun, S. Edwards and T. B. G. Egziabher (eds) *Important Bird Areas of Ethiopia*, Ethiopian Wildlife and Natural History Society, Semayata Press, Addis Ababa,.pp36–53

Descheemaeker, K., Muys, B., Nyssen, J., Poesen, J., Raes, D., Haile, M. and Deckers, S. (2006) 'Litter production and organic matter accumulation in exclosures of the Tigray highlands, Ethiopia', *Forest Ecology and Management*, vol 233, pp21–35

Dessie, G. and Kleman, J. (2007) 'Pattern and magnitude of deforestation in the south central rift valley region of Ethiopia', *Mountain Research and Development*, vol 27, pp162–168

Eastern Africa Bamboo Project (2007) 'Bamboo entrepreneurs in Addis Ababa', www.eabp.org.et

Edwards, S., Tadesse, M. and Hedberg, I. (eds) (1995) *Flora of Ethiopia and Eritrea, Vol 2, Part 2*, National Herbarium, Addis Ababa University, Addis Ababa, and University of Systematic Botany, Uppsala University, Uppsala, Sweden

Edwards, S., Demissew, S. and Hedberg, I. (eds) (1997) *Flora of Ethiopia and Eritrea, Vol 6*, National Herbarium, Addis Ababa University, Addis Ababa, and University of Systematic Botany, Uppsala University, Uppsala, Sweden

Edwards, S., Tadesse, M., Demissew, S. and Hedberg, I. (eds) (2000) *Flora of Ethiopia and Eritrea, Vol. 2, Part 1*, National Herbarium, Addis Ababa University, Addis Ababa, and University of Systematic Botany, Uppsala University, Uppsala, Sweden

EFAP (1994) *Ethiopian Forestry Action Program*, Ethiopian Forestry Action Program, Addis Ababa

Eldridge, K., Davidson, J., Hardwood, C. and Van Wyk, G. (1997) *Eucalyptus Domestication and Breeding*, Oxford Science Publication, Claredon Press, Oxford, UK

Embaye, K. (2000) 'The indigenous bamboo resources of Ethiopia', *Ambio*, vol 29, pp518–521

Embaye, K. (2003) *Ecological Aspects and Resource Management of Bamboo Forests in Ethiopia*, Doctoral thesis, Swedish University of Agricultural Sciences, Uppsala, Sweden

FAO (1979) *Eucalyptus for Planting*, Food and the Agricultural Organization of the United Nations Forestry Series No. 11, Rome

FAO (1988) *The Eucalypts Dilemma*, Food and the Agricultural Organization of the United Nations, Rome

FAO (2001) 'Global forest resources assessment 2000. Main report', Forestry Paper 140, Food and the Agricultural Organization of the United Nations, Rome

FAO (2004) 'Livestock sector brief: Ethiopia', Livestock Information Sector Analysis and Policy Branch, Food and the Agricultural Organization of the United Nations, Rome

FAO (2005) *Global Forest Resources Assessment 2005*, Food and the Agricultural Organization of the United Nations, Rome

FAO (2007) *State of the World Forests*, Food and the Agricultural Organization of the United Nations, Rome

Fetene, A. (2006) *Diversity and Socio-Economic Importance of Non-Timber Forest Products of Menagesha Suba Forest Area, Central Ethiopia*, MSc thesis, Wondo Genet College of Forestry and the Swedish University of Agricultural Sciences, Uppsala, Sweden

Friis, I. (1986) 'Zonation of forest vegetation on the south slopes of Bale Mountains, south Ethiopia', *SINET: Ethiopian Journal of Science*, vol 9, pp29–44

Friis, I. (1992) 'Forests and forest trees of north east tropical Africa', *Kew Bulletin Additional Series*, vol 15, pp1–396

Friis, I. and Tadesse, M. (1990) 'The evergreen forests of tropical northeast Africa', *Mitt. Inst. Allg. Bot. Hamburg*, vol 23a, pp249–263

Gebre-Egziabher, T. B. (1986) 'Ethiopian vegetation – Past, present and future trends', *SINET: Ethiopian Journal of Science*, vol 9 (supplement), pp3–13

Getaneh, T. (2008) *Remote Sensing and GIS Assisted Participatory Biosphere Reserve Zoning for Wild Coffee Conservation: Case of Yayu Forest*, MSc thesis, Addis Ababa University, Addis Ababa

Gobeze, T. (2008) *Impacts of Participatory Forest Management (PFM) on Forest Status and Livelihoods: Experience from the Bonga PFM Project in Ethiopia*, MSc thesis, Wondo Genet College of Forestry and Natural Resources, Awassa, Ethiopia

Gobeze, T., Bekele, M., Lemenih, M. and Kassa, H. (2009) 'Participatory forest management and its impacts on livelihoods and forest status: The case of Bonga forest in Ethiopia', *International Forestry Review*, vol 11, no 3, pp346–358

Hagos, F., Pende, J. and Nega, G. (2002) 'Land degradation and strategies for sustainable land management in the Ethiopian highlands, Tigray Region', International Livestock Research Institute Working Paper 25, ILRI, Nairobi

Haile, A., Bekele, M. and Ridgewell, A. (2009) 'Small and medium forest enterprises in Ethiopia', IIED Small and Medium Forest Enterprise Series No. 26, FARM-Africa and International Institute for Environment and Development, London

Hailu, M., Oba, G., Angassu, A. and Weladji, R. B. (2006) 'The role of area enclosures and fallow age in the restoration of plant diversity in northern Ethiopia', *African Journal of Ecology*, vol 44, pp507–514

Hedberg, I. and Edwards, S. (eds) (1989) *Flora of Ethiopia, Vol. 3*, Addis Ababa University, Addis Ababa, and Department of Systematic Botany, Uppsala University, Uppsala, Sweden

Hedberg, I. and Edwards, S. (eds) (1995) *Flora of Ethiopia and Eritrea, Vol 7*, Addis Ababa University, Addis Ababa, and Department of Systematic Botany, Uppsala University, Uppsala, Sweden

Hedberg, I., Edwards, S. and Nemomissa, S. (eds) (2003) *Flora of Ethiopia and Eritrea, Vol. 4(1)*, Addis Ababa University, Addis Ababa, and Department of Systematic Botany, Uppsala University, Uppsala, Sweden

Hedberg, I., Friis, I. and Edwards, S. (eds) (2004) *Flora of Ethiopia and Eritrea, Vol. 4(2)*, Addis Ababa University, Addis Ababa, and Department of Systematic Botany, Uppsala University, Uppsala, Sweden

Hedberg, I., Kelbessa, E., Edwards, S., Demissew, S. and Persson, E. (eds) (2006) *Flora of Ethiopia and Eritrea, Vol. 5*, Addis Ababa University, Addis Ababa, and Department of Systematic Botany, Uppsala University, Uppsala, Sweden

Hillman, J. C. (1993) *Ethiopia: Compendium of Wildlife Conservation Information. Vol 1*, New York Zoological Park, Bronx, New York

Hoekstra, D. A., Torquebiau, E. and Bishaw, B. (1990) *Agroforestry: Potentials and Research Needs for the Ethiopian Highlands*, Agroforestry Research Networks for Africa (AFRENA), ICRAF, Nairobi

Homann, S., Rischkowsky, B. and Steinbach, J. (2008) 'The effect of development interventions on the use of indigenous range management strategies in the Borana lowlands in Ethiopia', *Land Degradation and Development*, vol 19, pp368–387

ITC-P-Maps (2006) www.trademap.org/, accessed December 2006

Jagger, P. and Ponder, J. (2000) 'The role of trees for sustainable management of less favored lands: The case of *Eucalyptus* in Ethiopia', EPTD Discussion Paper No. 65, International Food Policy Research Institute, Washington, DC

Jenbere, D. (2009) *The Expansion of Eucalyptus Plantation by Smallholder Farmers and its Drivers: The Case of Arsi Negelle District, Southern Oromia, Ethiopia*, MSc thesis, Hawassa University, Wondo Genet College of Forestry and Natural Resources, Awassa, Ethiopia

Kebebew, Z. (2002) *Profitability and Household Income Contribution of Growing Eucalyptus globulus to Smallholder Farmers*, MSc thesis, Swedish University of Agricultural Sciences, Skinnskatteberg, Sweden

Kebede, B. (2006) *Land Cover/Land Use Changes and Assessment of Agroforestry Practices at Pawe Resettlement District, Northwestern Ethiopia*, MSc thesis, Wondo Genet College of Forestry and Natural Resources, Awassa, Ethiopia

Kebrom, T. and Hedlund, L. (2000) 'Land cover changes between 1958 and 1986 in Kalu District, Southern Wello, Ethiopia', *Mountain Research and Development*, vol 20, pp42–51

Kelbessa, E., Bekele, T., Gebrehiwot, A. and Handera, G. (2000) *The Socio-Economic Case Study of the Bamboo Sector in Ethiopia: An Analysis of the Production-to-Consumption System*, INBAR Working Paper No. 25, Addis Ababa, Ethiopia

Kvarnbäck, G. and Natvig, T. (1979) *A Proposal for Restructuring the Present Mills for Future Investments*, Forestry and Wildlife Development Authority (FAWDA), Addis Ababa

Lemenih, M. (2008) 'Current and prospective economic contributions of the forestry sector in Ethiopia', in T. Hechett and N. Aklilu (eds) *Proceeding of a Workshop on 'Ethiopian Forestry at Crossroads: On the Need for Strong Institutions'*, Addis Ababa University Press, Addis Ababa, pp59–82

Lemenih, M. and Bongers, F. (in press) 'Dry forests in Ethiopia and their silviculture', in S. Günter, B. Stimm, M. Weber and R. Mosandl (eds) *Silviculture in the Tropics*, Springer, Berlin, Germany

Lemenih, M. and Teketay, D. (2005) 'Effect of prior land use on the recolonization of native plants under plantation forests in Ethiopia', *Forest Ecology and Management*, vol 218, pp60–73

Lemenih, M. Gidyelew, T. and Teketay, D. (2004) 'Effects of canopy cover and understory environment of tree plantations on species richness, density and sizes of colonizing native woody species in southern Ethiopia', *Forest Ecology and Management*, vol 194, pp1–10

Lemenih, M., Tolera, M. and Karltun, E. (2008) 'Deforestation: Impact on soil quality, biodiversity and livelihoods in the highlands of Ethiopia', in I. B. Sanchez and C. L. Alonso (eds) *Deforestation Research Progress*, Nova Science Publishers, Hauppage, NY, pp21–39

Lemessa, D. (2006) *The Roles of Apiculture In Vegetation Characterization and Household Livelihoods in Walmara District, Central Ethiopia*, MSc thesis, Wondo Genet College of Forestry, Awassa, Ethiopia, and the Swedish University of Agricultural Sciences, Uppsala, Sweden

Logan, W. E. M. (1946) 'An introduction to the forests of central and southern Ethiopia', Imperial Forestry Institute Paper No. 24, Oxford Forestry Institute, Oxford, UK

Mamo, G., Sjaastad, E. and Vedeld, P. (2007) 'Economic dependence on forest resources: A case from Dendi District, Ethiopia', *Forest Policy and Economics*, vol 9, pp916–927

Mayaux, P., Holmgren, P., Achard, F., Eva, H., Stibig, H. J. and Branthomme, A. (2005) 'Tropical forest cover change in the 1990s and options for future monitoring', *Philosophical Transactions of the Royal Society B*, vol 360, pp373–384

McCann, J. C. (1995) *People of the Plow: An Agricultural History of Ethiopia, 1800–1990*, University of Wisconsin Press, Madison, WI

McCann, J. C. (1999) *Green Land, Brown Land, Black Land: An Environmental History of Africa, 1800–1990*, Heineman, Portsmouth, NH

Mekonnen, Z., Kassa, H., Lemenih, M. and Campbell, B. (2007) 'The role and management of *Eucalyptus* in Lode Hetosa District, Central Ethiopia', *Forests, Trees and Livelihoods*, vol 17, pp309–323

Mekuria, W., Veldkamp, E., Haile, M., Nyssen, J., Muys, B. and Gebrehiwot, K. (2007) 'Effectiveness of exclosures to restore degraded soils as a result of overgrazing in Tigray, Ethiopia', *Journal of Arid Environments*, vol 69, pp270–284

Mender, M., Emana, B., Asfaw, Z. and Badassa, B. (2006) 'Marketing of medicinal plants in Ethiopia: A survey of the trade in medicinal plants', research report prepared for Sustainable Use of Medicinal Plants Project, Institute of Biodiversity Conservation, Addis Ababa

Mengesha, B. (1996) *Natural Regeneration Assessment in Tiro Boter Becho Integrated Forest Development and Utilization Project*, MSc thesis, Swedish University of Agricultural Sciences, Skinnskatteberg, Sweden

Mengistu, T., Teketay, D., Hulten, H. and Yemshaw, Y. (2005) 'The role of exclosures in the recovery of woody vegetation in degraded dryland hillsides of central and northern Ethiopia', *Journal of arid Environments*, vol 60, pp259–281

Nedessa, B., Ali, J. and Nyborg, I. (2005) *Exploring Ecological and Socio-Economic Issues for the Improvement of Area Enclosure Management. A Case Study from Ethiopia*, Drylands Coordination Group Report No. 38, Oslo, Norway

Negarit Gazeta (2007) 'Forest Development, Conservation and Utilization', Federal Democratic Republic of Ethiopia (FDRE) Proclamation No 542/2007

Negash, M. (2002) *Socio-Economic Aspects of Farmers' Eucalyptus Planting Practices in the Enset–Coffee-Based Agroforestry System of Sidama, Ethiopia. The Case of Awassa and Shebedino Districts*, MSc thesis, Swedish University of Agricultural Sciences, Skinnskatteberg, Sweden

Negussie, A., Aerts, R., Gebrehiwot, K. and Muys, B. (2008) 'Seedling mortality causes recruitment limitation of *Boswellia papyrifera* in northern Ethiopia', *Journal of Arid Environments*, vol 72, pp378–383

Nigatu, A. (2004) *Sawnwood Market of Addis Ababa*, MSc thesis, Wondo Genet College of Forestry, Awassa, Ethiopia, and the Swedish University of Agricultural Sciences, Skinnskatteberg, Sweden

NRMRD (2001) *The Impact of Checkpoints Lifting on the Conservation of Forest Resources. Case Study on Addis Ababa and Two Other Prominent Forest Enterprises*, Natural Resources Management and Regulatory Department, Addis Ababa

Nune, S., Kassie, M. and Mungatana, E. (2009) *Forestry Resources Accounting: The Experiences of Ethiopia'*, Environmental Economic Policy Forum for Ethiopia (EEPFE), Addis Ababa

Nyssen, J., Poesen, J., Moeyerson, J., Deckers, J., Haile, M. and Lang, A. (2004) 'Human impact on the environment in the Ethiopian and Eritrean highlands – A state of the art', *Earth-Science Review*, vol 64, pp273–320

Nyssen, J., Frankl, A., Poesen, J., Haile, M., Haregeweyn, N., Moeyersons, J., Descheemaeker, K., Deckers, J. and Munro, N. (2009) 'Desertification? Northern Ethiopia re-photographed after 140 years', *Science of the Total Environment*, vol 407, pp2749–2755

Ogbazghi, W., Rijkers, T., Wessel, M. and Bongers, F. (2006) 'Distribution of the frank-incense tree *Boswellia papyrifera* in Eritrea: The role of environment and land use',

*Journal of Biogeography*, vol 33, pp524–535

Örlander, G. (1986) *Growth of Some Forest Trees in Ethiopia and Suggestions for Species Selection in Different Climatic Zones*, Swedish University of Agricultural Sciences, Umeå, Sweden

Pankhurst, R. (1995) 'The history of deforestation and afforestation in Ethiopia prior to World War I', *Northeast African Studies*, vol 2, pp119–133

PASDEP (Plan for Accelerated and Sustained Development to End Poverty) (2005) 'Ethiopia: Building on Progress: A Plan for Accelerated and Sustained Development to End Poverty, 2005/06–2009/10', Ministry of Finance and Economic Development, Addis Ababa

Pohjonen, V. and Pukkala, T. (1990) '*Eucalyptus globulus* in Ethiopia forestry', *Forest Ecology and Management*, vol 36, pp19–31

Reusing, M. (1998) *Monitoring Forest Resources of Ethiopia*, Ministry of Agriculture, Addis Ababa

Ritler, A. (1997) 'Land use, forests and the landscape of Ethiopia 1699–1865: An inquiry into the historical geography of Central/Northern Ethiopia', Research Report No. 38, Soil Conservation Research Project, Bern

RLALUP (2005) Rural Land Administration and Land Use Proclamation No. 456/2005, Federal Democratic Republic of Ethiopia (FDRE), Addis Ababa

Schmitt, C. (2006) *Montane Rainforest with Wild* Coffea arabica *in the Bonga Region (SW Ethiopia): Plant Diversity, Wild Coffee Management and Implications for Conservation*, Ecology and Development Series No. 27, Cuvillier Verlag, Göttingen, Germany

Seifu, A. (2002) *Farmers' Private Tree Planting and Management Tradition at Wondo Genet, Ethiopia*, MSc thesis, Wageningen University, Wageningen, The Netherlands

Seifu, K. (1998) *Estimating Land Cover/Land Use Changes in Munessa Forest Area Using Remote Sensing Techniques*, MSc thesis, Report No. 1998:32, Swedish University of Agricultural Sciences, Skinnskatteberg, Sweden

Senbeta, F. (2006) *Biodiversity and Ecology of Afromontane Rainforests with Wild* Coffea arabica L. *Populations in Ethiopia'*, PhD dissertation, Ecology and Development Series No. 38, Cuvillier Verlag, Göttingen, Germany

Senbeta, F. and Teketay, D. (2001) 'Regeneration of indigenous woody species under the canopy of tree plantations in Central Ethiopia', *Tropical Ecology*, vol 42, pp175–185

Senbeta, F., Teketay, D. and Näslund, B-Å. (2002) 'Native woody species regeneration in exotic tree plantations in Munessa-Shashemene Forest, Ethiopia', *New Forests*, vol 24, pp131–145

Senbeta, F., Woldemariam, T., Demissew, S. and Denich, M. (2007) 'Floristic diversity and composition of Sheko forests, southwest Ethiopia, *Ethiopian Journal of Biological Sciences*, vol 6, no 1, pp11–42

Tadesse, D. (2007) *Forest Cover Change and Socioeconomic Drivers in Southwest Ethiopia*, MSc thesis, Centre of Land Management and Land Tenure, Technische Universität München, Munich, Germany

Tadesse, D. and Woldemariam, T. (2007) 'Customary forest tenure in southwest Ethiopia', *Forests, Trees and Livelihoods*, vol 17, pp325–338

Tadesse, M. (1993) 'A survey of the evergreen forests of Ethiopia', in A. G. Berhanu (ed) *Proceedings of the National Workshop on Setting Forestry Research Priorities in Ethiopia*, Forestry Research Center, Addis Ababa, pp265–297

Teketay, D. (1999a) 'History, botany and ecological requirements of coffee', *Walia*, vol 20, pp28–50

Teketay, D. (1999b) 'Past and present activities, achievements and constraints in forest genetic resources conservation in Ethiopia', in S. Edwards, A. Demissie, T. Bekele and G. Haase (eds) *Proceedings of the National Forest Genetic Resources Conservation Strategy Development Workshop*, Institute of Biodiversity Conservation and Research and German Technical Cooperation (GTZ), Addis Ababa, pp49–72

Teketay, D. (2004) 'Forestry research in Ethiopia: Past, Present and future', in G. Balcha, K. Yeshitela and T. Bekele (eds) Proceedings of a National Conference on Forest Resources of Ethiopia: Status, Challenges and Opportunities, Addis Ababa, Ethiopia, 27–29 November 2002, pp1–39

Teketay, D., Anage, A., Mulat, G. and Eneyew, M. (1998) *Study on Forest Coffee Conservation*, Coffee Improvement Project, Addis Ababa

Teketay, D., Senbeta, F., Maclachlan, M., Bekele, M and Barklund, P. (in press) *Edible Wild Plants in Ethiopia*, Addis Ababa University Press, Addis Ababa

Teklay, T. (1996) *Problems and Prospects of Tree Growing by Smallholder Farmers. A Case Study in Feleghe-Hiwot Locality, Eastern Tigray*, MSc thesis, Swedish University of Agricultural Sciences, Skinnskatteberg, Sweden.

Teklu, G. (2003) *Expanse of Plantation Forest in Ethiopia. An Outcome of More Than Half a Century's Effort*, MoA, Addis Ababa

Temesgen, Z., Irwin, B., Jordan, G. and Mckee, J. (2007) 'Forests, use them or lose them, an argument for promoting forest-based livelihoods rather than alternative non-forest-based livelihoods within PFM programmes', in E. Kelbessa and C. de Stoop (eds) *Participatory Forest Management in Africa (PFM), Biodiversity and Livelihoods in Africa*, Government of Ethiopia in collaboration with other stakeholders, Addis Abeba, pp7–17

Terefe, B. (2009) Acacia drepanolobium *as Invasive Indigenous Species: Assessing Impact on Gum-Resin Resource, Range Land and Livelihood in Borana*, MSc thesis, Hawassa University, Wondo Genet College of Forestry and Natural Resources, Awassa, Ethiopia

Tesfaye, G. (2008) *Ecology of Regeneration and Phenology of Seven Indigenous Tree Species in a Dry Tropical Afromontane Forest, Southern Ethiopia*, doctoral dissertation, Addis Ababa University, Addis Ababa

Tesfaye, G. (2009) *Effects of Climate Variability on Species Diversity and Biomass Productivity of a Humid Afromontane Tropical Forest of Ethiopia: Evidence from Postfire Forest Regrowth*, Technical Report to the International START, Addis Ababa

Teshome, M. (2004) *Economics of Growing* Eucalyptus Globulus *on Farmers' Woodlot: The Case of Kutaber District, South Wollo, Ethiopia*, MSc thesis, Hawassa University, Wondo Genet College of Forestry, Awassa, Ethiopia

Tolera, M., Asfaw, Z., Lemenih, M. and Karltun, E. (2008) 'Woody species diversity in a changing landscape in the south-central highlands of Ethiopia', *Agriculture, Ecosystems and Environment*, vol 128, pp52–58

Tolossa, T. (2002) *Forest Dwellers' Association (WAJEB) as an Approach in Participatory Forest Management. The Case of Adaba-Dodola*, MSc thesis, Swedish University of Agricultural Sciences, Skinnskatteberg, Sweden

Toru, T. (2002) *Dependency on Forestry for Fuelwood and Prospects of Tree Growing by a Peasant Community. A Case Study of the Nejebesa Peasant Association, West Hararge, Ethiopia*, MSc thesis, Swedish University of Agricultural Sciences, Skinnskatteberg, Sweden

Turnbull, J. W. (1999) 'Eucalypt plantations', *New Forests*, vol 17, pp37–52

Van Muysen, W., Brusselmans, A., Abate, T., Saleem, M., Rampelberg, S. and Deckers, J. (1998) 'Land use sustainability in the Tero Jemjem catchment of the Ethiopian highlands', *The Land*, vol 2, no 1, pp29–38

Vivero, J. L. P. (2002) 'Forest is not only wood: The importance of non-wood forest products for the food security of rural households in Ethiopia', in D. Teketay and Y. Yemshaw (eds) *Forests and Environment, Proceedings of the Fourth Annual Conference of Forestry Society of Ethiopia, 14–15 January 2002*, Forestry Society of Ethiopia, Addis Ababa, pp16–31

Wassie, A., Teketay, D. and Powell, N. (2005) 'Church forests in North Gonder Administrative Zone, northern Ethiopia', *Forests, Trees and Livelihoods*, vol 15, pp349–373

Wassie, A. (2007) *Ethiopian Church Forests: Opportunities and Challenges for Restoration*, PhD dissertation, Wageningen University, Wageningen, The Netherlands

Wassie, A., Sterck, F. J., Teketay, D. and Bongers, F. (2009) 'Tree regeneration in church forests of Ethiopia: Effects of microsites and management', *Biotropica*, vol 41, no 1, pp110–119

Watson, E. E. (2003) Examining the potential of indigenous institutions for development: A perspective from Borana, Ethiopia', *Development and Change*, vol 34, pp287–309

WBISPP (1995) *Towards a Strategic Plan for the Sustainable Development and Conservation of the Woody Biomass Resources*, Woody Biomass Inventory and Strategic Planning Project, Addis Ababa

WBISPP (2004) *Forest Resources of Ethiopia*, Woody Biomass Inventory and Strategic Planning Project, MoARD, Addis Ababa

WCMC (World Conservation Monitoring Centre) (1994) *Biodiversity Data Sourcebook*, World Conservation Press, Cambridge, UK, http://openlibrary.org/b/OL917455M/Biodiversity_data_sourcebook

WHO (1998) *Regulatory Situation of Herbal Medicines: A Worldwide Review*, WHO/TRM/98.1, Geneva

Wøien, H. (1995) 'Deforestation, information and citations: A comment on environmental degradation in highland Ethiopia', *Geojournal*, vol 37, no 4, pp501–512

Woldemariam, T. (2003) *Vegetation Ecology of the Yayu Forest in SW Ethiopia: Impacts of Human Use and Implications for In Situ Conservation of Wild Coffea arabica L. Populations*, Ecology and Development Series No. 10, Cuvillier Verlag, Göttingen, Germany

Woldemariam, T., Borsch, T., Denich, M. and Teketay, D. (2008) 'Floristic composition and environmental factors characterizing coffee forests in southwest Ethiopia', *Forest Ecology and Management*, vol 255, pp2138–2150

Woldu, K. (1999) *Natural Regeneration of Juniperus procera (Mocht) in Eucalyptus globulus (Labill) Plantation at Entoto Mountain, Central Ethiopia*, MSc thesis, Swedish University of Agricultural Sciences, Skinnskatteberg, Sweden

Woldu, Z. and Namomissa, S. (2006) 'The influence of banning bushfire on the vegetation of the Borana drylands in southern Ethiopia', in A. M. Nikundiwe and J. D. L. Kabigumila (eds) *Drylands Ecosystems: Challenges and Opportunities for Sustainable Natural Resources Management. Proceedings of the Regional Workshop held at Hotel Impala, Arusha, Tanzania, June 2006*, University of Dar es Salaam, Dar es Salaam, Tanzania

Workaffess, W. and Kassu, G. (2000) *Coffee Production System in Ethiopia: Proceedings of the Workshop on Control of Coffee Berry Disease (CBD), 13–15 August 1999*, Addis Ababa University Press, Addis Ababa, Ethiopia

Worku, A. (2006) *Population Status and Socio-Economic Importance of Gum and Resin Bearing Species in Borana Lowlands, Southern Ethiopia*, MSc thesis, Department of

Biology, Addis Ababa University, Addis Ababa

Yami, M. (2005) *Impact of Area Exclosures on Density and Diversity of Large Wild Mammals in Douga Tembien, Tigray, Ethiopia*, MSc thesis, Mekelle University, Mekelle, Ethiopia

Yebeyen, D. (2006) *Population Status of Acacia senegal (Linne) Willd. and its Gum Quality in the Central Rift Valley of Ethiopia*, MSc thesis, Wondo Genet College of Forestry, Awassa, Ethiopia

Yirdaw, E. (2002) *Restoration of the Native Woody-Species Diversity, Using Plantation Species as Foster Trees, in the Degraded Highlands of Ethiopia*, PhD dissertation, Viikki Tropical Resources Institute, University of Helsinki, Helsinki

Zeleke, G. and Hurni, H. (2001) 'Implications of land use and land cover dynamics for mountain resource degradation in the northwestern Ethiopian highlands', *Mountain Research and Development*, vol 21, pp184–191

Zemede, A. and Mesfin, T. (2001) 'Prospects for sustainable use and development of wild food plants in Ethiopia', *Economic Botany*, vol 55, pp47–62

# 3

# Forests and Forestry in Uganda

## Joseph Obua and Jacob Godfrey Agea

## INTRODUCTION

Uganda is a land-locked country measuring 241,500km². It is located in the eastern region of Africa and lies between latitude 1°30 south and 4° north and longitude 29°30 and 35° west. It is bordered by Kenya in the east, Tanzania and Rwanda in the south, Democratic Republic of Congo in the west and Sudan in the north. Most of Uganda forms part of the interior plateau of the African continent. It is characterized by flat-topped hills in the central, western and eastern parts of the country. The rise of the plateau in the eastern and western parts of the country is represented by mountainous topography found along the borders, for example the Rwenzori mountains and Mufumbira volcanoes in the west and Mt Elgon and Mt Kadaru in the east. About 20 per cent of the country is covered by water bodies such as Lake Victoria, Lake Albert, Lake Edward, Lake Kyoga and the River Nile (AFRENA, 1988).

The climate of Uganda is influenced by the Inter-Tropical Convergence Zone (ITCZ). In most parts of the country, the seasons are fairly well marked as rainy and dry seasons. Annual rainfall ranges from 625 to 2250mm, with most parts of the country receiving 1000 to 1500mm. In the northeastern parts, the annual rainfall is as low as 625mm while in the islands of Lake Victoria it may reach 2250mm. The mean temperatures over the whole country show great variations depending on elevation and landscape. Mean annual maximum temperatures in most parts of the country vary from 27.5 to 32.5°C; mean annual minimum temperatures from 15 to 20°C.

Uganda lies in the transition zone between the east African savanna vegetation systems and the moist tropical forests of the Congo basin. It contains 7 of Africa's 18 biogeographic regions (Howard, 1991). The vegetation classification and description used in Uganda are still based on Langdale-Brown et al (1964). There are 10 main vegetation types: high montane moorland and heath; medium altitude forest/savanna mosaic; moist thicket; woodland; wooded savanna; grass savanna; steppe; bushland and dry thicket; swamp (wetlands); and cultivation communities (NEMA, 1996).

Over 96 per cent of all energy consumed is derived from biomass fuel and of this 82 per cent is used for domestic purposes, principally cooking (Watts and Otiti, 1992). Charcoal use and inefficient methods of charcoal production are leading to deforestation in the areas surrounding urban settlements. Lack of fossil fuel resources, scarcity of foreign exchange and inadequate infrastructure make it impracticable to encourage the switch to kerosene, gas or electricity. Added to this, Uganda has an annual population growth rate of over 3 per cent. This implies that the demand for wood fuel resources is likely to grow further.

According to a 1986 World Bank report, Uganda's biomass resource base was being replenished at an annual rate of 15.6 million m$^3$, but at the same time was being consumed at a rate of 18.3 million m$^3$ per year, clearly a recipe for environmental disaster (Watts and Otiti, 1992). These valuable forest resources are disappearing rapidly. The 1992 Uganda National Environmental Action Plan (NEAP) estimated that deforestation was occurring in Uganda at an annual rate of 500km$^2$, while the Food and Agriculture Organization (FAO) of the United Nations (1993) estimated it to be 650km$^2$. If the rate of deforestation were to continue unabated, most of the forested area of Uganda would disappear by the end of the 21st century. The socio-economic indicators of the country are given in Table 3.1. Basically, Uganda is an agrarian country, dominated by peasant (subsistence) farmers. It is on this basis that the government has formulated the Plan for the Modernization of Agriculture and the National Forestry Plan to transform the economy.

In this chapter, an overview of the status, structure and distribution of forests in Uganda, the historical profile of forest resource management, evolution and landmarks of forest policy, and the institutional set-up for forest management in Uganda are presented. The contribution of forests to socio-economic development, forest revenue systems, causes of forest degradation, invasive species and their influence on forest management, and restoration attempts for degraded forests in Uganda are highlighted.

## STATUS, STRUCTURE AND DISTRIBUTION OF FORESTS IN UGANDA

The National Biomass Study carried out in the late 1990s and early 2000s, and published in 2003, is the most comprehensive analysis of land use in Uganda (Winterbottom and Eilu, 2006). Uganda has a total estimated area of

**Table 3.1** *Socio-economic data for Uganda for the period 2000–2007*

| Type of data | 2000 | 2005 | 2006 | 2007 |
|---|---|---|---|---|
| Population total (millions) | 24.69 | 28.95 | 29.90 | 30.93 |
| Population growth (annual %) | 3.0 | 3.2 | 3.2 | 3.4 |
| GNI, Atlas method (current US$) (billions) | 6.39 | 7.93 | 8.90 | 10.47 |
| GNI per capita, Atlas method (current US$) | 260 | 270 | 300 | 340 |
| GNI, PPP (current international $) (billions) | 16.22 | 23.84 | 25.90 | 28.46 |
| GNI per capita, PPP (current international $) | 660 | 820 | 870 | 920 |
| Life expectancy at birth, total (years) | 46 | 50 | 51 | - |
| Mortality rate, under-5 (per 1000) | 145 | 136 | 134 | - |
| Prevalence of HIV, total (% of population aged 15–49) | – | 6.4 | – | 5.4 |
| Forest area (km²) (thousands) | 40.6 | 36.3 | – | – |
| Agricultural land (% of land area) | 62.3 | 64.5 | – | – |
| GDP (current US$) (billions) | 5.93 | 8.74 | 9.50 | 11.21 |
| GDP growth (annual %) | 5.6 | 6.7 | 5.1 | 6.5 |
| Inflation, GDP deflator (annual %) | 3.8 | 7.9 | 8.6 | 8.2 |
| Agriculture, value added (% of GDP) | 37 | 33 | 31 | 29 |
| Industry, value added (% of GDP) | 20 | 18 | 18 | 18 |
| Services etc., value added (% of GDP) | 42 | 49 | 51 | 53 |
| Time required to start a business (days) | – | 34 | 28 | 28 |

*Note:* GNI = gross national income; PPP = purchasing power parity; GDP = gross domestic product.
*Source:* World Development Indicators Database (2008)

24,155,058 hectares (241,500km²), of which farmland (35 per cent) is the most extensive land use (Table 3.2). Woodland and bushland account for about 16.7 per cent and 5.9 per cent respectively of the total area of Uganda. Tropical high forests and plantation forests account for about 4 per cent and 0.14 per cent of the total area. A significant area under tropical high forests, however, has been degraded, and this has serious implications for the sustainability of tropical high forests in Uganda.

Uganda's forests are also structured into central forest reserves (CFRs) and those forests outside protected areas (forests on private and customary lands). Of much importance are the CFRs categorized as production and protection forests. Production forests, which include savanna bushland and grassland areas, were gazetted for supply of forest products and future development of industrial plantations. The protection forests include all the tropical high forests, savanna woodlands and/or grasslands that protect watersheds and water catchments, biodiversity, ecosystems and landscapes that are prone to degradation under uncontrolled human use (Winterbottom and Eilu, 2006).

CFRs of ecological and biodiversity importance includes CFRs whose main functions are to protect biodiversity, water catchments, riverbanks, lakeshores and stabilization of steep slopes. A total of 1,073,983 hectares in 353 CFRs

**Table 3.2** *Land use and land cover of Uganda*

| Land use | km² | % | Gazetted forest (km²) | Non-gazetted forest (km²) | National park (km²) |
|---|---|---|---|---|---|
| | | | Area of Uganda's forest | | |
| Plantations | 345 | 0.14 | 306 | 19 | 20 |
| (broadleaved) | 189 | 0.08 | 186 | 3 | |
| (conifer) | 156 | 0.06 | 120 | 16 | |
| Tropical high forest | 8847 | 3.66 | 4170 | 1467 | 3210 |
| (fully stocked) | 6039 | 2.50 | | | |
| (degraded) | 2808 | 1.16 | | | |
| Woodland | 40,278 | 16.67 | 7200 | 33,078 | |
| Bushland | 14,199 | 5.88 | | | |
| Grassland | 51,119 | 21.16 | | | |
| Swamps | 4831 | 02.00 | | | |
| Farmland | 84,617 | 35.03 | | | |
| (small-scale subsistence) | 83,931 | 34.75 | | | |
| (large-scale subsistence) | 686 | 0.28 | | | |
| Build-up areas | 364 | 0.15 | | | |
| Open water | 36,909 | 15.28 | | | |
| Impediments | 39 | 0.02 | | | |
| **Total area** | **241,548** | **100** | **11,982** | **34,563** | **3230** |
| | | | 23.6% | 69.9% | 6.5% |

*Source:* Uganda Forest Department (1999)

(Figure 3.1) has been categorized as of ecological importance and designated for strict protection. The whole cattle corridor covering most of western through central to northeastern Uganda would be rendered unviable for cattle grazing and agriculture if these reserves were destroyed. The 'water for production' programme in support of the Poverty Eradication Action Plan (PEAP) requires these natural regulators and reservoirs of water flow. There are over 1.3 million people in Mubende, Kiboga and Kibaale Districts of western Uganda. They depend on the water that is trapped by the forests in the Mubende–Kiboga hills that constitute an important part of the Kafu and Katonga river systems. The forests in these hills and valleys ensure that the boreholes, wells and dams in the area are constantly refilled. The forests and wetlands ringing Lake Victoria stand between the survival and extinction of the fish in the lake, on which millions of people depend for their livelihoods. In 2002 fish exports earned Uganda US$87.9 million (Winterbottom and Eilu, 2006).

The forests and wetlands along the River Nile are a buffer between the pollutants generated by human activities and the river's fresh waters. Millions of people live along the Nile and their wellbeing is threatened when the water becomes polluted. In the Eastern Region the 21,870 hectares of West Bugwe, Igwe-Luvunya and South Busoga CFRs are the only natural forests remaining in the region. These have now been degraded by encroachment. As the harsh Karamoja weather creeps onto the rest of the country, people could be adversely affected if the forests are converted into agricultural land. The Forest Nature

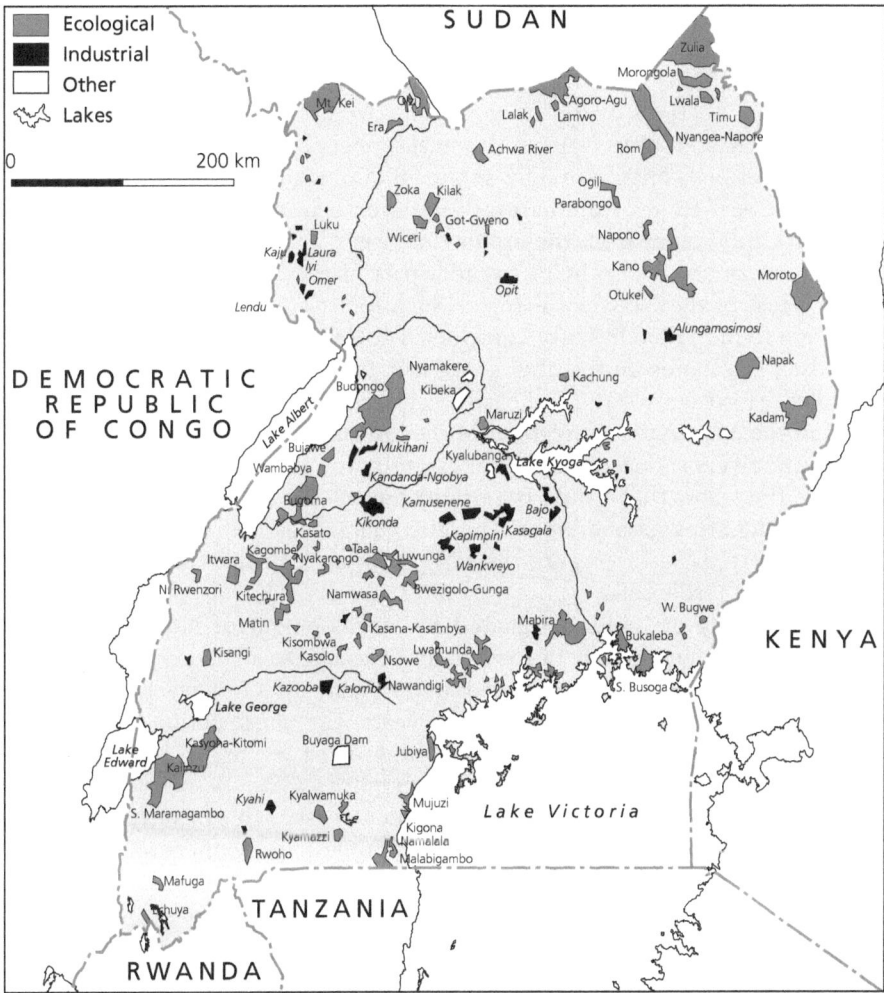

**Figure 3.1** *Map of Uganda with forest reserves*

Conservation Master Plan (FNCMP) of 2002, which was produced after a comprehensive biological inventory, categorizes the conservation importance of forest reserves as prime, core and secondary. Some 840,100 hectares (over 70 per cent of the total CFR area) have been categorized as such. Most of the tropical high forests in the Albertine rift fall in one or other of these categories. The CFRs of Karamoja, Kitgum, Moyo and Yumbe in northern Uganda constitute another stretch of important biodiversity areas located mainly in the mountains/hills that have been war ravaged and denuded of tree cover. The Lake Victoria crescent also constitutes another belt stretching from the wetland forests of Sango Bay in the south, through the lakeshore forests of Masaka, Mpigi and Mukono in the central region and tapering off with the natural forests in Mayuge and Bugiri Districts in the east.

Although the CFRs are important for biodiversity, the FNCMP has zoned them into 351,900 hectares for production zones, 220,800 hectares of strict nature reserve (SNR) zones for preservation of biodiversity and, in between the two, 267,400 hectares of buffer zones that are used to provide non-timber forest products. Out of an estimated annual timber consumption of 250,000m$^3$, forest plantations contribute only about 50,000m$^3$ (Winterbottom and Eilu, 2006). The rest comes from natural forests on private lands and CFRs. The timber in CFRs comes from the production zones, and contributes to the local domestic incomes. It also helps to meet part of the costs for protecting the forests. Most of the CFRs with tropical high forest are either rich in biodiversity or are found in biodiversity corridors. They also protect watersheds, river valleys and lakeshores. In the drier areas, lush tropical high forests are found in the valleys (for example the mahogany-rich forests in Gulu, Kitgum and West Nile and the *Markhamia* forests of Mubende and Kiboga. These forests also protect the rivers and streams that originate in the hills. Their 'sponge' effect soaks up the water that would otherwise run off and get lost and releases it slowly to the streams and wells, thereby guaranteeing all-year-round flow. Unfortunately these forests are the targets for agricultural land encroachers because of the fertile soils.

The majority of CFRs designated for development of industrial forest plantations are found mainly in Luwero and Nakasongola Districts. In the late 1960s and early 1970s, plans were initiated to scale up forest plantation establishment in the area. For more than 10 years, the construction industry in Uganda has been growing at a rate of 5–8 per cent and uses a lot of timber, plywood and poles. Major government programmes that have been consuming large amounts of wood products in the recent past include construction of power dams, schools and resettlement programmes. Today, there are less than 3000 hectares of industrial plantations in the country (Winterbottom and Eilu, 2006).

Drastic changes in the forest cover have taken place in Uganda during the past century. In 1890 forests covered approximately 10,800,000 hectares or 52 per cent of Uganda's surface area. By 1996, forest cover had declined to about 20 per cent. Tropical high forest cover declined from 12.7 per cent of total land area in 1900 to 3.6 per cent by 2000 (Forest Department, 2003). The country's annual deforestation rate has increased by 21 per cent since the end of the 1990s. The country lost an average of 86,400 hectares of forest or 2.27 per cent of its forest cover per year between 2000 and 2005 (FAO, 2007). On a generational timescale, Uganda lost 26.3 per cent of its forest cover (1.3 million hectares) between 1990 and 2005 (FAO, 2007). This forest loss is directly threatening some of the highest concentrations of biodiversity in Africa: Uganda is home to more than 5000 plant species, 345 species of mammals and 1015 types of birds. About 18 per cent of Uganda is presently forested (FAO, 2007). The Ministry of Water, Lands and Environment (MWLE, 2001) estimated that over one-third of the original forest area in forest reserves has disappeared and another third of the remaining tropical high forests are seriously degraded. In woodlands, particularly those on public land, conversion and degradation are even more pronounced, and of the once established forest plantations little remain.

# HISTORICAL PROFILE OF FOREST RESOURCE MANAGEMENT IN UGANDA

Formal forest management in Uganda started over 100 years ago with the establishment of the Scientific and Forestry Department in 1898 (Troup, 1922). The department had a mandate to carry out research, but did not charge fees for forest products or collected revenue (Kamugisha, 1993). By 1917, the name was changed to the Forestry Department and its mandate was expanded to include the production of timber and wild rubber and to engage in commercial forest exploitation and to supply the government with its own sawn-wood requirements. These functions were later transferred to the private sector in 1926 (Kamugisha, 1993). Like many countries within the region, Uganda's first forest reserves were gazetted in the 1930s and were facilitated by policies and laws formulated by the colonial government. The main objective of creating an elaborate network of forest reserves was to ensure that there was adequate supply of the country's wood needs, particularly for industrial purposes. By then the increasing forest frontier population was perceived as a serious threat to forest conservation. Until 1940, the Forest Department managed Uganda's forests through the process of command and control (Kamugisha, 1993). The department's focus was on the establishment of industrial forest plantation and maintenance of watershed protection areas. This system lacked incentives that could encourage the local communities to perceive forest resources being managed for the common good of all. Instead, the approach was marked with constant conflicts between the department and local communities. In the long term the command and control approach to forest conservation did not adequately achieve the objectives it was set up for.

Before independence, advocacy for incentive-based approaches were intensified by various stakeholders, leading to promotion of local community involvement in management of forest resources, which in some cases led to the creation of village forest reserves (declared and controlled by the local authorities), local forest reserves (declared by the central government but managed and controlled by local authorities) and central forest reserves (declared and managed by the central government) (Kamugisha, 1993). However, after independence, village and local forest reserves were abolished and put in the hands of the central government, with all the revenues going to the central treasury. This over-centralization of forest resources management, which was in place until the early 1990s, had an adverse effect on the relationship between the local communities and the Forest Department. However, since the early 1990s incentive-based approaches have been increasingly promoted in forest management in Uganda.

The policy framework for Uganda has gradually shifted to one which is supportive of community involvement in forest management and the use of incentive-based measures such as revenue sharing. The Constitution of 1995 explicitly recognizes the significance of the environment sector (including forestry) in promoting communities' livelihoods and health. Similarly, the National Environment Statute of 1995 emphasizes the importance of involving and empowering local councils and local communities in environmental management. In addition, the Wildlife Statute 1996 recognizes the need to collaborate with different stakeholders in wildlife management.

# EVOLUTION AND LANDMARKS OF FOREST POLICY IN UGANDA

Forest policy in Uganda has a long history, dating back to 1929. Four revisions were made in 1948, 1970, 1987 and 2001. The revisions reflected the distinct changes in the perceived role of forestry in Uganda. The first policy of 1929 was developed when the colonial state was seeking to gain formal control over much of the land (Kamugisha, 1993). The main justification for scheduling forest reserves was to ensure water catchments protection. This was a far-sighted policy that looked ahead to a time when the water catchments would be threatened by agricultural expansion. Timber production forests were also gazetted.

By the time of revising the policy in 1948, Uganda was beginning to change rapidly: there was growth in population and more awareness of the importance of national economic development in the post-war era. In addition to emphasis on retaining forests for their climatic and other indirect values, the 1948 policy stressed the need 'to foster among the people of Uganda a real understanding of the value of forests', the need for effective extension services and acquisition of land for planting new forests (Kamugisha, 1993; Karani, 1993). Under this policy, some forest reserves were converted to plantation, in others logging intensified, sawmills flourished, and forest refinement and other silvicultural works were encouraged. This reflected the realization of forests for economic development. Other national forests were cleared for agriculture, which was believed to be a better economic use of land than forestry in some well-wooded areas.

The forest policy was further revised in 1970; although it maintained the main provisions of the 1948 policy, a new addition was provision for efficient conversion of wood and wood products (Kamugisha, 1993; Karani, 1993). A third revision came in 1988 with a greater focus on environmental aspects of forestry, for example conservation of biodiversity and rare species, and the need for more active protection of forest resources, for research in silviculture and tourism, for promotion of agroforestry, and an overall emphasis on environmentally sustainable forestry (Kamugisha, 1993; Karani, 1993). 20 per cent of all natural forests were to be turned into 'strict nature reserves' in which no human activity was permitted except scientific studies. 30 per cent was to be become 'buffer zone' with 'limited' forest harvesting being permitted, and the remain-

---

**BOX 3.1 PILLARS OF UGANDA'S FOREST POLICY OF 2001**

- Forestry on government land
- Forest on private land
- Commercial forest plantations
- Forest products processing industries
- Collaborative forest management
- Farm forestry
- Forestry biodiversity conservation
- Watershed management
- Urban forestry
- Education, training and research
- Supply of tree seed and planting materials

---

ing 50 per cent was to be left for management for sustainable utilization (Kamugisha, 1993; Karani, 1993). These proportions, however, applied only to forests that were managed by the Forest Department and the management options did not consider forests on private landholdings (Kamugisha, 1993).

In 2001, the government approved a new forest policy (Box 3.1) aimed at developing an integrated forest sector and achieving sustainable increases in the economic, social and environmental benefits of forests and trees by all the people of Uganda, especially the poor and vulnerable (MWLE, 2001). The policy institutionalizes community forestry and addresses the need to manage forests on private land.

# INSTITUTIONAL SET-UP FOR FOREST MANAGEMENT IN UGANDA

There has been institutional restructuring in Uganda since the promulgation of the Constitution of Uganda in 1995. The Forest Department was restructured into a semi-autonomous body – the National Forest Authority (NFA). This is the lead agency for forest management in Uganda. It was established in 1998 to replace the Forest Department. By the 1960s Uganda's forest sector had a worldwide reputation, particularly for research into tropical high forest management. The political and economic upheavals of the 1980s, however, precipitated a general decline in all its aspects of operation. According to Hamilton (1984), forest management planning was far-sighted and the forest sector was an effective organization with a high degree of control over its land. This has changed during the last 10 to 15 years, when the forest policy has become short term and restricted in its aims, all forest working plans have become out of date, and many management systems designed to control activities in forest reserves have become ineffective. In 1991, the government transferred the Bwindi, Mgahinga, Elgon, Kibale and Semliki forest reserves to the Uganda National Park (now the

Uganda Wildlife Authority). The transfer of these forest reserves to the Uganda Wildlife Authority (UWA) was not well received by some staff from the forest sector as it meant reduced sources of earning and control.

The NFA strives to be financially viable and to operate in a business like manner, while leaving forest sector policy, planning and legislation to the relevant ministry and its cross-sectoral coordination structures. The National Forest Plan of 2002 stipulates the need to manage the central forest reserves on a sustainable basis to optimize the economic, environmental and social functions of the forest estate and to reduce poverty through the active involvement of the private sector and local communities. The plan also mandates the NFA to supply high-quality forestry-related products and services to government and the private sector on a contractual basis. However, within the Ministry of Water and Environment, there is a Forest Sector Support Division responsible for formulating national policies, standards, legislation and plans for the management of forests; mobilizing resources for forest management; coordinating and supervising national projects of forest management; monitoring the performance of the NFA; and inspecting, monitoring and coordinating the activities of local government in forest management.

# CONTRIBUTION OF UGANDA'S FORESTS TO SOCIO-ECONOMIC DEVELOPMENT

## Forests' contribution to GDP

The forest sector accounts for less than 2 per cent of the national income in official statistics, with a slightly declining tendency (Kazoora, 2001). However, a significant part of the income generated from forests is in the informal sector, such as non-traded products and services. Fuelwood collected and used by nearly 98 per cent of households is an example. GDP and other standard economic calculations refer only to traded outputs and do not capture the socio-economic and environmental importance of forests. Hence, the real value of forestry is insufficiently reflected in GDP calculations. If GDP were to be adjusted to include the unvalued forest goods and services, the contribution of forestry would rise to about 6 per cent (Table 3.3).

## Forests' contribution to energy use and rural livelihoods

Woody biomass contributes almost 90 per cent of the energy used in Uganda (Kazoora, 2001). Generally, the woodfuel demand is higher than the demand for any other fuel. In a recent study (Energy for Sustainable Development, 1996), it was reported that aggregate consumption of solid woody biomass is 30 million $m^3$ (about 17.2 million metric tonnes) per year, based on household consumption; beer brewing; fish drying; tobacco and tea curing; lime, tiles and bricks production; domestic heating; bakeries; and educational, prison, medical

Table **3.3** *Adjusted contribution of forestry to Uganda's GDP in 1998*

| Item | Non-adjusted GDP Uganda shilling billions | %GDP | Adjusted GDP Uganda shilling billions | %GDP |
|---|---|---|---|---|
| **Formal sector/monetary sector** | | | | |
| Sawn timber | | | 40.0 | 0.5 |
| Poles | | | 5.4 | 0.225 |
| Firewood | | | 21.0 | 0.26 |
| Charcoal | | | 57.0 | 0.7 |
| Tourism | | | 2.7 | 0.33 |
| Other non-wood forest products (NWFPs) | | | 20.0 | 0.25 |
| Total formal forest sector contribution | 61.2 | 0.8 | 146.1 | 19 |
| **Informal sector/non-monetary sector** | | | | |
| Poles | | | 6.0 | |
| Firewood | | | 160.0 | |
| Fodder | | | 4.0 | |
| Other non-wood forest products (NWFPs) | | | 40.0 | |
| Total in-formal forest sector contribution | 55.9 | 0.7 | 210.0 | 2.75 |
| **Non-marketable outputs** | | | | |
| Watershed benefits | | | 20 | |
| Carbon sequestration | | | 26.1 | |
| Biodiversity option value | | | 3.5 | |
| Erosion control | | | 60.0 | |
| Groundwater | | | 2.0 | |
| Total non-marketable outputs | N/A | | 112.3 | 1.45 |
| Grand total sector | | 1.5 | 468.4 | 6.1 |

*Source:* Background to the budget 2001/2002

and military institutions. Some of this biomass is used to produce 400,000 tonnes (80 million bags) of charcoal per annum, used mainly in urban centres. Per capita (unweighted) consumption is estimated at 157kg per year. Besides this, local communities depend on forests for medicinal plants, building poles, fruits and honey, and in some places game meat.

## Forests' contribution to industry and construction

The volume of sawn logs cut annually is today estimated at 100,000m³, 75,000m³ and 50,000m³ from forest reserves (natural forest), plantations (mainly conifers) and private/public land respectively. These estimates were based on timber that was cleared for movement from districts in 1997 and was captured by the Natural Forest Management and Conservation Project database (Kazoora, 2001). An adjustment of 50 per cent on these estimates has been added to cater for timber that is used within the districts and which is illegally transported outside.

Table 3.4 *Uganda's forest biodiversity*

| Classification | Specific examples |
|---|---|
| General composition | 427 species of trees, 329 species of birds, 12 species of diurnal primates and 7 butterfly species |
| Globally threatened with extinction | 4 primate species, 2 other mammal species, 6 bird species and 2 butterfly species |
| Endangered | Mountain gorilla |
| Vulnerable | Chimpanzee, Hoest monkey, elephant, leopard, Grauer's rush warbler and cream-banded shallow butterfly |
| Rare | Nahan's francolin, African green broadbill, flycatcher and forest ground thrush |
| Intermediate | The Uganda red colobus monkey and Kibale ground thrush species |

*Source:* NEMA (2002)

## Ecological functions of the forests

Forests provide a wide range of environmental services, which unfortunately are not monetized in the estimation of GDP. Such services include protection of watersheds and soil, carbon sequestration, microclimate regulation, and acting as habitat for wildlife and biodiversity (Table 3.4). Today, with the rampant negative political influence on forest conservation, like in the recent case of attempts to give away part of the Mabira forest to a private investor for sugar cane growing and parts of Kalangala forest reserves having been given to an investor for growing oil palm, these non-monetized values of the forests are being undermined in favour of agriculture.

## FOREST REVENUE SYSTEMS IN UGANDA

### Charges for round-wood production

Revenue charges for wood production in Uganda are classified into timber royalty, forest produce fees and licence fees. The levels of these charges are contained in the Forest Produce Fees and Licence Order of 2000 (Government of Uganda, 2000) and these charges are levied both in forest plantations and natural forests, but not in private forests. Timber royalty is charged on the basis of volume ($m^3$) of round wood taken by sawmills and pit-saws. The rate charged per cubic metre varies with species and is based on the value of different timber species and market demand. These charges are currently grouped into three classes: class one representing the species with the most value and highest demand, followed by class two and class three (Arumadri, 2001).

In addition to timber royalty, sawmills and pit-saws also pay registration fees of USh1,400,000 and USh350,000 respectively for their forest concessions (Government of Uganda, 2000). For sawmills consuming an average of

2300–4200m³ per year, a concession of five years is awarded. For sawmills consuming 500–2300m³ per year, a concession of two years is awarded and for hand-saws, a one-year concession is offered (Arumadri, 2001). The registration fee is paid every year until the concession expires, after which a new application is required. After the registration fees have been paid, a licence is issued to the sawmiller or pit-sawyer, which shows the area of operations and describes the products that may be taken, states the annual allowable cut (AAC) and species that may be harvested, and gives the minimum cutting diameter for each species. The licence is only issued for harvesting; management of the forest remains the responsibility of the National Forestry Authority.

The forest produce fees are charges levied on smaller sizes of round wood. Poles are classified according to diameter classes and a distinction is made between poles from forest plantations and poles from the natural forest. Charges are levied per pole for the smaller poles (0.5–14cm DBH) and per running metre for the larger poles (15–24cm DBH) (Government of Uganda, 2000). A cutting charge per pole is also levied on every pole that is harvested. Faggots (withes) are charged per head-load and fence posts are charged per running metre. Similarly, palms are charged per running metre. Firewood from forest plantations and natural forest reserves is charged per stacked cubic metre. For firewood from other types of public land, firewood is charged per stacked cubic metre or a monthly licence is issued, depending on the interest of the person. In most cases, monthly licences are issued to people who harvest firewood regularly (for example firewood traders). The amount charged for the licence varies with the scale of production, with large-scale firewood producers paying more than small traders. A charge is levied on the haulage or conveyance of firewood, which is based on the size of the vehicle used. A lorry is charged the most and a bicycle is charged the least. Timber grading fees (payable per cubic metre of sawn wood) are also contained in the Forest Produce Fees and Licences Order of 2000 (Government of Uganda, 2000). However, they are not charged at the moment because the National Forestry Authority has not resumed this service.

## Charges for production of non-wood forest products and services

In Uganda, the most common non-wood forest products include charcoal, bamboo, Christmas trees, seeds and seedlings, palms, rattan canes, minor forest produce (MFP), and minerals extracted from the forest. Forest services include ecotourism, grazing and hunting. Licences to produce and sell charcoal are charged per person per month and licences for conveyance are charged according to the size of the vehicle (lorries, pickups, canoes and bicycles) (Arumadri, 2001). Bamboo and Christmas trees are charged per pole and per tree respectively. Seeds are charged per kilogram and seedlings are charged per seedling. For MFP (for example materials for beds, chairs, mats and baskets, fruits, honey, herbal medicines, ornamentals, woodcarvings, walking sticks, drum frames, and brewing troughs), casual trade licences are issued and a fee is paid per person per year (Government of Uganda, 2000). Similarly, licences are issued per month per person for harvesting wild coffee. Gum arabic and resins are charged per kilogram. Licences for extracting forest minerals (for example brick-making materials and sandstone) are charged per person per month and higher charges are levied on the production of minerals near cities than in other areas (Arumadri, 2001).

## Charges on processed product production

Another charge paid by sawmillers and pit-sawyers is 17 per cent value added tax (VAT), which is charged on sawn wood and is paid to the Uganda Revenue Authority (URA). In Uganda, all businesses must charge VAT unless their annual gross income is less than USh20 million (Arumadri, 2001). Below this threshold, businesses are not required to charge VAT. Small sawmillers and pit-sawyers do not usually have to pay this tax because their gross income does not reach the threshold. However, in these cases, the sawmillers and pit-sawyers have to pay a tax of 15 per cent (of the value of timber sold) to the National Forestry Authority. Currently, all producers in private forests and all pitsawyers are below the VAT threshold and have to pay this tax (Arumadri, 2001). In fact, the bulk of timber production in Uganda comes from small producers and traders who do not reach the VAT threshold, so only a small fraction of the potential revenue in the form of VAT is currently collected. The National Forestry Authority does not have a mandate to levy charges on any other type of production (for example production of wood-based panels, pulp and paper). However, the URA collects general taxes from the forest industry (for example income and corporation taxes). The URA uses a broad classification of all businesses and divides them into large, medium and small taxpayers, depending on their income (Arumadri, 2001). Each class has a flat rate of income tax that is paid annually. Therefore, the amount of income tax paid by a business in the forestry sector depends on the level of income it generates. Other general taxes, fees and licence fees (for example for the export or production of processed forest products) are the responsibility of the Ministry of Commerce.

## Charges on forest products trade

The National Forestry Authority does not collect any charges on forest product trade. However, trading licences are required to sell processed forest products and these are issued by urban or local authorities, depending on the location. These licences are issued for one year and the revenue from the charges for them goes to local government (Arumadri, 2001). In terms of international trade, there are no taxes or inspection fees on exports. This is the current policy of the government of Uganda and is intended to promote exports and increase foreign exchange earnings. The only tax on international trade is a re-export tax, which is charged on forest products that are imported and then re-exported to other countries. At the moment, all round wood used in Uganda is produced from domestic sources (Arumadri, 2001). There are no imports of round wood, and round-wood exports were banned in the 1980s. Most forest product exports are exports of wooden handicrafts and woodcarvings. The number of people exporting these products is quite small, so the amount of exports is small and is unlikely to have a significant impact on forest resources.

## Fines and penalties

Apart from all of the above charges, there are also fines and penalties for breaking forestry laws and regulations. When a person is convicted of an offence under the provisions of the Forests Act or any other rules and regulations, they may be imprisoned for up to six months and/or fined up to USh2000 (Government of Uganda, 2000). All of the equipment and forest products associated with the offence may be seized by the court and sold, with the revenue going to the National Forestry Authority. A convicted person may also pay compensation equal to the value of any products lost through illegal acts. The problem with the fine described above is that it is not high enough to discourage unlawful practices. The effect of such low fines, especially low fines for offences committed in the forest, was that they were encouraging illegal activities, with the result that offences were on the increase. For example, if an offender knows that they may only lose 10 per cent or less of the value of their production, they would often prefer to act illegally. Such uncontrolled and, in most cases, unprofessional harvesting leads to forest degradation. In addition, harvesting more than the AAC results in overexploitation and reduces the resource base.

Having realized this, the National Forestry Authority has started to enforce higher penalties (Arumadri, 2001). For example, when an offence has been committed, all forest products are now seized and are sold by the National Forestry Authority at a public auction. This is organized on a regular basis and these auctions now account for approximately 15 per cent of monthly revenue collection. A timber-monitoring task force has been given the responsibility to collect illegal forest products from timber depots, building sites, transit vehicles, backyards and so forth and to bring this to the National Forestry Authority headquarters for auctioning.

Other charges, such as ground rent, are paid by farmers who use government forest reserve land to plant trees. Rent is charged per hectare per year and, until recently, this rent was fixed at USh1500 (Government of Uganda, 2000). Rental payments for other uses of the forest (for example radio masts and electricity power lines) are not currently included in the forest revenue system. However, with the current self-financing National Forestry Authority all of these potential sources of revenue might be exploited in the near future. Innovative revenue-raising schemes might also be examined, such as payments for carbon storage and the development of privately owned ecotourism projects.

# CAUSES OF FOREST DEGRADATION IN UGANDA

As already mentioned above in the section on the status, structure and distribution of forests in Uganda, the forest situation in the country is currently bad. Many forces of degradation have affected much of the country's forest and the forest cover is still dwindling. The causes of forest degradation are described below.

## Encroachment, land grabs and changes in land use (degazettement)

Encroachment, degazettement and grabbing of forested land/reserves has a long history in Uganda, largely driven by a combination of factors, including population growth, inequitable land and income distribution, and development policies (Uganda Wildlife Society, 2005). There is a growing trend of change of land use of protected areas to agriculture (sugar cane, palm oil, tea, tobacco and so on) or industrial expansion (Uganda Wildlife Society, 2005). The protected areas are perceived by politicians and investors as a land bank for future appropriation for investment. This trend is worrying and has already claimed the Bugala Islands for palm oil plantation, Namanve CFR for an industrial park and part of Pian Upe Wildlife Reserve for large-scale agriculture, and is likely to affect the South Busoga forests, which are among the few remaining forests on the shores of Lake Victoria.

## High population pressure

Despite the high incidence of disease, including HIV/AIDS, Uganda's population is growing fast and is over 80 per cent rural. The human population growth rate for Uganda currently is well over 3.4 per cent, while the average world population growth rate is about 1.3 per cent. Human density estimates are equally astonishing, with Uganda's national average of 102 people/km$^2$ compared to the world's average of 42 people/km$^2$. Annually, more land must be brought under cultivation to feed the increased number of people and hence there is forest habitat clearance.

## Overexploitation of forest resources

There is overexploitation of forest resources to meet the ever-increasing demands for firewood, charcoal, and construction materials such as timber and building poles, most of which is commercial driven. For example, the high demand for bamboo poles from Echuya Forest Reserve and from Bwindi and Mgahinga National Parks has led to habitat destruction. The majority of the forest loss has occurred outside of protected areas. Approximately 25 million tonnes of wood are consumed annually in Uganda, which translates to about 1.1 tonnes per capita per year. The majority of that wood is used as household firewood (65 per cent), charcoal (16 per cent), and commercial and industrial firewood (14 per cent) (Winterbottom and Eilu, 2006). The trend in loss of forest cover shows an accelerated rate of deforestation in Uganda compared to a number of other countries. The National Biomass Study Project (Forestry Department, 2003) estimates that per capita forest area will decline from 0.3 ha in 1991 to 0.1 ha in 2025, if there is no serious investment in forestry. Today, while 50 per cent of all the tropical high forest on private land is degraded, only 15 per cent in forest reserves is degraded. Given that most of the forest loss takes place on private lands, it is more than ever necessary to protect CFRs.

## Armed conflicts, civil unrest and refugees

Armed conflicts have contributed to deforestation and the abandonment of the management of protected areas. The insecurity in northern and southwestern Uganda have made it difficult for managers to be effective custodians of the protected areas in the region (Nampindo et al, 2005). In the early 1980s, many peri-urban plantation forests were cleared for security reasons. This has in turn led to greater pressures on the surrounding natural forests for fuelwood, poles and timber. For example, in northern Uganda, the LRA (Lord's resistance Army) conflict has had unequal impact on woody biomass. Forest areas adjacent to the internally displaced persons (IDP) camps have experienced significant losses. The total woody cover changes noted between 1985 and 2002 demonstrate a decrease of 8750km$^2$ (Nampindo et al, 2005). At the same time, civil unrest in neighbouring countries has resulted in influxes of refugees into Uganda. These refugees need land on which to settle, cut poles with which to build settlements, and collect fuelwood for cooking and heating.

## Weak capacity of the forest sector

In Uganda, the forest sector has been faced with several problems that explain its inability to meet the challenges of today. Low budgetary allocations undermine its ability to monitor illegal activities and to ensure compliance. Political interference has also undermined the technical authority of the staff. About 800,000m$^3$ of logs are cut each year, a rate that exceeds sustainable cutting levels by a factor of about four. In terms of enforcement, the penalties laid down for infringement of forest land are very low compared to the potential gains from illegal activities. Those in charge of enforcing the legislation are lax as they not well motivated to do so and often work together with those destroying the forests.

## Policy failures

The main source of policy failure has been the Forest Act 1964 and its related forest rules. According to Kamugisha (1993), Uganda's forest policy has been evolving since 1929. However, three weaknesses have continued to exist: failure to institutionalize sustainable community participation, neglect of forests outside the gazette system, and absence of detailed and adequately responsive guidelines on interpretation of the policy, especially from a legal point of view.

## Political influence

Important decisions regarding forests lie in the hands of top politicians rather than the National Forestry Authority. For instance, the former Ugandan President Idi Amin encouraged Ugandans to settle in the Mabira Forest Reserve and grow crops in what was termed 'double production'. Encroachers were evicted by the Forest Department in the late 1980s. Recently, there was a request by the Sugar Corporation of Uganda Limited (SCOUL) to government to use about 7100 ha of Mabira forest land for sugar cane growing.

## Market failures

Forest degradation is partly a result of decisions by agents such as private entre-preneurs, corporations, ordinary farmers and communities. Generally, the main agents of forest degradation are in the private sector. An underlying cause of deforestation is the discrepancy between the values of these private agents and those of society. Because of this, the satisfaction of the private agents' objec-tives may be in conflict with the satisfaction of society's objectives. Much of the forest degradation in Uganda is traceable to the malfunctioning of the markets. Because some forest services are not traded on the market (for example watershed protection, soil erosion control, capturing carbon, maintain-ing scenic beauty and preserving biodiversity resources) (Harrison and Goldman, 2002), individuals (especially private business owners) do not have incentives to protect forests for these services because they do not privately benefit from them.

In general, where an individual does not obtain the full value of social and non-monetary benefits provided by forests, there will be less incentive to maintain lands under forest cover. The market fails to generate the signals that would lead private operators in the direction of satisfying social objectives. In many cases, Adam Smith's 'invisible hand' fails and signals the wrong priorities to private-sector decision-makers. Frequently, for the reasons already set out, these signals lead to forest degradation. In fact Contreras-Hermosilla (2000) noted that if private agents are not compensated for the values of forests that do not have a financial, marketable dimension, they will be less interested in managing forests. If they do not have to pay for some of the costs of depleting forests, they are more likely to convert forested lands to other uses.

Other causes of forest degradation are rampant fires, especially in the case of dryland forests, greed and corruption, absence of formal employment for a large section of the populace, who therefore resort to illegal activities in the forest reserve in order to meet their livelihood demands, and invasive species.

# FOREST RESTORATION INITIATIVES IN UGANDA

Concern about the degradation of the Uganda's forests, and in particular tropical high rainforests, has grown considerably and has resulted in various initiatives to reverse this trend and to develop strategies and actions for sustainable forest management.

## Landscape approach

Although there are several initiatives which have been implemented to restore degraded forests in the country, one cannot say that they have consciously adopted a landscape approach. Common among them is that they have been site-specific. Most of these initiatives have been carried out in government-gazetted areas. Furthermore, the Forestry Nature Conservation Master Plan (March 1999) observes that the nature forest reserves established in Uganda in 1950s and 1960s were inadequate in assuring minimum landscape scale for sustaining minimum viable populations. This is even without assessing existing forest reserves for other forest functions. The concentration on government forest reserves for a long time also dictated the scale of the interventions. Although some initiatives, especially agroforestry, may have started as site-specific activities by AFRENA (Agroforestry Research Networks for Africa) in Kabale and the Vi Agroforestry Programme in Rakai, they have spread out. Furthermore, they have also broadened their focus on functions. Agroforestry in Kabale initially focused on restoring soil productivity (Kazoora, 2001), although it later took on biodiversity as well. The other factor that may have caused delay to adopting a landscape approach to forest restoration in the country is the government's putting off making a land-use plan.

## Recognition of forest functionality

In 1981, Hamilton, working with his students at Makerere University, tried to establish the environmental changes they had observed between 1966 and 1981, particularly with regard to forest. They found out that the climate was becoming more arid: crop yields per unit area declined, probably because of reduced soil fertility, the number of trees had decreased, the quantity and quality of grass fodder had declined, non-piped water supplies had become dirtier and less reliable, fuel had become scarce, the large wild animals had become rare and, lastly, the area of cultivated land had increased (Kazoora, 2001). It is not surprising that since then, and both within and outside government forest reserves, there has been a shift in focus of public and private investment to restoring some of the lost forest functions, and to flag the emerging prominence of other functions.

The government's tree-planting campaign initiated in 1992 was in response to the above problems. Several NGOs, working with communities, also started tree planting, mainly at the time to deal with the looming energy crisis. Soon the private sector followed too, particularly after it had been shown that tree growing was a paying activity. The same campaign is backed by the country's National Forestry and Tree Planting Act that was enacted in 2004 (Government of Uganda, 2004). In southwestern Uganda, where soil fertility had declined owing to high human population and not giving land fallow periods, the intro-duction of agroforestry was willingly received. Moreover, some of the tree species grown for soil fertility improvement are also a source of firewood and fodder for the popular zero-grazing in the area (Agea et al, 2007).

Around forest resources (and parks) such as Mt Elgon, Bwindi, Kibale and Semliki, projects that have been implemented since the late 1980s have taken an integrated conservation and development project (ICDP) approach. Products that communities derived from these ecosystems are being substituted outside them on people's own land. This has been developed as a mechanism to reduce pressure on forests. It is also a recognition that forests can no longer be looked at purely for their environmental values. They must also be looked at for the social and economic values in a broad context of understanding and sustaining sustainable development.

The Forests Absorbing Carbon Emissions (FACE) Project, which started in Kibale and Mt Elgon National Parks with funding support from the Norwegian Agency for Development Cooperation (NORAD) and the government of The Netherlands, has highlighted the significance of the carbon sequestration function. The Plan for Modernization of Agriculture has stressed that agroforestry and farm forestry will be given special attention in recognition of their potential to restore land degradation. In response to declining water quality, the National Environment Management Authority (NEMA) has put in place guidelines for the protection of riverbanks in compliance with National Environment Statute 1995 which reiterated the need to protect hilltops, water-sheds and riverbanks. To protect biodiversity outside the protected areas, NEMA has formulated the National Biodiversity Strategy and Action Plan (Republic of Uganda, 2001). The eviction of 40,000 Bakiga in the Kibale Forest Game Corridor in 1992 was to restore the function of the reserve as a corridor for wildlife and biodiversity between Queen Elizabeth National Park and Kibale Forest (Kazoora, 2001).

A European Union-funded project, 'Forest Restoration in East Africa, Indian Ocean Islands and Madagascar' (FOREAIM) involving three European countries (France, the UK and Norway) and three African countries (Madagascar, Kenya and Uganda), is trying to develop innovative tools and management strategies to enable restoration of degraded humid forest ecosystems by advancing under-standing of the mechanisms of forest degradation/restoration and their potential impacts on local populations, policymakers, governments and markets. In Uganda, the project is being implemented in Mabira Forest Reserve. The project is focused on the impact of forest degradation on economic, societal, biological and biophysical factors, evaluating the potential of targeted native species such as *Albizia coriaria* to increase incomes, developing rehabilitation strategies based on trees that are ecologically important and/or produce valuable wood and non-wood products, developing strategies to ensure supplies of planting stock for species whose seeds are recalcitrant, evaluating genetic robustness of the redeveloping vegetation and its capacity to withstand change, and studying vegetation dynamics in fallows and regenerating secondary forests to determine the need for enrichment planting.

## Tradeoffs among forest functions

Forest conservation competes with other land uses like agricultural production and settlement. The initiatives have generally been weak in taking a deliberate focused approach to balance the tradeoffs. As a consequence, conflict among uses has taken even political dimensions. People had to be evicted from Kibale Game Corridor in 1992, Mt Elgon Forest Reserve in the early 1990s, Mabira Forest Reserves late 1980s and Mgahinga Forest Reserve. Evictions had been carried out for a range of purposes including biodiversity conservation. However, realizing the effect of conflict on community relations, some initia-tives adopted an integrated conservation and development project (ICDP) approach. Initiatives testifying to this approach are the Mt Elgon Conservation Development Project (Kanyesigye and Muramira, 2001) and the Kibale and Semliki Development and Conservation Project (Kazoora, 2001). These projects have strived to satisfy the conservation objectives while at the same time enabling communities around the protected areas to use their land for economic development.

## Devolution of forest management

Available information suggests that there are different scenarios of decision-making in support of restoration of degraded forests in Uganda. One scenario is where the communities, through a negotiated effort, agree on the rights, respon-sibilities and benefits with the NFA or the Uganda Wildlife Authority (UWA) for some forest functions. This is usually under the collaborative forest manage-ment agreements as in the case of Mabira Collaborative Forest Management. The second scenario is where clan leaders lead in decision-making and defining rules, which are followed by the rest of the community. This is the case in Karamoja, where patches of forests have been conserved under stringent rules of the *Akiriket* (Kazoora, 2001). The *Akiriket* is the traditional 'shrine' of the

Karamojong elders where they go to hold meetings to discuss the affairs of the communities.

# CONCLUDING REMARKS

The restoration and management of Uganda's degraded forests should be based on the priorities and objectives of all concerned stakeholders including the private sector. If properly restored and managed, they can play an important role in the production of timber, wood and non-wood forest products for local and national use and international trade and, as such, can directly help reduce poverty. The environmental and socio-cultural benefits of restoring and managing degraded forests should, therefore, be fully recognized and endorsed at the national level.

One fundamental challenge in the restoration of Uganda's degraded forest landscape is to ensure the internalization of externalities. Individuals investing in forest landscape restoration must be compensated for those investments that generate social benefits, and which they would not undertake because of long-term perceived benefits. This calls for political commitment on the use of economic instruments. The government must henceforth come out boldly and makes a financial commitment from public budgetary provisions to support incentives. The command approaches cannot work to restore some functions in hotspots. However, as a necessity, a position paper has to be prepared to justify the need to restore degraded forest landscapes from environmental/ecological, economic and social perspectives.

As Uganda prepares to put in place its own adequate incentives, building on the Forest Policy of 2001 and the National Forestry and Tree Planting Act of 2004, it should take advantage of existing international incentives and financing mechanisms like the Global Environment Facility (GEF), joint implementation, debt relief for environmental programmes and the carbon sequestration fund.

A number of activities are responsible for the high level of deforestation and forest degradation in Uganda, for example sugar cane, palm oil, tea and tobacco growing. It is not surprising, for instance, that the tea and tobacco companies in Uganda have, as part of their corporate responsibility, programmes to establish woodlots and/or support local communities in afforestation programmes. Therefore, government needs to tie enforcement of forest landscape restoration to economic activities or corporate bodies that are causing enormous forest degradation and loss of tree cover in Uganda. In addition, the NFA or the District Forestry Services should be helped to license certain activities such as brick-making, lime-making and charcoal-burning, after being shown evidence of their providing their own sources of energy.

Finally, there is a need for research on the impacts of sectoral and macroeconomic policies and legislations on forest degradation in Uganda, on socio-economic evaluation of successfully rehabilitated/restored forest areas, on nursery and field trials of single and mixed tree/shrub species for degraded land planting, on harmonization of demands on land resources, notably agriculture, on forest production, especially where these are in conflict, as in the case of Mabira Forest Reserve, and on integrated and holistic approaches, including industrial and other off-farm livelihood opportunities to reduce pressure on forest resources. Sustainable agroforestry production systems must be designed in a way that is affordable by the resource-poor.

# REFERENCES

Agea, J. G., Namirembe, S., Bukenya, M., Zziwa, A and Waiswa, D. (eds) (2007). *Agroforestry In-Service Training Manual: Design of Appropriate Agroforestry Interventions in Uganda*, Fountain Publishers, Kampala

AFRENA (Agroforestry Research Networks For Africa) (1988) *Agroforestry Potentials for the Land-Use Systems in the Bimodal Highlands of Eastern Africa: Uganda*, ICRAF, Nairobi

Arumadri, J. (2001) 'The forest revenue system and government expenditure on forestry in Uganda', paper prepared for the FAO work-programme component on financing sustainable forest management, Working Paper FSFM/WP/08, FAO, Rome

Contreras-Hermosilla, A. (2000) 'The underlying causes of forest decline', CIFOR Occasional Paper No. 30, ISSN 0854-9818, CIFOR, Jakarta

Energy for Sustainable Development (ESD) (1996) 'A study of woody biomass derived energy supplies in Uganda', report for the EU Natural Forest Management and Conservation Project, Forestry Department, Ministry of Environment, Kampala

FAO (1993) 'Forest resources assessment 1990: Tropical countries', FAO Forestry Paper No. 112, Rome

FAO (2007) *State of the World's Forests*, Food and Agricultural Organization of the United Nations, Rome

Forest Department (2003) *National Biomass Study Technical Report*, Ministry of Water Lands and Environment, Kampala

Government of Uganda (2000) 'The Forest Produce Fees and Licences Order 2000', Statutory Instruments 2000 No. 16, Government of Uganda, Kampala

Government of Uganda (2004) *National Forestry and Tree Planting Act*, Ministry of Water, Lands and Environment, Kampala

Hamilton, A. C. (1984) *Deforestation in Uganda*, Oxford University Press, Nairobi

Harrison, M. and Goldman, I. (2002) *A Livelihoods Approach to Redesign of Forestry Services in Uganda*, Livelihoods Connect and DFID, London

Howard, P. C. (1991) *Nature Conservation in Uganda Tropical Forest Reserves*, IUCN, Gland, Switzerland

Kamugisha, R. J. (1993) *Management of Natural Resources and Environment in Uganda. Policy and Legislation Landmarks, 1890–1990*, RSCU/SIDA, Nairobi

Kanyesigye, J. and Muramira, E. (2001) 'Decentralization, participation and accountability: Analysing collaborative management models from Mt Elgon National Park and Mabira Forest Reserve in Uganda', paper prepared for the World Resources Institute, Centre for Basic Research, Kampala

Karani, P. (1993) *Sustainable Management of Tropical High Forests in Uganda*, Commonwealth Secretariat, London

Kazoora, C. (2001) 'Forest landscape restoration: Uganda country report', compiled by IUCN-EARO and WWF-EARPO, Sustainable Development Centre, Kampala

Langdale-Brown, I., Osmaston, H. A. and Wilson, J. G. (1964) *The Vegetation of Uganda and its Bearing on Land-Use*, Government Printers, Entebbe, Uganda

MWLE (2001) 'The Uganda Forestry Policy, 2001', Ministry of Water, Lands and Environment, Kampala

Nampindo, S., Guy, P. P. and Plumptre, A. (2005) *The Impact of Conflict in Northern Uganda on the Environment and Natural Resource Management*, Wildlife Conservation Society, New York

National Biomass Study Project (NBS) (1999) Forestry Department unpublished data

NEAP (1992) 'National Environmental Action Plan, National Environmental Management Policy Framework (draft)', Ministry of Environment, Kampala

NEMA (National Environment Management Authority) (1996) State of the Environment Report for Uganda 1996, NEMA, Kampala

NEMA (2002) State of the Environment Report for Uganda 2002, NEMA, Kampala

Republic of Uganda (2001) 'Draft Uganda National Biodiversity Strategy and Action Plan', National Environment Management Authority (NEMA), Kampala

Troup, R. S. (1922) *Report on Forestry in Uganda*, Crown Agents for the Colonies, London

Uganda Forestry Department (1999) unpublished data, National Biomass Study Project (NBS), Forestry Department, Kampala

Uganda Wildlife Society (2005) *Nature Watch Magazine for Advocacy on Environment and Wildlife Conservation*, November

Watts, P. and Otiti, T. (1992) 'Household energy activities in Uganda', in *Household Energy Developments in Southern and East Africa, Boiling Point 29*

Winterbottom, B. and Eilu, G. (2006) 'Uganda Biodiversity and Tropical Forest Assessment', EPIQ II Task Order No. 351, USAID Contract EPP-I-00-03-00013-01, International Resources Group (IRG), Washington, DC

World Bank (2008) World Development Indicators, Volume 1, World Bank Publications, Washington, DC

# 4

# Forests and Forestry in Tanzania

*Shabani A. O. Chamshama and Vincent G. Vyamana*

## INTRODUCTION

The Tanzania mainland has forest resources covering about 38.8 million hectares, 41 per cent of Tanzania's $94.5 \times 10^6$ha land area (United Republic of Tanzania, 2007a). Less than 2 per cent comprises tropical high forest or other closed forest cover, and the bulk is open woodland of the miombo type, dominated by *Brachystegia* spp. About 80,000ha is under industrial pine, cypress and teak government plantations, important for sustainable production of industrial round wood, and there are 25,000ha (United Republic of Tanzania, 2001) of *Tectona grandis* L.f., *Acacia mearnsii* De Wild, pines and eucalyptus plantations owned by three private companies – the Kilombero Valley Teak Company (KVTC), Tanganyika Wattle Company (TANWATT) and Forest Escarpment Company (FES). About $13 \times 10^6$ha (34 per cent) of the total forest area is gazetted as forest reserves, of which $1.6 \times 10^6$ha are under natural forests for water catchment protection and conservation of biodiversity. In addition, $23.8 \times 10^6$ha (61 per cent) of the forest and woodland are unreserved, and $2 \times 10^6$ha (5 per cent) are in national parks (United Republic of Tanzania, 2007a). Most unreserved forest occurs on, or adjacent to, village lands, for which village councils are legally the land and forest managers (United Republic of Tanzania, 2001). Often, however, the treating of these forests as open-access resources results in severe deforestation and degradation (Luoga et al, 2000; United Republic of Tanzania, 2001). Figure 4.1 shows key locations in Tanzania referred to in this chapter.

Forests and woodlands are important for the livelihoods of people and the economy of the country. It is estimated that forests provide 92 per cent of fuel

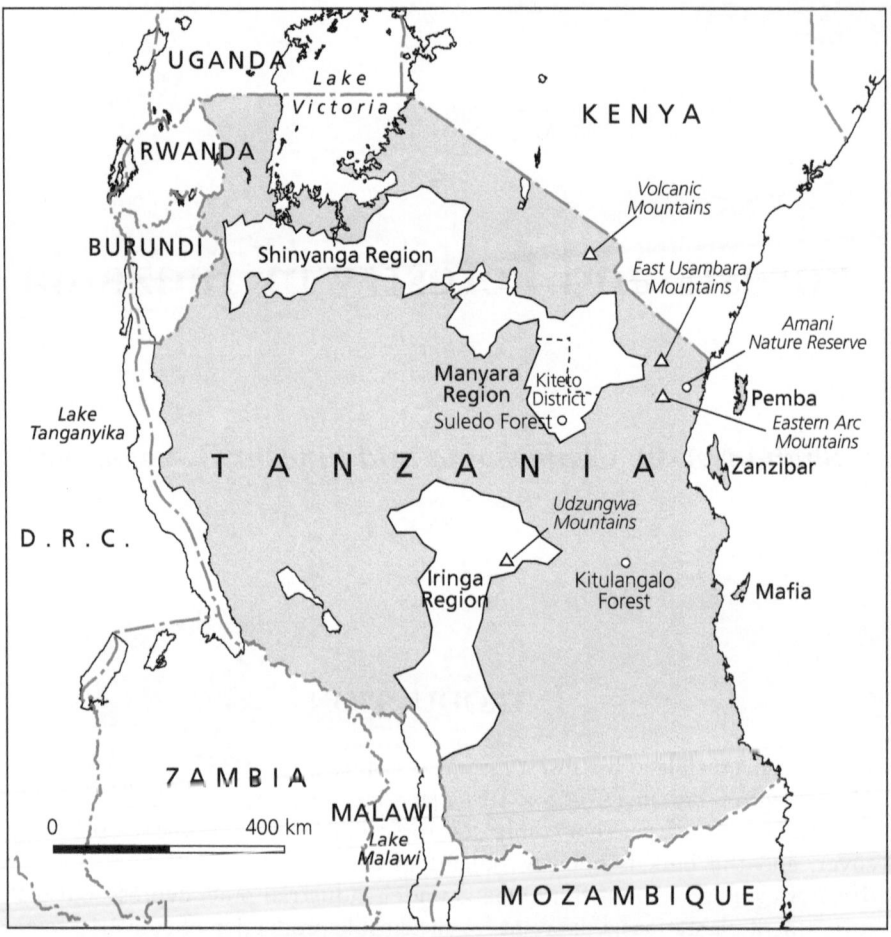

**Figure 4.1** *Map of Tanzania*

energy. They also protect watersheds that are sources of water for power generation and irrigation, and conserve soil and add nutrients to the soil for agricultural production. The forest sector contributes 2–3.4 per cent of gross domestic product per annum (United Republic of Tanzania, 2001). In addition, forests provide important subsistence goods such as fruits, mushrooms, medicinal herbs and vegetables for the majority of the rural population, and at least some cash income for up to 54 per cent of rural Tanzanians (Luoga et al, 2000). Despite the importance of forests to the national economy and the livelihoods of the people, Tanzania experiences a high rate of deforestation: from 130,000 to 500,000 hectares per year. Most deforestation occurs in miombo woodlands, particularly in general or unreserved land adjacent to villages, which constitute 57 per cent of the total forest land. Negative impacts of deforestation and forest degradation include loss of ecological services (biodiversity and watersheds), the loss of goods, such as timber, fuelwood, charcoal and non-timber forest products (Lamb et al, 2005), and the loss of livelihood sources for more than 80 per cent of rural Tanzanians (United Republic of Tanzania, 2005). Thus deforestation and forest degradation constitute a huge opportunity cost to Tanzania and her people.

From a conservationist perspective, the United Republic of Tanzania (2001) and Kilahama (2006) have identified the main challenges that Tanzanian forests and forestry face: threats from human activities in the form of encroachment, shifting cultivation, illegal logging, charcoal-making and wildfires. However, the implications of deforestation and forest degradation are more complex (Kaoneka and Solberg, 1997). With few exceptions (Scrieciu, 2007), they are indicative of technological, policy and market inefficiencies, and failure to provide alternative technologies or incentives for livelihood activities that do not jeopardize the integrity of the environment (Andreoni and Levinson, 2001). Human activities represent efforts to earn livelihoods; the resulting land degradation is an inescapable outcome of resource utilization (Luoga et al, 2000). In addition, uncoordinated or inadequate extension services may encourage land uses that are not sustainable. In fact, communities are willing and able to restore the degraded forests if they are involved and their indigenous knowledge and interests are considered (Blomley et al, 2008). Taking people's interests on board entails reconciliation of household goals of food security and improved income with the public goal of forest conservation through flexible policies that allow reasonable tradeoff between the two (Kowero et al, 2005). In addition, the need for coordination and cooperation among different sectors cannot be overemphasized, because human activities that cause deforestation and forest degradation are complex and intertwined through sector interactions – as among fisheries, wildlife and agriculture.

In this chapter, we give an account of Tanzanian forest resources and their importance, and then outline the legal status of forests in the country, highlighting current provisions for forest management as relevant context. We conclude with brief reports on two encouraging initiatives taken to restore the forest resource role of degraded dryland, the biome where the need for restoration has become most urgent.

# TANZANIAN FORESTS

From a management perspective, it is convenient to distinguish five Tanzanian forest situations, three arising in the relatively moist coastal and mountain areas and two which represent dryland ecosystems. There are two coastal situations – the strictly maritime mangrove forests and the more variable forest areas of the coastal plain. The mountain areas are the Eastern Arc mountains in the east (Lovett and Wasser, 1993; Burgess et al, 2007), the high ground along the Albertine Rift in the west (Burgess et al, 2007) and the volcanic mountains in the north (White, 1983). The dryland situations are the miombo woodlands of the semi-arid areas and the *Acacia-Commiphora* bushland and thicket of the most arid region.

# Mangrove

In contrast with the other Tanzanian vegetation of forestry interest, mangrove vegetation does not represent a distinctively African flora, but one also present on the tropical coasts of other continents, and is treated by White (1983) as 'azonal'. The constituent species are almost all tolerant of some salinity and exclusive to this specialized vegetation and absent from freshwater swamp forests and dryland forests. In Tanzania, mangrove forests occur on the sheltered shores of deltas, alongside river estuaries and in creeks where there is an abundance of fine-grained sediment (silt and clay) in the upper part of the inter-tidal zone. As well as on the mainland, mangroves are well represented on the coasts of Mafia, Pemba and Zanzibar. The most luxuriant development of mangrove forest is as closed pure stands of *Rhizophora mucronata* Lam., 25m tall.

Low tree diversity and general vascular plant diversity and the absence of noteworthy national or regional endemism make mangrove forest of limited biodiversity conservation significance. However, an important role is played in the protection of the sensitive estuarine environments where mangrove forests are most extensive. A more widely recognized role is as a source of building poles, from harvesting of *R. mucronata*, and demand for these has prompted Tanzania's Forestry and Beekeeping Division to plan a revival of mangrove management, which has been in abeyance for several decades. Past management activity involved enrichment planting and harvesting controls through licensing of annual coupes. When a new management programme commences, it will be able to benefit from extensive relevant experience in the Pacific and Indian Oceans, where floristically very similar forest has been intensively and successfully managed for many years (Holmes, 1995).

# Coastal forests

Tanzania's coastal plain extends inland from 50km to as much as 200km from the seaboard, and is mostly <400m elevation except where it adjoins the mountain blocks of the Eastern Arc and, at ca. 800m, grades into White's (1983) 'transitional rainforest' (included for present purposes with mountain area forests). These are the forests typical of the Tanzania section of White's Zanzibar–Inhambane regional mosaic. Coastal forests encompass vegetation mosaics of lowland evergreen (mainly southern) or semi-deciduous (mainly northern) forest, scrub forest and swamp forest, interspaced with areas of cultivation and fallow. In Tanzania, as in other countries within the Zanzibar–Inhambane region, the coastal forest cover surviving today is very fragmented. Tanzania's 69,200ha of this forest type is in 179 separate forest patches (Burgess and Clarke, 2000; WWF-EARPO, 2002). The most luxuriant of the coastal forest types is structurally complex rainforest with a closed, floristically mixed canopy reaching 40m with a range of smaller tree and shrub species beneath it.

Past forestry interest in indigenous species has centred on a small number of economic timber trees – particularly *Milicia excelsa* (Welw.) Berg. (widespread), *Brachylaena huillensis* O.Hoffm. (very characteristic of the north of the Zanzibar–Inhambane region in northern Tanzania) and, to a lesser extent, *Afzelia quanzensis* Welw. (present throughout the Tanzania coastal plain but also as smaller individuals in woodland formations across much of the country). Some active management has been undertaken with *M. excelsa* and *B. huillensis* (Holmes, 1995). Raising pure stands of *M. excelsa* has long been known to be problematic, because of the gallfly *Phytolyma* and, as alternatives, protocols for promoting natural regeneration and for enrichment have been developed. In Tanzania, silvicultural attention to *B. huillensis* has been limited to tending routines aiming to free seedlings and saplings from competition for soil moisture while preserving shelter from a canopy. Silvicultural attention in Kenya has been wider, with success in raising plantations of the species. In recent years, however, interest in coastal forest silviculture has been overtaken by interest in safeguarding their rich biodiversity and in activities which will achieve this whilst still enabling local communities to benefit from forest products and benefits. The high biodiversity importance of the coastal forests arises from both the high species richness, despite the small total area, and significant numbers of vascular plant species endemic to the forests of the Zanzibar–Inhambane region (White, 1983; Burgess and Clarke, 2000; WWF-EARPO, 2002).

## Mountain area forests

The non-coastal closed forests of Tanzania consist of forest formations at low altitudes (lowland forest) and higher altitudes (submontane, montane and upper montane forests). These can all be further divided according to their degree of wetness into moister and drier forests (Lovett and Wasser, 1993). Further complicating the situation are the driest and most floristically distinct of all the forests of the mountains at elevations >1500m and subject to 800–1200mm mean annual rainfall, often typified by gregarious *Hagenia abyssinica* (Bruce) J.F.Gmel. and *Juniperus procera* Engl., species which occur very sparsely, if at all, in the other forests. Floristically, the forests at the lower altitudes are referable to the forests of the Zanzibar–Inhambane region. However, with elevation increasing from 800m to 1200m, or a little higher, the species of the coastal forests decline in variety and numbers. Certain families well represented in the lowlands illustrate this change, including Annonaceae and the legumes. Conversely, Afromontane tree species (species essentially restricted to areas of mountain terrain) become more and more prominent, in what White (1983) terms 'transitional rainforest'. Above 1200–1500m the dominance among the trees of species found primarily in mountain forests has prompted the designation Afromontane forest. There is a wide variety of species, some locally present in relatively high concentrations, but no strongly dominant families among the larger trees, although the presence of the gymnosperm family Podocarpaceae is noteworthy.

In Tanzania, most Afromontane forest is in the Eastern Arc. This is the chain of ancient crystalline Precambrian basement mountain blocks uplifted more than 30 million years ago and under the direct climatic influence of the Indian Ocean, which stretches from the Taita Hills in Kenya to Tanzania's Udzungwa Mountains (Burgess et al, 2007). There are differences between forested areas in terms of distance from the seaboard, prevailing terrain gradient, rainfall regime, and incidence and extent of cloud cover. Because of these differences, the altitudinal limits of the submontane, mountain and upper montane forma-tions vary with mountain block and there is also much variation in forest structure. Where growing conditions are favourable, the tallest trees often reach 40–50m in height, forming a broken upper canopy of wide-crowned emergents some 10–15m above a more continuous middle stratum, and additional layers of smaller tree species and shrubs may also be present. The structure of the adjoining transitional forest is similar. Forest structure and complexity are reduced as growing conditions become less favourable with higher altitude, steeper terrain offering shallower soil and increasing exposure.

There has been a long history of exploiting both transitional and Afromontane forests for timber from a range of species. From the transitional forest the economic timber species include *Beilschmiedia kweo* (Mildbr.) Robyns and R.Wilczek and *Cephalosphaera usambarensis* (Warb.) Warb. Those exploited from the Afromontane forests are *Afrocarpus falcatus* (Thunb.) C.N.Page, *Entandrophragma excelsum* (Dawe and Sprague) Sprague, *Ficalhoa laurifolia* Hiern, *Ocotea usambarensis* Engl., *Olea capensis* L., *Parinari excelsa* Sabine, *Podocarpus latifolius* (Thunb.) Mirb. and *Prunus africana* (Hook.f.) Kalkm. Although forestry has long recognized the value of the mountain forests as a source of timber, there has been little progress in the development and introduc-tion of management measures for the economic species of these difficult terrains and working conditions. The species for which useful progress in developing management procedures was achieved were *Ocotea usambarensis* and *Cephalosphaera usambarensis*. At one time, the gregariousness and value of O. *usambarensis* were thought to justify developing the species for plantations. When this proved not to be feasible, attention switched to measures to ensure the development of a new crop where stands had been harvested – through a combination of advance actions to weaken potential competitors whose growth would be favoured by removal of harvested stems and post-exploitation libera-tion of regenerating root suckers (Holmes, 1995). *Cephalosphaera usambarensis* is simply introduced as enrichment to planted stands of *Maesopsis eminii* Engl., which nurses it. Forestry's main activity in the mountains was to establish planta-tions, mainly of pines and eucalypts above 1500m and of teak and *Maesopsis eminii* at lower levels. Areas that have retained a cover of natural forest, often exploited at an early date, are currently considered to have soil and catchment protection as their key role and management inputs are made only to maintain their integrity by preventing encroachment and illicit tree felling (Figure 4.2). There is also high biodiversity value in these forests. The closed forests of the Eastern Arc, in particular, are generally species-rich: Burgess and co-workers report 2000 taxa of plants (including over 800 endemic species).

**Figure 4.2** *Agriculture–forest boundary in the Uluguru Mountains, Tanzania*

*Source:* J. R. Healey

## Miombo woodland

In Tanzania, miombo woodlands account for a little more than half of the land with natural vegetation (Table 4.1). They occupy the southeast and the central plateau of the north, but a non-miombo corridor 500km long and 60–120km wide separates these areas. Typical miombo woodland extends from sea level up to 1600m, but the woodlands also occur in attenuated form between 1600m and 2000m. The wide range of mean annual rainfall (500–1200mm) is associated with a gradation from dry (mainly central Tanzania) to wet (mainly southwestern Tanzania), each with distinct floristic and functional features (White, 1983). Wet miombo occurs in areas with >1000mm of mean annual precipitation (Frost, 1996a). Miombo woodlands are generally frost-sensitive

**Table 4.1** *Areas of Tanzania referable to different forest types*[*]

| Subdivision | Mangrove | Coastal forest | Mountain area forests | | Miombo woodlands | | Bushland/ thicket |
|---|---|---|---|---|---|---|---|
| | | | Low levels | Higher levels | Moist | Dry | |
| Area (km²) | 108 | 692 | 9140 | 9560 | 139,121 | 75,961 | 165,529 |

*Note:* * The areas indicated are those contributing to the 41 per cent of Tanzania's mainland area retaining a natural vegetation cover.
*Source:* Adapted from Holmes (1995) and Nshubemuki (1998)

and associated with monthly means of minimum daily temperature $\geq 0°C$, although withstanding occasional frosts as severe as $-4°C$ at the altitudinal limit (ca. 2000m).

Miombo woodland occurs with other vegetation types in a mosaic related to drainage characteristics. Woodland is present on the well-drained soils of the middle and upper parts of ridges, while seasonally waterlogged valley bottoms are occupied by hydromorphic grassland on vertisols, often interspersed with patches of bushland or wooded grassland with *Combretum* spp. and other species. These grassy depressions, locally known as *dambos* or *mbuga*, are important sites for cultivation and livestock grazing. Scattered termite mounds within the *mbugas* bear a distinctive tree cover, including muninga (*Pterocarpus angolensis* DC.) and African blackwood (*Dalbergia melanoxylon* Guill. and Perr.). Also common in the grasslands, and sometimes in farmed woodlands, are species of *Acacia*: *A. abyssinica* Benth., *A. robusta* Burch., *A. tortilis* (Forssk.) Hayne, *A. xanthophloea* Benth., and others. In riverine communities, palm grassland with *Borassus aethiopum* Mart. and *Hyphaene* spp. (doum palm) may occur. In northern and central Tanzania, and in the Iringa region in the southern highlands, at the upper altitudinal limit, miombo woodlands give way to *Protea-Dombeya* highland grassland (Iddi and Sjöholm, 1997).

Mature miombo woodland is typically single-storeyed, with the larger trees 10–20m tall, and mostly single stemmed. The canopy is rarely closed and typically dominated by leguminous trees of the subfamily Caesalpinioideae: particularly of the genera *Brachystegia, Isoberlinia* and *Julbernardia* (White, 1983). Trees of other genera also occur, notably *Afzelia, Albizia, Burkea, Combretum, Dalbergia, Erythrophleum, Monotes, Pericopsis, Pterocarpus, Swartzia, Strychnos, Sterculia, Uapaca* and *Xeroderris* (Holmes, 1995). Miombo trees are typically deciduous, the length of leafless period varying from year to year and according to site. Normally the trees become leafless during the dry season and flushing of new shoot growth occurs in September and October, just before the rains. The ground layer varies from a dense and tall (1–2m) grass stand to a sparse but continuous cover of herbaceous plants. Most of the grass species are $C_4$-plants and perennial, sometimes tussock-forming.

Generally, dry season fire occurs regularly in the grassy miombo under-storey, although with variation in frequency and timing within the dry season. The complete absence of fire from areas of miombo woodland is rare except in those dense miombo forests which have an evergreen understorey and little grass. In any year, a proportion of the woodland estimated at around 40–60 per cent (Chidumayo, 1995; Frost, 1996b) escapes burning. Fire-return intervals at any location depend on the fuel accumulation and the potential sources of ignition. As miombo fires are fuelled largely by grass, fire intensity is influenced by grass biomass production. Fires tend to be more frequent and intense in areas where canopy cover and grazing intensity are low and mean annual rainfall high. Late dry season fires are more intense and destructive, while late wet/early dry season fires are less intense, because the fuel material is still relatively moist (Frost, 1996b). If fires are prevented, litter accumulation increases the risk of accidental intensive fires. Woody material contributes little to the main fire front, but may continue burning long afterwards, creating localized, deep and sterile ash beds (Frost, 1996b). In general, fire is more destructive to living woody plants than to grasses, the damage suffered depending on fire intensity, the size and phenological state of the plant, and the plant species. Physiologically active plants are at higher risk of damage than dormant ones.

The soils supporting miombo are typically acidic, with low cation exchange capacities, and low contents of organic matter, nitrogen, exchangeable cations and extractable phosphorus (Frost, 1996b). Within the woodlands, however, termite mounds represent nutrient-rich microsites in the generally nutrient-poor landscape, having higher levels of total nitrogen, acid-extractable phosphorus and cations than surrounding soils (Jones, 1989). These characteristics are considered to arise from the concentration and decomposition of organic matter in the mounds, and the concentration of minerals through the evaporation of soil water in the mounds' chimneys (Montgomery and Askew, 1983).

Fire effects complement termite effects on nutrient cycling, with burning accelerating carbon and nutrient flow. If fire occurs regularly, much of the grass and litter is burned before removal by termites. Regular burning releases nutrients in a single pulse, resulting in more rapid nutrient recycling. In the absence of fire, more material is available for termites to transport to their mounds. Within the mounds, this material is protected from fires, and its nutrient contents are only slowly released (Frost, 1996b).

*Brachystegia* and the other legumes characteristic of the miombo woodland canopy are not valued timber species. Further, strongly seasonal dryland conditions prevail, combined with poor soils and frequent fires. As a consequence, forestry has mostly lacked incentive for devising and introducing specialized miombo management programmes, or for converting miombo to plantations of exotic timber on a significant scale. With two miombo tree species, in particular, however, Tanzania has supplied a wood product to overseas markets. These species are *Dalbergia melanoxylon* (the wood of which commands very high prices as the material from which high-quality woodwind instruments are made) and *Pterocarpus angolensis* (a fine hardwood valued for parquet flooring, veneers and furniture). It has been in areas with *P. angolensis* present in harvestable quantities that, with the involvement of the local community, the limited miombo management so far applied has been concentrated. The approach is a simple one, with emphasis on protection from the most severe fire effects – generally by exposing the population to controlled late wet/early dry season burning. Trees other than *P. angolensis* are cleared by farmers, who then cultivate the area. If crop cultivation ceases, the increased incidence of severe (late) fires in the fallow that develops is countered with fire control measures – creation of firebreaks and prescribed early burning of vegetation between the retained *P. angolensis* trees. Within patches of high *P. angolensis* stocking, provision may be made for hoeing or discing to prevent fire completely and to minimize competition from other species (Holmes, 1995). Exceptionally, enrichment with transplanted suffrutices may be undertaken and established individuals of the species are tended. In addition to being the source of these specialized woods, miombo woodlands are of significance for apiculture based on wild bees and fuelwood supply. For apiculture, disturbance of the woodland should be minimal and the key management role is to preserve their integrity; areas designated for valuable hardwoods or fuelwood production are inappropriate. Sustainable fuelwood production operates under a management process of coupe clear-felling at intervals of around 10 years, yielding a harvest of small diameter wood, much of which is harvested coppice shoots. As with miombo hardwood production, fire control through imposition of managed early burning, to prevent injury to the crop from high intensity late fires, is necessary.

Miombo is the main natural vegetation of the vast Zambezian region and many of the miombo plants occur widely through the seven main countries within it. Biodiversity values in miombo are more commonly attached to the fauna than the flora and in Tanzania it is the Wildlife Division which plays the lead role in faunal conservation – sometimes with inputs from forestry expertise if vegetation problems arise. Nevertheless, the flora is also of interest and distinctive, with very few of the constituent species present in typical stands of any of the closed Afromontane and Zanzibar–Inhambane forest vegetation considered above. Among the tree genera, *Brachystegia* (with most of its 30 or so species restricted to the Zambezian region, making it the most characteristic) and *Uapaca* (with about a dozen species) are noteworthy. Much less attention has been given to woodland biodiversity than forest biodiversity in Tanzania, but closer analysis is likely to demonstrate additional plant biodiversity value. Burger (2001) provided some evidence that the miombo areas of central Tanzania are among the richest in plant species of the Zambezian region.

## *Acacia-Commiphora* bushland and thicket

This is the natural vegetation of the low (400–700mm) mean annual rainfall area, the arid Somalia-Masaai region, of northern Tanzania. Beyond Tanzania, this is vegetation extending northwards through Kenya into the Horn of Africa and across the Red Sea to the Arabian Peninsula. There is generally a fairly continuous layer of herbaceous plants, especially grasses, although where soils are very shallow and stony this herbaceous cover may be thin and sparse. Woody plants mostly occur singly and well separated from neighbouring individuals, although locally there may be areas of dense woody plant thicket from which herbs are almost absent. Areas of exposed rock tend to support more individuals and species of woody plants than their surroundings. Thickets are up to 5m (occasionally 7m) high, while the scattered individual trees are generally from 2m (*Acacia* spp.) to 4m (*Commiphora* spp.) tall, although occasional trees of other genera may be considerably taller – particularly baobabs (*Adansonia digitata* L.) and mng'ong'o (*Sclerocarya birrea* (A.Rich.) Hochst.), which may reach 8–10m. As the name indicates, the genera *Acacia* and *Commiphora*, each represented by numerous species, are particularly characteristic components of the woody vegetation. Within the Somalia-Masaai landscape, however, the prominence of *Acacia* spp. is greater on alluvial soils and where habitat degradation has been a result of overgrazing.

In the arid conditions of today's Somalia-Masaai region, water availability is by far the most dominant influence on tree regeneration and growth, in contrast with the miombo, where moisture availability, soil infertility and fire all have considerable significance. The soils of the *Acacia-Commiphora* bushland and thicket have not shared the Cretaceous/Tertiary era impoverishment through leaching of the miombo soils (King, 1978), and present rainfall levels are too low for dissolved nutrients to be lost from the rooting zone by downward percolation. Moisture availability governs closely the onset and cessation of the short periods of active plant growth, and when rainfall is erratic there may be more than one period of growth within the same annual seasonal cycle. Year-to-year rainfall variation is also important, with trees at risk of high drought mortality in years of exceptionally low rainfall and prolific regeneration in years of unusually high precipitation. Fires are irregular, and rarely intense, because little grass fuel accumulates and areas may escape fire for several years in succession.

Because of the challenging climate and low potential productivity, this part of the country has been largely disregarded in terms of formal forestry activity. However, some of the tree species present, such as gum arabic (*Acacia senegal* (L.) Willd.), produce marketed gums or other exudates or locally appreciated fruits (for example *Balanites* spp.). Also, the Somalia-Masaai woody plants serve fodder and fuelwood roles; in some areas there has been significant, although little-regulated, charcoal production and site degradation. As an African phytogeographical unit, the Somalia-Masaai region, despite its extent $(1.9 \times 10^6 \text{km}^2)$, is species-poor: White (1983) suggests 2500 species, half of which are found in the *Acacia-Commiphora* bushland. In Tanzania, the noteworthy biodiversity aspects of the *Acacia-Commiphora* bushland and thicket are plant species which are not known elsewhere in the country and the associated fauna, particularly arid-land birds and large mammals.

# FOREST MANAGEMENT AND MANAGEMENT PLANS

Documented forest management in Tanzania originated in the German colonial period with the introduction of a forest administration and the creation of the first forest reserves in the country over the period from 1902 to the outbreak of World War I (Schabel, 1990). However, it was not until 1957 that forestry activities were required to comply with an official national Forest Ordinance. The Forest Ordinance of 1957 formalized state dominance in the management of forest resources, covering both industrial plantations and natural forests. By the mid-1990s a global shift towards decentralized forest management was taking place, with delegation of forest management rights and responsibilities to a local level as a strategy to achieve sustainable forest management and development. In Tanzania, as elsewhere, this led to a major review of forest policy and legislation (Wily and Mbaya, 2001). A new Forest Policy (United Republic of Tanzania, 1998) took effect in 1998 and led to a new Forest Act. The Forest Act (United Republic of Tanzania, 2002) covers the creation and declaration of forest reserves owned by central or local government. In addition, the Forest Act gives the Ministry of Natural Resources and Tourism (MNRT) the power where necessary to alter the status of a National or Local Authority Forest Reserve 'to become a Village or Community Forest Reserve'. While this provision gains direct support from a land law (United Republic of Tanzania, 1999) defining how land may be removed from the general or government class of land into village land, there are no records of its use by the MNRT for any alteration of forest legal status.

## Centralized forest management

The Tanzanian Forestry and Beekeeping Division (FBD) provides overall policy guidance for the nation's forestry sector, including setting fees and royalties that have to be paid (United Republic of Tanzania, 2007b) and technical oversight and supervision. The FBD holds primary responsibility for the management of state-owned forests, including industrial forest plantations and natural forests in forest reserves (United Republic of Tanzania, 2002). However, in practice, the decentralized system of government places much responsibility for forest conservation and management with FBD district administrations. Exceptions are made for industrial forest plantations, for several major catchment forests and for forests with high biodiversity values; all these remain under the direct management of the FBD.

## Participatory forest management

The recent legislation thus makes transfers of forest resource ownership and management responsibilities to local communities feasible. Consequently, a community-based approach to securing and managing forests, generally referred to as participatory forest management (PFM), has emerged as a central element in the FBD's strategy for ensuring the sustainable management and conservation of Tanzania's forests (United Republic of Tanzania, 2001; Blomley and Ramadhani, 2006). In Tanzania, the two major approaches to the implementation of PFM are community-based forest management (CBFM) and joint forest management (JFM). The FAO (2004) noted that when conventional forest management plans developed primarily for timber production are used for PFM they impose financial and technical demands farmers cannot meet. Simpler forest management plans drawn up by communities with technical assistance from the forest sector have been adopted in Tanzania (United Republic of Tanzania, 2007c).

CBFM and JFM approaches differ in terms of forest ownership and cost/benefit flows. CBFM, where trees are owned and managed by a village government through a Village Natural Resources Committee, applies on village land or private land. In this case, the owner carries most of the costs and accrues most of the benefits related to management and utilization, with a minimal role for central government, and the district authorities have a role only in monitoring. For CBFM, the Forest Act requires a management plan to be in place before any village or private forestry status is recognized. Preparation of the plan is a village council responsibility but must involve relevant government officials, the local authority in the vicinity of the forest, and the local communities granted the rights of occupancy or lease from the private entity. Approval from the director of the FBD has to be obtained before it is operational. After approval, Village Natural Resources Committees oversee implementation. In practice, a good number of villages protect their forests only through by-laws or forest management plans (Blomley and Ramadhani, 2006).

JFM is currently a strongly favoured approach to the management of state-owned forests, with management responsibilities and returns divided between the state and the communities adjacent to the forest (Blomley et al, 2008). It takes place on 'reserved land' owned and managed by either central or local government. Villagers typically enter into agreements to share management responsibilities with the forest owner. The Forest Act requires a joint management agreement prepared by the central government, or designated district authority, to be formally made with local communities adjacent to the state forests before any JFM initiative starts.

CBFM and JFM operate in accordance with the Village and Ujamaa Villages Act of 1975, and the Local Government (District Authorities) Act of 1982, which empower the village councils to make rules in the form of by-laws, recognized in courts of law, which facilitate management of village land and village forests (Kihiyo and Kajembe, 2000). Since the Forest Act was enacted, forest area under PFM has increased considerably, but there has been no increase in forest area under private ownership (United Republic of Tanzania, 2007a). By 2006, the area under CBFM was 1,641,000 hectares in 670 villages, and the area under JFM was 1,386,000 hectares in 568 villages (Blomley and Ramadhani, 2006).

## Management plans for forests in Tanzania

In principle, Tanzania's forests are managed according to forest management plans under which the needs and interests of forest-adjacent communities are taken into account where appropriate, always consistent with overall aims of resource sustainability and poverty reduction (United Republic of Tanzania, 2001, 2002 and 2007a). The plans are based primarily on silvicultural rules and prescribe management procedures, both for natural and industrial plantation forests, setting out relevant management objectives, silvicultural measures and, where applicable, marketing arrangements for forest products (United Republic of Tanzania, 2007a). In the case of the silvicultural provisions, any pressing research needs are indicated.

Institutional responsibilities for the preparation and implementation of forest management plans are clearly defined (United Republic of Tanzania, 2002 and 2007a). Finance severely constrains action on the ground, however. In practice, as a result, recent plans tend to have been prepared and implemented with development partners' support, particularly in natural forests with global biodiversity values. For example, Amani Nature Reserve was initiated with support from the government of Finland (Amani Nature Reserve, 1998), while preparation of the strategy for conservation of Eastern Arc mountain forests has been developed with the support of the Global Environmental Facility (CMEAMF, 2008). Until now, PFM has also been heavily donor-dependent (Blomley and Ramadhani, 2006), despite the government's indication of commitment to PFM as a key strategy (United Republic of Tanzania, 2001 and 2007c). Under few PFM management plans are silvicultural rules implemented: Vyamana et al (2008), studying poverty impacts of PFM in the Eastern Arc area of Tanzania, found that from stipulated rules concerning enrichment planting, resource monitoring, patrols and boundary clearing, only patrolling and boundary clearing rules were implemented. In Iringa, a comparable situation arose from lack of interest and incentive for forest conservation among communities (Topp-Jørgensen et al, 2005). In the southern highlands of Tanzania, neglect of resource monitoring has been estimated to reduce potential forest revenue by 80 per cent (Topp-Jørgensen et al, 2005).

# FOREST LANDSCAPE RESTORATION

Forest landscape restoration is a process for re-establishing ecological integrity and enhancing human wellbeing in deforested or degraded landscapes (WWF, 2007). It covers both forest restoration and forest rehabilitation. Forest restoration is the process of restoring a forest to its original structure, composition and function, while forest rehabilitation seeks only to revive capacity for providing goods and services (FAO, 2003). Natural regeneration, assisted natural regeneration, enrichment planting, plantations, agroforestry, and various soil and water conservation techniques are all used in forest landscape restoration (Lamb et al, 2005; WWF, 2007). In Tanzania, techniques already in use include plantations, natural regeneration, agroforestry, and various soil and water conservation techniques (WWF, 2007). Plantations are too restricted in extent to provide sustainable livelihoods and environmental services for the large land areas demanding restoration (Lamb et al, 2005), while assisted natural regeneration and enrichment planting have been tried only in research activity (Mugasha, 1996; Mbwambo and Nshubemuki, 2007). Kaale (2001) concluded that natural regeneration through active involvement of local communities promoted under PFM, and supported by the new forestry legislation and programme, was by far the most promising option for restoration of the large areas of degraded land in Tanzania. A number of PFM studies have since reported improved forest regeneration, biodiversity, forest growth and wellbeing of community members (see, for example, Topp-Jørgensen et al, 2005; Blomley and Ramadhani, 2006;

Blomley et al, 2008). CBFM is regarded as the most appropriate way to achieve forest landscape restoration, and is successful because local communities are allocated clear forest-land rights, and traditional knowledge and practices are taken into account. In contrast, the contributions of JFM to people's livelihoods will remain questionable as long as benefit-sharing mechanisms are not clear and limitations on income generation potential persist (Topp-Jørgensen et al, 2005; Blomley and Ramadhani, 2006; Vyamana et al, 2008).

## Pastoral *Masaai* communities in Manyara Region

This case study is based on Sjöholm and Luono's (2003) study centred on Suledo Forest, which is miombo woodland (167,000ha) in Kiteto District, southern Manyara Region. Since pre-colonial times the area has been inhabited by the *Masaai* pastoral communities with grazing as the dominant land use. Historically, *Masaai* communities maintained a sustainable, traditional, extensive and flexible grazing system. After independence, the *Masaai* traditional grazing pattern was progressively disrupted by an increasing population and a diversification of society as a result of immigration of people from different, mostly agricultural, tribes. As a result, pressure on the woodland increased in the form of pit-sawing and agriculture expansion, leading to deforestation and forest degradation.

In 1993, the government, without consulting the local communities, expressed an intention to gazette the area as a central government forest reserve. This implied denying local access to the forest for grazing, which would disrupt local people's livelihoods. The result was a conflict, with the local people and their leaders opposing the government, such that the local community became very hostile to foresters and prevented them doing their work. A year later, however, through a donor-funded project, the government introduced community-based forest management taking account of local knowledge and practices. The scheme involved nine villages bordering the forest. Simple participatory resource assessment and land-use planning was conducted involving all of the nine villages. As a result, each village's land was divided into specific management zones, and use rules were developed for each of the zones. This constituted a simple forest management plan. In addition, common by-laws were formulated, and Village Natural Resources Committees established to enforce the by-laws and oversee the overall forest management, based on the terms of reference drawn up by all community members.

After eight years of community-based forest management at Suledo, the forest cover was improved and natural regeneration was similar to the case of *ngitili* in the Shinyanga region (Sjöholm and Luono, 2003). In addition, the forest continued to provide goods and services for the livelihoods of local people, including firewood, medicines, water, fruits and forest foods.

# The *ngitili* system of agro-pastoral communities in Shinyanga Region

*Ngitili* means enclosure and refers to an ancient and traditional range (bushland and thicket) management system of *Wasukuma* agro-pastoralists for ensuring the availability of dry season fodder reserves. Areas of standing vegetation (trees, shrubs, forbs and grasses) are fallowed and enclosed from the onset to the end of the wet season and opened for livestock grazing at the peak of the dry season (Kamwenda, 2002; Iddi, 2003). Protection of *ngitilis* is a responsibility of traditional village guards (*sungusungu*) and community assemblies known as *Dagashida*, led by the council of elders, who enact and enforce by-laws.

The *ngitili* concept as currently understood came into being through 'Hifadhi Ardhi Shinyanga (HASHI)', the Kiswahili designation of a soil conservation project in Shinyanga Region, which began in 1986. In the 1960s the Shinyanga Region was covered with dense *Acacia* scrub and miombo woodlands but by 1985 the region had been turned into semi-desert, mainly as a result of tree felling to control livestock-disease-carrying tsetse flies and clearing land for cash crop production. HASHI was initiated by the government, its objective being to devolve as much as possible the control of natural resources management activities to the local communities of the region, thereby improving their livelihoods (Iddi, 2003). Previous government efforts to control land and forest degradation in the Shinyanga Region, including compulsory destocking and conventional exotic tree planting schemes, had failed because they were top–down and bureaucratic (Monela et al, 2005). The HASHI project actively involved local communities in the whole process of landscape restoration and built on traditional knowledge and practices. The revival of *ngitili* had been proposed by the local communities and since it was revived the system has helped to restore the woodland vegetation and the associated services (Monela et al, 2005) through natural regeneration, which is cheap to establish and therefore easily adoptable (Iddi, 2003).

Monela et al (2005) found that more than 350,000 hectares of land was occupied by restored or newly established *ngitili*, of which about 50 per cent was owned by groups and the other 50 per cent by individuals. Benefits from *ngitili* were estimated at US$14 per person per month, which is much higher than the average monthly spending per person in rural Tanzania (US$8.5). The benefits gained were from products harvested from *ngitili*, including fuelwood, timber, medicinal plants, fodder, thatch-grass for roofing, and wild foods such as bushmeat, edible insects, fruit, vegetables and honey (Barrow and Mlenge, 2004; Monela et al, 2005). In addition, restored vegetation through *ngitili* has been reported to maintain water storage in reservoirs for domestic and livestock use. The free access to communal *ngitili* for poor people provides a safety net for the poorer households who have no or very little individual *ngitili* (Monela et al, 2005).

# CONCLUDING REMARKS

Tanzania is still endowed with extensive forest resources, but their sustainability is threatened by the human activities that cause deforestation and forest degradation, and occur when communities strive to earn their livelihoods. Thus it is important to recognize a forest–livelihoods linkage if sustainable forest management is to be achieved. As people's livelihoods are embedded in many sectors (for example agriculture, fisheries and forestry), halting deforestation must be approached from a multi-sector perspective. Similarly, sustainable forest management can only be realized if the forestry sector aims to optimize the dual objective of improving forest condition and conserving the environment, while at the same time improving livelihoods of the people, particularly the poor, who largely depend on forest resources for their livelihoods.

Although the science of landscape restoration may be new, efforts to restore degraded landscapes in Tanzania are not. The success stories on forest landscape restoration have always been associated with situations where communities were actively involved, and their interests, local knowledge and practices taken into account. This notion is already part of the current policies and legislation in almost all sectors, which are providing the necessary enabling environment for restoration of degraded lands. The initial positive impacts of landscape restoration provide guidance and encouragement for wider success in the future.

# REFERENCES

Amani Nature Reserve (1998) *General Management Plan*, Forestry and Beekeeping Division, Ministry of Natural Resources and Tourism, Dar es Salaam

Andreoni, J. and Levinson, A. (2001) 'The simple analytics of the environmental Kuznets curve', *Journal of Public Economics*, vol 80, pp269–286

Barrow, E. and Mlenge, W. (2004) *Ngitili for Everything: Woodland Restoration in Shinyanga, Tanzania*, United Republic of Tanzania Ministry of Natural Resources and Tourism/International Union for the Conservation of Nature and Natural Resources, Eastern Africa Regional Office, Nairobi

Blomley, T. and Ramadhani, H. (2006) 'Going to scale with Participatory Forest Management: Early lessons from Tanzania', *International Forestry Review*, vol 8, pp93–100

Blomley, T., Pfliegner, K., Isango, J. and Zahabu, E. (2008) 'Seeing the wood for the trees: An assessment of the impacts of participatory forest management on forest condition in Tanzania', *Oryx*, vol 42, pp380–391

Burger, A-M. (2001) 'A study – using WORLDMAP – of distributions of African savanna plants', *Systematics and Geography of Plants*, vol 71, pp301–312

Burgess, N. D. and Clarke, G. P. (eds) (2000) *The Coastal Forests of Eastern Africa*, International Union for the Conservation of Nature and Natural Resources, Cambridge, UK, and Gland, Switzerland

Burgess, N. D., Butynski, T. M., Cordeiro, N. J., Doggart, N. H., Fjeldså, J., Howell, K. M., Kilahama, F. B., Loader, S. P., Lovett, J. C., Mbilinyi, B., Menegon, M., Moyer, D. C., Nashanda, E., Perkin, A., Rovero, F., Stanle, W. T. and Stuart, S. N.

(2007) 'The biological importance of the Eastern Arc Mountains of Tanzania and Kenya', *Biological Conservation*, vol 134, pp209–231

Chidumayo, E. N. (1995) *Handbook of Miombo Ecology and Management*, Stockholm Environment Institute, Stockholm

CMEAMF (Conservation and Management of the Eastern Arc Mountain Forests) (2008) *Eastern Arc Mountains Conservation Strategy and Action Plan. Draft Report*, Forest and Beekeeping Division, Ministry of Natural Resources and Tourism, Dar es Salaam

FAO (2003) Second Expert Meeting on Harmonization of Forest Related Definitions for Use by Different International Processes and Instruments, FAO, Rome

FAO (2004) *Simpler Forest Management Plans for Participatory Forestry*, FAO, Rome

Frost, P. (1996a) 'Miombo: Savannah, woodland or forest?', in B. Campbell (ed) *The Miombo in Transition: Woodlands and Welfare in Africa*, Center for International Forestry Research, Bogor, Indonesia, p7

Frost, P. (1996b) 'The ecology of miombo woodlands', in B. Campbell (ed) *The Miombo in Transition: Woodlands and Welfare in Africa*, Center for International Forestry Research, Bogor, Indonesia, pp11–57

Holmes, J. (1995) *Natural Forest Hand Book for Tanzania: Forest Ecology and Management*, Faculty of Forestry, Sokoine University of Agriculture, Morogoro, Tanzania

Iddi, S. (2003) 'Community participation in forest management in the United Republic of Tanzania', in *Proceedings of the Second International Workshop on Participatory Forestry in Africa. Defining the Way Forward: Sustainable Livelihoods and Sustainable Forest Management through Participatory Forestry*, 18–22 February 2002, FAO, Rome and Ministry of Natural Resources and Tourism, Dar es Salaam, Tanzania, pp59–67

Iddi, S. and Sjöholm, H. (1997) 'Managing natural forest at the village level: Reaching the ultimate development goal', paper presented at XI World Forestry Congress, Antalya, Turkey, 13–22 October, Volume 5, Topic 26, in *Proceedings of the XI World Forestry Congress, Antalya, Turkey*

Jones, J. A. (1989) 'Environmental influences on soil chemistry in central semiarid Tanzania', *Soil Science Society of America Journal*, vol 53, pp1748–1758

Kaale, B. K. (2001) *Forest Landscape Restoration: Tanzania Country Report*, WWF/IUCN, Dar es Salaam

Kamwenda, G. J. (2002) 'Ngitili agrosilvipastoral systems in the United Republic of Tanzania', *Unasylva*, vol 53, pp46–50

Kaoneka, A. R. S. and Solberg, B. (1997) 'Analysis of deforestation and economically sustainable farming systems under pressure of population growth and income constraints at the village level in Tanzania', *Agriculture, Ecosystems and Environment*, vol 62, pp59–70

Kihiyo, V. B. M. S. and Kajembe, G. C. (2000) 'The Tanzanian Ujamaa policy: Its impact on community based forest management', in W. S. Gombya-Ssembajjwe and A. Y. Banana (eds) *Community Based Forest Resource Management in East Africa*, Makerere University, Kampala, pp34–48

Kilahama, F. B. (2006) 'Human activities and environmental conservation in Tanzania', in L. D. B. Kinabo and W. S. Abeli (eds) *Transforming Livelihoods of Smallscale Farmers: Contribution of Agricultural and Natural Resources Research. Proceedings of the First Annual PANTIL Research Workshop*, Sokoine University of Agriculture, Morogoro, Tanzania, 25–27 September, pp3–12

King, L. (1978) 'The geomorphology of central and southern Africa', *Monographiae Biologicae*, vol 31, pp3–17

Kowero, G., Nhantumbo, I. and Tchale, H. (2005) 'Reconciling household goals in southern African woodlands using weighted goal programming', *International Forestry Review*, vol 7, pp294–304

Lamb, D., Erskine, P. D. and Parrotta, J. A. (2005) 'Restoration of degraded tropical landscapes', *Science*, vol 310, pp1628–1632

Lovett, J. C. and Wasser, S. K. (eds) (1993) *Biogeography and Ecology of the Rain Forests of Eastern Africa*, Cambridge University Press, Cambridge, UK

Luoga, E. J., Witkowski, E. T. F. and Balkwill, K. (2000) 'Economics of charcoal production in miombo woodlands of eastern Tanzania: Some hidden costs associated with commercialization of the resources', *Ecological Economics*, vol 34, pp243–257

Mbwambo, L. and Nshubemuki, L. (2007) 'Principal ways to promote amelioration of degraded miombo ecosystems in semi-arid Tanzania', *Working Papers of the Finnish Forest Research Institute*, vol 50, pp88–93

Monela, G., Chamshama, S. A. O., Mwaipopo, R. and Gamassa, D. (2005) *A Study on the Social, Economic and Environmental Impacts of Forest Landscape Restoration in Shinyanga Region, Tanzania*, United Republic of Tanzania Ministry of Natural Resources and Tourism/International Union for the Conservation of Nature and Natural Resources, Eastern Africa Regional Office, Nairobi

Montgomery, R. F. and Askew, G. P. (1983) 'Soils of tropical savannas', in F. Bourlière (ed) *Ecosystems of the World, 13: Tropical Savannas*, Elsevier, Amsterdam, pp63–78

Mugasha, A. G. (1996) *Compendium of Silviculture in the Tropical Natural Forests with Special Reference to Tanzania*, Faculty of Forestry, Sokoine University of Agriculture, Morogoro, Tanzania

Nshubemuki, L. (1998) *Selection of Exotic Tree Species and Provenances for Afforestation in Tanzania*, Faculty of Forestry, University of Joensuu, Joensuu, Tanzania

Schabel, H. G. (1990) 'Tanganyika forestry under German colonial administration', *Forest and Conservation History*, vol 34, pp130–141

Scrieciu, S. S. (2007) 'Can economic causes of tropical deforestation be identified at a global level?' *Ecological Economics*, vol 62, pp603–612

Sjöholm, H. and Luono, S. (2003) 'Traditional pastoral communities securing green pastures through participatory forest management: A case study from Kiteto District, United Republic of Tanzania', in *Proceedings of the Second International Workshop on Participatory Forestry in Africa. Defining the Way Forward: Sustainable Livelihoods and Sustainable Forest Management through Participatory Forestry*, 18–22 February 2002, Arusha, Tanzania, pp131–151

Topp-Jørgensen, E., Poulsen, M. K., Lund, J. F. and Massao, J. J. (2005) 'Community-based monitoring of natural resource use and forest quality in montane forests and miombo woodlands of Tanzania', *Biodiversity and Conservation*, vol 14, pp2653–2677

United Republic of Tanzania (1998) *National Forest Policy*, Forestry and Beekeeping Division, Ministry of Natural Resources and Tourism, Dar es Salaam

United Republic of Tanzania (1999) *Tanzania Land Act (No. 4 of 1999)*, Government Printer, Dar es Salaam

United Republic of Tanzania (2001) *National Forest Programme in Tanzania 2001–2010*, Ministry of Natural Resources and Tourism, Dar es Salaam

United Republic of Tanzania (2002) *Forest Act, 2002*, Ministry of Natural Resources and Tourism, Dar es Salaam

United Republic of Tanzania (2005) *Tanzania Poverty and Human Development Report, 2005*, Research and Analysis Working Group of the Poverty Monitoring System, Dar es Salaam

United Republic of Tanzania (2007a) *Guidelines for Preparation of Management Plans for Natural Forests in Tanzania: A Synthesis for Forest Managers*, Forestry and Beekeeping Division, Ministry of Natural Resources and Tourism, Dar es Salaam

United Republic of Tanzania (2007b) *Forest Harvesting Guidelines*, Ministry of Natural Resources and Tourism, Dar es Salaam

United Republic of Tanzania (2007c) *Community-Based Forest Management Guidelines for the Establishment of Village Forest Reserves and Community Forest Reserves*, Forestry and Beekeeping Division, Ministry of Natural Resources and Tourism, Dar es Salaam

Vyamana, V. G., Chonya, A. B., Sasu, F. V., Rilagonya, F., Gwassa, F. N., Kivamba, S., Mpessa, I. and Ndowo, E. A. (2008) 'Participatory forest management in the Eastern Arc mountain area of Tanzania: Who is benefiting?', paper presented to symposium 'Who benefits from community forestry? Insights from North and South', 12th Biennial Conference of the International Association for the Study of Commons on 'Governing shared resources: Connecting local experience to global challenges', Cheltenham, UK, 14–18 July

White, F. (1983) 'The vegetation of Africa', *UNESCO Natural Resources Research*, vol 20, pp1–356

Wily, L. A. and Mbaya, S. (2001) *Land, People and Forests in Eastern and Southern Africa at the Beginning of the 21st Century. The Impact of Land Relations on the Role of Communities in Forest Future*, IUCN-EARO, Nairobi

WWF (2007) *Five Years of Implementing Forest Landscape Restoration Lessons to Date*, WWF International, Gland, Switzerland

WWF-EARPO (East Africa Regional Programme Office) (2002) *Coastal Forests of Eastern Africa: Action Plan*, WWF-EARPO, Nairobi

5

# Composition, Structure and Regeneration of Miombo Forest at Kitulangalo, Tanzania

*John A. F. Obiri, John B. Hall and John R. Healey*

## INTRODUCTION

The miombo woodlands of Africa cover over 2.7 million km$^2$ and are among the largest dry deciduous forests in the world. In Tanzania, dry forests of the miombo type cover about 390,000km$^2$ (Ministry of Natural Resources and Tourism, 1998). In the last two decades these forests have been threatened by frequent fires and uncontrolled human utilization (Campbell, 1996) leading to their current state of high degradation. Most degradation follows slash and burn activities where trees are cut down for fuelwood and charcoal, and the fallow land is used for subsistence farming. Subsequently the exposed ground is vulnerable to soil erosion and invasion by pioneer species.

This study was undertaken in the miombo of Kitulangalo Forest (Figure 5.1) in Tanzania's Morogoro Region. Kitulangalo Forest (6°34'–6°45'S, 37°53'–38°04'E) is adjacent to the main highway from Morogoro town (65km) to Dar es Salaam (140km). It covers about 2600ha, of which 2100ha is a reserve under government control and 500ha is under the control of Sokoine University of Agriculture and managed as a private forest. These two forests are hereafter distinguished as 'Government Forest Reserve' and 'University Forest' respectively. Legally, both are protected and with limited public access but, in practice, strict protection (especially harvest restriction) is only observed in the

University Forest, which has a visible presence of forest guards and well-demarcated borders with regularly maintained fire lines. An open-access public forest (400ha), hereafter termed 'Public Land Forest', is adjacent to the Government Reserve and University forests. From the 2100ha Government Forest Reserve, an area of 800ha was included in this study, along with the entire University Forest and all the Public Land Forest. These areas under three different forest management responsibilities, and associated levels of disturbance, constitute the basis for this comparative study, where we hypothesize that their species composition differs. All three forests are subject to additional disturbance which takes place along the forest margin adjoining the highway, via which bags of charcoal are conveyed for sale in urban centres, and to a lesser extent at the other edges that are adjacent to farmed land. We further hypothesize that the floristics of tree regeneration varies along a disturbance gradient from the forest edge towards the forest core, with invasive species dominating the more disturbed areas.

We have taken a broad view of the category of species considered 'invasive' to include species matching a number of different criteria. The first is species that dominate sparsely vegetated disturbed sites and are little represented in undisturbed vegetation. The second is incoming species present in the region but previously unrecorded in Kitulangalo Forest, which has previously been subjected to extensive floristic inventory, for example by Nduwamungu (1996). The third is species previously rare but increasing markedly in abundance, and the fourth is introduced species that are strictly alien to Morogoro District. Our study has also sought to identify species strongly but negatively affected by forest degradation (for example those showing low levels of regeneration in disturbed forest). Lambeck (1997 and 2002) suggests that such species are vulnerable and should be used as focal species in restoration programmes.

Through a transect-based survey, species of tree and shrub (hereafter referred to simply as 'tree') occurring in the Kitulangalo Forest were quantitatively assessed. The survey was designed to indicate:

- the composition and population size-class distribution of species within the three forests under different management responsibilities (University Forest, Government Forest Reserve and Public Land Forest); and
- any differences in patterns of occurrence of well-represented species along the disturbance gradient from the less-disturbed forest core to the highly disturbed highway edge and to the moderately disturbed opposite edge of the forest.

# METHODS

## Sampling area and transect layout

A systematic distribution of sample plots was used. Straight-line transects were established perpendicular to the highly disturbed forest edge at the Morogoro to

**Figure 5.1** *Mature miombo forest near Kitulangalo, Tanzania*

*Source:* J. R. Healey

Dar es Salaam highway. The transect lines were at 300m intervals, the first being 150m from one edge of the forest perpendicular to the highway. Of the 32 transect lines, 13, 11 and 8 were laid in the Public Land Forest, Government Forest Reserve and University Forest respectively. The transect lines varied in length from 3km to 0.5km according to the distance to the opposite (and moderately disturbed) boundary of the forest along the Sangasanga river. Along each transect line an inter-plot spacing of 150m between centres was used, so 48, 82 and 117 samples were assessed in Public Land Forest, Government Forest Reserve and University Forest respectively. As well as trees, stumps encountered in samples were identified as far as possible and their diameters recorded.

## Sample recording

Circular, concentric sample plots were the basis for the census:

- radius = 2m ($12.6m^2$ area): identification and enumeration of all trees with basal diameter <1cm and height ≤1m (the seedling and sapling size class, comprising individuals of both seed and sprout origin);
- radius = 5m ($78.5m^2$): identification and enumeration of all trees (diameter at breast height, DBH, ≥1cm) and measurement of their DBH;
- radius = 10m ($314.2m^2$) identification and enumeration of all trees (DBH ≥10cm) and measurement of their DBH;
- radius = 15m ($706.9m^2$): identification and enumeration of all trees (DBH ≥15cm), measurement of their DBH, height measurement of dominant canopy and sub-canopy trees; identification of grasses and assessment of their height and percentage ground cover.

Plant voucher specimens were collected as necessary and identified by C. Ruffo (Tanzania National Tree Seed Project).

## Data analysis

The composition of tree species recorded was collated separately for each of the three forests under different management responsibility. Grass abundance (in terms of height and cover) was compared amongst the three forests using analysis of variance (ANOVA). Relationships between the seedling/sapling density of tree species and grass abundance were investigated using linear regression separately in each forest. To investigate spatial variation in seedling/sapling density along a gradient of forest disturbance, data from the transect lines in only the University Forest were used (as this forest had the most undisturbed core, providing the maximum contrast with the disturbed edges). Distribution of stem densities across the full size range by diameter class was used to characterize the structure of species' populations separately for each of the three forests.

# RESULTS

## Species composition

In total, 104 tree species, representing 30 families, were identified. The number of species recorded was highest in the Government Forest Reserve (n = 84) and lowest in the Public Land Forest (n = 59, Figure 5.2). 43 species were recorded in all three forests. 13 species were recorded only in the Government Forest Reserve, with fewer in only the Public Land Forest (7 species) or the University Forest (9 species). The occurrence restrictions of species suggest a degree of ecological differentiation based on site environment or disturbance tolerance within the miombo forest community.

The vegetation of the Government Forest Reserve and University Forest can be divided into three strata: canopy, sub-canopy and understorey. The canopy is discontinuous, 14–20m in height, and dominated by two Caesalpiniaceae: *Brachystegia boehmii* Taub. and *Julbernardia globiflora* (Benth.) Troupin. Less abundant canopy-level species include *Diospyros kirkii* Hiern (Ebenaceae), *Pterocarpus angolensis* DC. (Fabaceae), *Sclerocarya birrea* (A.Rich.) Hochst. (Anacardiaceae), *Sterculia quinqueloba* (Garcke) K.Schum. (Sterculiaceae) and *Xeroderris stuhlmannii* (Taub.) Mendonça and Sousa (Fabaceae). The sub-canopy, at 8–15m, consists of scattered trees, with *Acacia nigrescens* Oliv. (Mimosaceae), three species of Combretaceae, *Combretum*

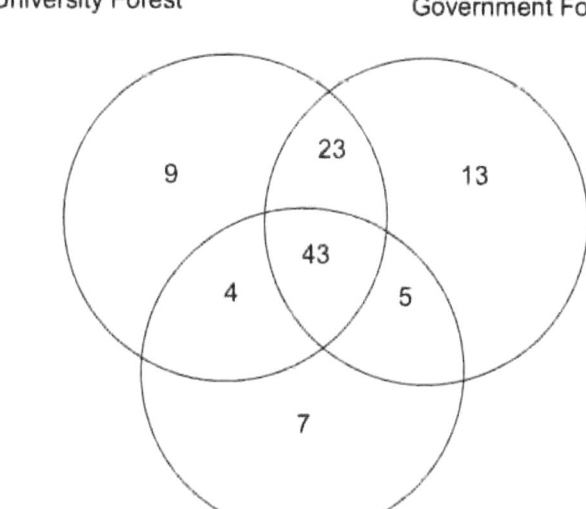

University Forest      Government Forest Reserve

9    23    13

43

4    5

7

Public Land Forest

**Figure 5.2** *Distribution of recorded tree and shrub species amongst the University Forest, Government Forest Reserve and Public Land Forest at Kitulangalo, Tanzania*

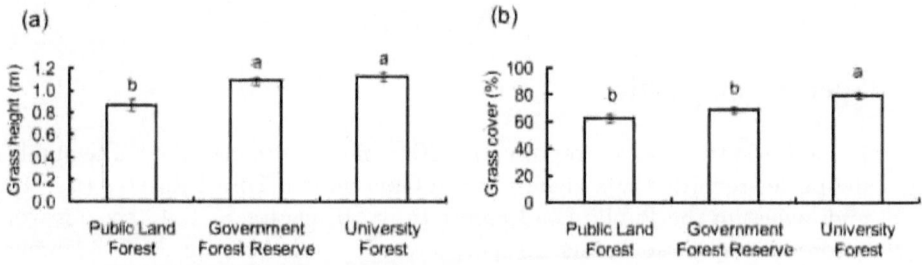

**Figure 5.3** *Grass abundance in the Public Land Forest, Government Forest Reserve and University Forest at Kitulangalo, Tanzania*

*Notes:* Mean and standard error of (a) grass height and (b) grass cover. Bars with different letters indicate forests with significantly
(P < 0.01) different grass abundance.

*adenogonium* A. Rich., *C. molle* G.Don and *C. zeyheri* Sond., and *Diospyros kirkii* being dominant. The understorey, below 8m, is composed largely of small trees and shrubs (for example *Dichrostachys cinerea* (L.) Wight and Arn. (Mimosaceae), *Diplorhynchus condylocarpon* (Muell.Arg.) Pichon (Apocynaceae), *Dombeya rotundifolia* (Hochst.) Planch. (Sterculiaceae) and *Lonchocarpus bussei* Harms, in the Fabaceae). A fourth stratum, impermanent and changing with season and fire events, is the grass and herbaceous layer. Overall, mean grass height and cover percentage were 1.06m (SE = 0.02m; n = 241) and mean grass cover was 71.5 per cent (SE = 1.4 per cent; n = 241) respectively. The dominant Poaceae were *Hyparrhenia* spp., *Panicum maximum* Jacq. and *Themeda triandra* Forssk.

Grass height and grass percentage cover differed amongst the three forests (ANOVA: *P* < 0.01, see Figure 5.3) and declined in a trend from University Forest to Government Forest Reserve to Public Land Forest (corresponding to the trend of decreasing protection). In the University Forest the number (all species) of tree seedlings/saplings per plot was strongly inversely related to percentage grass cover ($F_{[1, 75]}$ = 15.5, *P* < 0.001) but was unrelated to grass height ($F_{[1, 76]}$ = 0.53, *P* > 0.05). In the Public Land and Government Reserve forests the density of tree seedlings/saplings (of all species) varied independently of both percentage grass cover and grass height.

## Tree regeration along a disturbance gradient

Overall, the density of tree seedlings/saplings fluctuated slightly around a mean close to one per m$^2$ (Figure 5.4a), but for individual species there were different patterns along the transect lines in the University Forest. Seedlings/saplings of *Dichrostachys cinerea* had a higher density near the disturbed forest edges than in the forest core (Figure 5.4b); this species tended to colonize newly disturbed sites often unoccupied by other species. The rapid establishment of

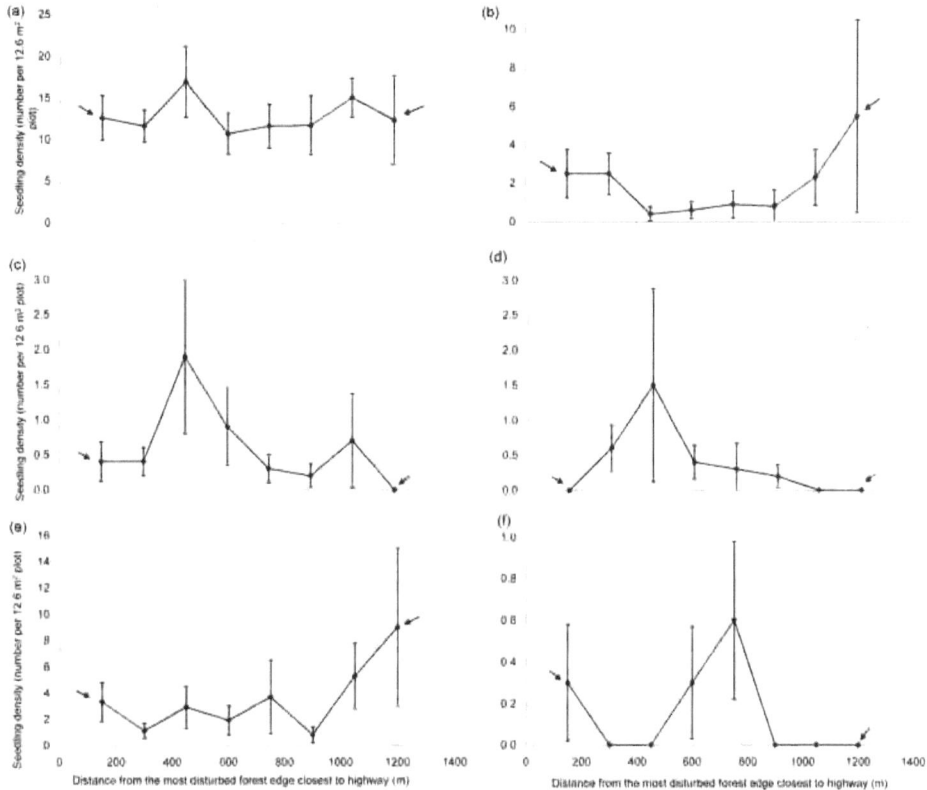

**Figure 5.4** *Density of tree seedlings and saplings across a disturbance gradient in the University Forest at Kitulangalo, Tanzania, from the most disturbed forest edge adjacent to the highway, through the less disturbed forest core to the moderately disturbed forest edge adjacent to the Sangasanga River*

*Notes:* (a) all species, (b) *Dichrostachys cinerea*, (c) *Brachystegia boehmii*, (d) *Dalbergia melanoxylon*, (e) *Julbernardia globiflora* and (f) *Acacia nigrescens*. Arrows indicate the positions of forest edges and standard error bars for means are shown.

this species in early successional sites may inhibit colonization by other species. A second pattern was displayed by *Brachystegia boehmii* and *Dalbergia melanoxylon* Guill. and Perr. (Fabaceae), species having higher densities of seedlings/saplings in the forest core (300–600m from the highway) and lower levels near the forest edges (Figure 5.4c and d). The limited regeneration of these two species may be reduced by forest disturbance. For a further species, *Julbernardia globiflora*, the high density of seedlings/saplings showed no clear trend along the transect lines (Figure 5.4e), suggesting an effectively stable and continually regenerating population across a wide range of disturbance conditions, and indicating the potential importance of this species in the resilience of the forest ecosystem as a whole. Finally, *Acacia nigrescens* had a low and patchy seedling/sapling density, also with no clear trend across the forest (Figure 5.4f).

## Tree population structure and species dynamics

Nine species were represented sufficiently to merit examination of their population size-class distributions. With *Julbernardia globiflora*, *Dichrostachys cinerea*, *Dombeya rotundifolia* and *Acacia nigrescens*, stem densities progressively declined with increasing tree diameter in all three forests (Figure 5.5a–d), indicating stable regeneration under the range of current conditions. Tolerance of disturbance by these species is indicated by their maintenance of this population structure in the Public Land Forest. For a fifth species, *Pterocarpus angolensis*, there was a very sharp reduction in densities of individuals ≥1cm DBH in the Public Land Forest and ≥10cm DBH in the University Forest (Figure 5.5e), and in neither of these forests were any individuals recorded ≥20cm DBH (despite this species typically reaching diameters in excess of 50cm). In two species, *Acacia goetzei* Harms (Mimosaceae) and *Annona senegalensis* Pers. (Annonaceae), their overall densities of seedlings/saplings (<1cm DBH) were similar to the densities of individuals in the 1–10cm and 10.1–20cm DBH classes, though not reaching a level of 100/ha in any size class (Figure 5.5f and g). However, it was notable that the abundance of *Acacia goetzei* was greatest in the Public Land Forest whereas that of *Annona senegalensis* was least in that highly disturbed forest. *Acacia polyacantha* Willd. (Mimosaceae) was noteworthy for the large differences in its population size-class distributions among the three forests (Figure 5.5h): in the Public Land Forest the trend approximates to that of *Acacia nigrescens*, but in the University Forest no individuals <10cm DBH were recorded (as was the case for *Sterculia africana*) and in the Government Forest Reserve there was a deficit of individuals >10cm DBH (as seen in *Acacia goetzei*). *Sterculia africana* completely lacked seedling/sapling (<1cm DBH) individuals but between the Government Reserve and University forests it was represented with an even density in all the size classes 1–50cm DBH (Figure 5.5i); it occurred only as very rare larger trees in the Public Land Forest.

# DISCUSSION

The forests under the three systems of management responsibility differed in the number and composition of species recorded, which could reflect the different selective exploitation pressures that they face. The low number of species in the open-access area of the Public Land Forest can be attributed to the high level of human disturbance emanating from the adjacent communities, a view consistent with those of Hanna et al (1995) and Mwase et al (2007), who concluded that common property regimes encourage exploitation and natural resource destruction. One suggestion for countering this effect is to increasingly shift the land tenure system from open access and government control to a more management-defined leasehold status, or one decentralized to communities with promotion of longer-term sustainable management (Mwase et al, 2007). Participatory forest management (either community-based or jointly

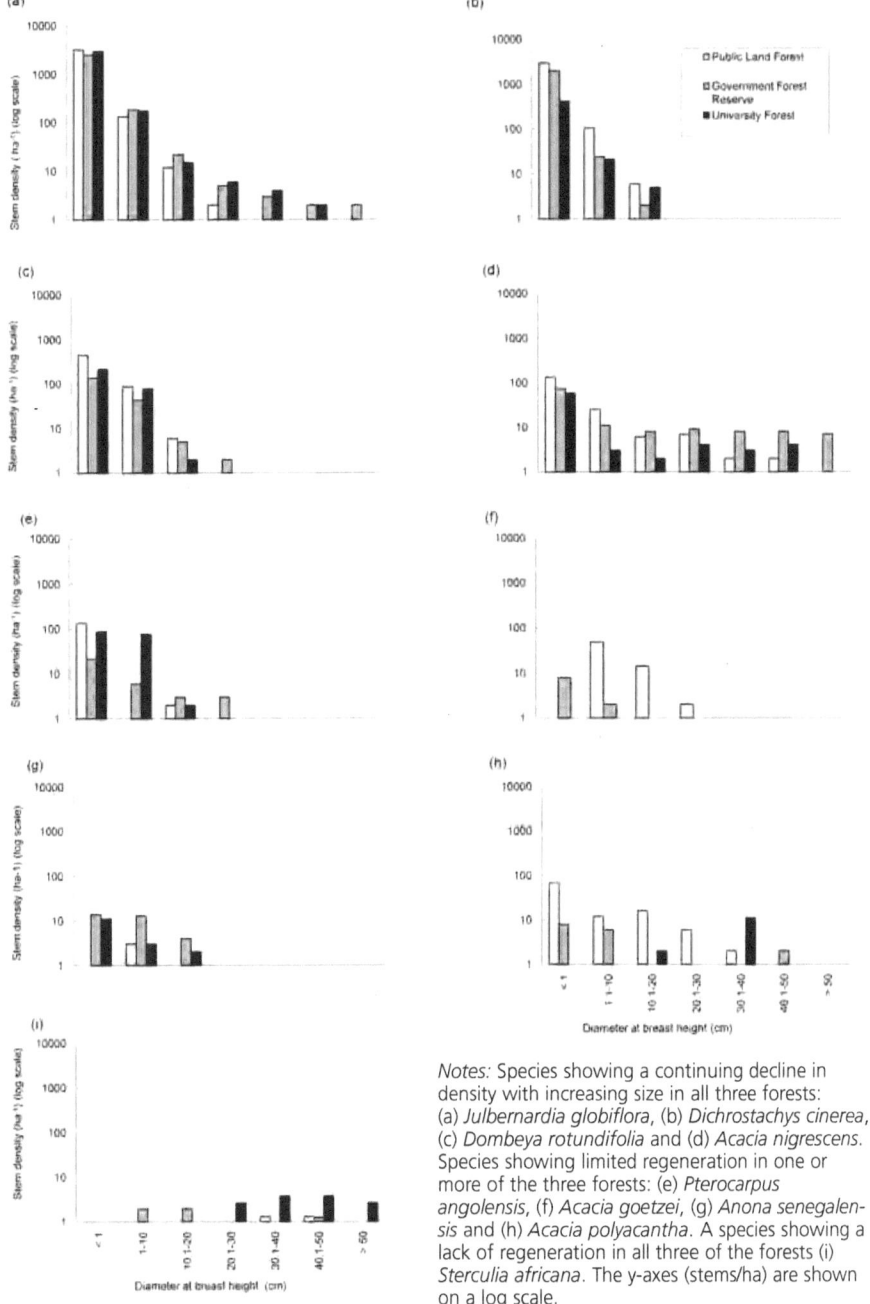

Notes: Species showing a continuing decline in density with increasing size in all three forests: (a) *Julbernardia globiflora*, (b) *Dichrostachys cinerea*, (c) *Dombeya rotundifolia* and (d) *Acacia nigrescens*. Species showing limited regeneration in one or more of the three forests: (e) *Pterocarpus angolensis*, (f) *Acacia goetzei*, (g) *Anona senegalensis* and (h) *Acacia polyacantha*. A species showing a lack of regeneration in all three of the forests (i) *Sterculia africana*. The y-axes (stems/ha) are shown on a log scale.

**Figure 5.5** *Population diameter size-class distributions of tree and shrub species in the University Forest, Government Forest Reserve and Public Land Forest at Kitulangalo, Tanzania*

between communities and government) is now promoted under Tanzanian forest policy (Blomley and Ramadhani, 2006). A possible mechanism for its implementation is advocated by Lund and Treue (2008), designating forests as village land forest reserves (VLFR) put under village committee management. Decentralization through VLFR increases the efficiency and control of forest resource use and allows local taxation of forest products. There are precedents such as the reduction of forest exploitation and degradation achieved through taxation incentives encouraging more efficient use of wood for tobacco curing by local users (Sauer and Abdallah, 2007). In our study area, the main contributor to degradation is unregulated and illegal charcoal production. For a nearby area facing a comparable problem, Lund and Treue (2008) noted positive consequences of the decentralization of charcoal production when the trade was legalized and regulated.

If human disturbance pressure can be reduced, the capacity for recovery of the populations of the dominant tree component of the forest ecosystem will be a key factor in its resilience, affecting both its value for ecosystem services (for example habitat for biodiversity) and potential to maintain a sustainable yield of forest products. Luoga et al (2004) found that 90 per cent of the 39 tree species harvested in the Public Land Forest had the capacity to recover by coppice resprouting, which occurred in 74 per cent of the cut stumps overall. The present study also indicates a high density of tree seedlings/saplings with a mean value close to $1/m^2$ across a large forest disturbance gradient; a high proportion of these may also be sprouts from below-ground organs (Box 5.1). Nonetheless, the negative correlation found between tree seedling/sapling density and grass cover in the least disturbed University Forest indicates the importance of competition with grasses as a limiting factor in tree regeneration, which appears to be a fundamental ecological characteristic of this dry miombo forest.

The response of the populations of the well-represented tree species to disturbance in the miombo forest at Kitulangalo was assessed to compare their resilience and to predict potential changes in tree community composition under continuing high disturbance levels. Whilst it is a useful source of evidence, population size-class distribution may not accurately predict capacity to recover from disturbance in dry tropical forest tree species (McLaren et al, 2005); therefore it is valuable that the present study could supplement this with information about the variation in populations with forest disturbance level. On the basis of both its higher seedling/sapling densities at the most disturbed forest edges and the dominance of its population in the most disturbed Public Land Forest by small-sized trees, *Dichrostachys cinerea* is considered to be a pioneer species, with a high potential to increase its population size following disturbance. A similar population structure in the disturbed Public Land Forest of a further four species (*Julbernardia globiflora, Dombeya rotundifolia, Acacia nigrescens* and *Acacia polyacantha*) also provided evidence of their resilience following disturbance and potential pioneer status. Whilst large-sized individuals of *Acacia polyacantha* were found in the least disturbed University Forest, there was no evidence that it was regenerating there. Luoga

---

**BOX 5.1 SOIL SEED BANK UNDER TWO INVASIVE TREE SPECIES IN MIOMBO FOREST**

In order to test the potential sources of tree species regeneration under the canopy cover of invasive tree species, a separate study of the soil seed bank was carried out in Kitulangalo forest. Soil was collected down to 12cm depth from beneath the crowns of trees of Acacia polyacantha and A. nigrescens in forest patches which they had each heavily invaded (continuous canopy) and lightly invaded (scattered trees), and in adjacent areas from which they were absent. The soil was spread in trays in a greenhouse and 1320 germinants were observed. As a result, viable seed of 19 species was found to be present, but only four of these were tree species (in declining abundance, Combretum molle, Albizia versicolor Oliv., Acacia polyacantha and Pseudolachnostylis maprouneifolia Pax); together they accounted for only per cent of the germinated seeds. By far the most abundant woody species was the shrub Phyllanthus reticulatus Poir. Solanum incanum L. was also present and the remaining 13 species were herbs and graminoids. No germinants of A. nigrescens were found. The soil seed bank density under A. polyacantha ($77/m^2$) was significantly greater than under A. nigrescens ($9/m^2$) or under uninvaded forest ($29/m^2$); however, there were no notable differences in its species composition. In contrast, there was no significant difference in soil seed bank density (or species richness) under A. nigrescens trees and under adjacent uninvaded forest. There was also no marked difference in the density or species richness of the seed bank between soil from under lightly and heavily invaded stands of either tree species.

The canopy of these two invasive trees may have directly affected the soil seed bank, through the rate of seed input and effect of shade on germination and, indirectly, through reducing grass density, the resulting fuel load, and thus the frequency and intensity of fires. All of these factors may have contributed to the high soil seed bank density under crowns of A. polyacantha but do not explain the much lower density under A. nigrescens. However, the over-riding consideration is the very low abundance of tree seed observed in the soil seed bank of this ecosystem. While this may be insufficient as a source of forest recovery after disturbance, there is considerable evidence that resprouting from below-ground stems and roots may instead be the principal form of regeneration for many miombo tree species. After disturbance so intensive or frequent that it kills existing trees/seedlings below ground, given the paucity of the soil seed bank, tree regeneration may be dependent on seed rain and thus be much slower; in this case dispersal distance (partly determined by the patch area of forest that is disturbed) may be a crucial limiting factor.

---

et al (2004) found that disturbance, particularly to the root systems, is necessary for successful regeneration of this species. However, the other four species in this proposed pioneer group did show a good capacity to regenerate in the University Forest. This may indicate that they have a broader regeneration niche (Grubb, 1977) than a strict dependence on highly disturbed conditions, or that here even in the well-protected University Forest, disturbance (for instance due to fire and browsing) is sufficient to sustain a high density of regeneration of strongly pioneer species (Box 5.2).

> **BOX 5.2** EXPERIMENTAL TESTING OF THE EFFECTS OF
> FIRE AND WEEDING DISTURBANCE ON TREE SPECIES
> REGENERATION IN MIOMBO FOREST
>
> In order to test the environmental factors influencing the rate of tree regeneration, a separate split-plot experiment was undertaken in Kitulangalo forest. It showed that neither the stem density nor the species richness of seedlings/saplings differed significantly between 45 plots that were placed under tree canopies and 45 paired control plots placed (within 10m distance) in the open. However, after one year, burning treatment significantly increased seedling/sapling density (by 81 per cent, $P = 0.001$) and species richness (by 68 per cent, $P = 0.023$). Manual weeding of grass had an even greater positive impact, increasing density by 138 per cent ($P = 0.0001$) and species richness by 93 per cent ($P = 0.004$).
>
> Under the crowns of trees of two pioneer *Acacia* species the density of their own seedlings/saplings was positively correlated with tree size for *A. polyacantha* ($r^2 = 0.917$, $P = 0.001$) and for *A. nigrescens* ($r^2 = 0.434$, $P = 0.008$). However, there was no such relationship for the generalist species *Julbernardia globiflora* ($r^2 = 0.04$, $P = 0.53$).
>
> Out of 25 tree species that are important components of the miombo forest, 15 increased and 10 showed a decline or no change in seedling/sapling density under undisturbed conditions over one year of monitoring, 20 increased after burning alone, 21 under weeding alone, and 22 under the weeding and burning treatment. This experiment provides evidence that, up to a threshold of intensity and frequency, disturbance which reduces the dominance of the ground layer by grasses is beneficial to the regeneration of miombo forest.
>
> 9 of the 11 species whose results are reported in the main study were included in this experiment. Of these, 6 increased in seedling/sapling density over the year in all four treatments, whereas 2 species (*Acacia nigrescens* and *A. goetzei*) showed a decline or no change (respectively) in the undisturbed treatment while they increased markedly in all three disturbance treatments. *Dichrostachys cinerea* had the greatest increase in seedling/sapling density of all 25 species in each of the four treatments. This provides support for the conclusions about the ecology of these 3 species (disturbance-benefiting and potentially invasive) drawn from the main study.
>
> *Julbernardia globiflora* and *Combretum molle* (all widespread in their distribution) had an increase in their seedling/sapling stem density irrespective of disturbance and have been identified as resilient 'framework' species. While *Dichrostachys cinerea* also has the resilience to potentially play this role, further checks should be made of its invasive risk before it is promoted for this purpose. On the basis of the intolerance of its regeneration to disturbance, *Spirostachys africana* Sond. is considered vulnerable and may require active management to maintain its population.

The species with the highest relative abundance in the most disturbed Public Land Forest was *Dichrostachys cinerea*. Its capacity to establish so freely as a pioneer in severely degraded environments suggests that it has an attractive potential for playing a positive role in forest restoration as a 'nurse' or 'framework' species. We observed that the typical main canopy species of miombo are able to grow through the open *D. cinerea* crowns and overtop them; therefore it does not appear to have a capacity to competitively exclude regeneration of tree species below its low shrub crown. Nonetheless, until further research has been carried out to ascertain whether *D. cinerea* carries an unacceptable risk of becoming invasive, it may be safer to promote the use of *Dombeya rotundifolia* and *Julbernardia globiflora* as nurse species.

Other tree species showed a lower capacity for population resilience in disturbed forest areas: *Brachystegia boehmii* and *Dalbergia melanoxylon* had relatively low seedling/sapling densities near the forest edges and the highly valued timber species *Pterocarpus angolensis* showed a much lower abundance of trees >1cm DBH in the Public Land Forest than in the other two forests. However, the high density of seedlings/saplings of *P. angolensis* in this disturbed forest did indicate a potential for population recovery should sufficient protection be implemented to enable these stems to grow and survive beyond that size.

The miombo forests in Tanzania are suffering heavy exploitation and degradation, reflected in negative effects on species composition and ecosystem services. In much of the miombo of Morogoro Region, pioneer tree and shrub species are increasing in abundance at the expense of the large canopy trees typical of the undisturbed community. While reversing this trend will require intervention to reduce the current rate of disturbance, this action is only likely to be successful if local stakeholders are empowered with more secure tenure over forest resources as a component of more sustainable livelihoods. Improved awareness would need to be based on enhanced monitoring of forests to reveal which important components of their biodiversity have been declining as a result of current exploitation and disturbance. A desirable option would be for the open-access Public Land Forest areas and at least part of the Government Forest Reserve to come under participatory forest management. To enable this, changing their status to VLFR managed through committees of village representatives involved in monitoring and familiar with the conservation status of the important species is an option worthy of consideration.

# REFERENCES

Blomley, T. and Ramadhani, H. (2006) 'Going to scale with Participatory Forest Management: Early lessons from Tanzania', *International Forestry Review*, vol 8, pp93–100

Campbell, B. M. (1996) *The Miombo in Transition: Woodlands and Welfare in Africa*, Center for International Forestry Research, Bogor, Indonesia

Grubb, P. J. (1977) 'Maintenance of species richness in plant communities – Importance of regeneration niche', *Biological Reviews of the Cambridge Philosophical Society*, vol 52, pp107–145

Hanna, S., Folke, C. and Maler, K. G. (1995) 'Property rights and environmental resources', in S. Hanna and M. Munasinghe (eds) *Property Rights and the Environment: Social and Ecological Issues*, The World Bank, Washington, DC, pp15–29

Lambeck, R. J. (1997) 'Focal species: A multi-species umbrella for nature conservation', *Conservation Biology*, vol 11, pp849–856

Lambeck, R. J. (2002) 'Focal species and restoration ecology: Response to Lindemayer et al', *Conservation Biology*, vol 16, pp549–551

Lund, J. F. and Treue, T. (2008) 'Are we getting there? Evidence of decentralized forest management from the Tanzanian miombo woodlands', *World Development*, vol 36, pp2780–2800

Luoga, E. J., Witkowski, T. F. and Balkwill, K. (2004) 'Regeneration by coppicing (resprouting) of miombo (African savanna) trees in relation to land use', *Forest Ecology and Management*, vol 189, pp23–35

McLaren, K. P., McDonald, M. A., Hall, J. B. and Healey, J. R. (2005) 'Predicting species response to disturbance from size class distributions of adults and saplings in a Jamaican tropical dry forest', *Plant Ecology*, vol 181, pp69–84

Ministry of Natural Resources and Tourism (1998) *Natural Forest Policy*, The United Republic of Tanzania, Government Printer, Dar es Salaam

Mwase, W. F., Bjornstand, A., Bokosi, K. M., Kwapata, M. B. and Stedje, B. (2007) 'The role of land tenure in conservation of tree and shrub species diversity in miombo woodlands of southern Malawi', *New Forests*, vol 33, pp297–307

Nduwamungu, J., (1996) 'Tree and shrub diversity in miombo woodlands: A case study of SUA Kitulanghalo Forest Reserve, Morogoro, Tanzania', MSc dissertation, Faculty of Forestry, Sokoine University of Agriculture, Morogoro, Tanzania

Sauer, J. and Abdallah, J. M. (2007) 'Forest diversity, tobacco production and resource management in Tanzania', *Forest Policy and Economics*, vol 9, pp421–439

6

# Church Forests – Relics of Dry Afromontane Forests of Northern Ethiopia: Opportunities and Challenges for Conservation and Restoration

*Alemayehu Wassie, Frans Bongers,
Frank J. Sterck and Demel Teketay*

## Introduction

Forests all around the world have been fragmented into small patches, and forest structure and species composition have been influenced by this fragmentation and habitat loss. Reduction of habitat has been described as one of the main causes of diminishing biological diversity in the tropics (Hill and Curran, 2001; Ross et al, 2002; Santos et al, 2007). Dry Afromontane forests in the northern Ethiopian highlands represent a particular case in this respect. Extensive deforestation in the Ethiopian highlands has led to small and isolated forest patches around churches and in poorly accessible mountain areas and these patches are located within a landscape matrix of intensively used agricultural land. The thousands of church forests in the dry Ethiopian mountains can be considered a special case of sacred groves, traditionally managed small forest patches with considerable potential for conservation (Bhagwat and Rutte, 2006). While the highly fragmented Ethiopian highland landscape is centuries

**Figure 6.1** *Church forests in South Gondar*

*Note:* Forests enveloping (a) Debresena Mariam, (b) Zagua Rafael and (c) Korata Wolete Petros.
*Source:* Alemayehu Wassie

old (McCann, 1997; Boerma, 2006), in some areas fragmentation is still increasing (Bekele, 2003). Zeleke and Hurni (2001), for instance, used remote sensing combined with field verification to show that in Gojam at 1800–2800m altitude the Afromontane forest cover reduced from 27 per cent in 1957 via 2 per cent in 1982 to 0.3 per cent in 1995. Most conversion was to agricultural land. Many remaining church forest patches are also currently declining in area and have been degrading over the last decades (Bingelli et al, 2003; Wassie et al, 2005a and b; Aerts et al, 2006b). Understanding the factors that influence patterns of species and structural composition in fragmented systems may be critical to conserving the remaining forests (Turner, 1996; Ewers and Didham, 2006).

The forests around the Ethiopian Orthodox Tewahido Churches (EOTC) are visible from a great distance and are usually built on small hills overlooking the surrounding villages (Figure 6.1). The local people refer to these churches

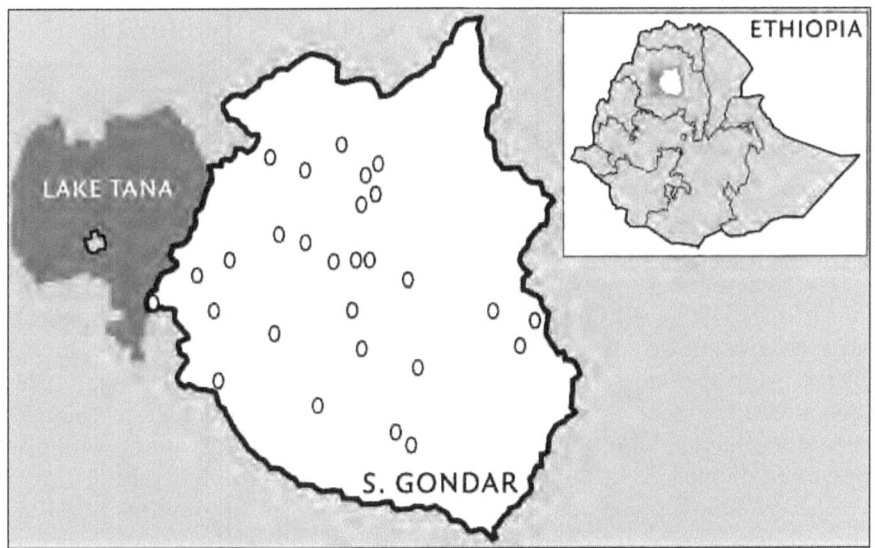

**Figure 6.2** *Map of studied church forests in the South Gondar Administrative Zone, Ethiopia*

*Note:* The locations of the church forests studied are indicated (n=28).

with the surrounding forest patch as *debr* or *geddam*, a religiously holy place as well as a respected and powerful social institution. The EOTC has over 38 million followers, 500,000 clergy and 35,000 churches in Ethiopia. In addition to its religious activities, it has also a long history of conservation of its forests, which usually surround the churches. Although the main purpose of churches is as places for worship, burials and meditating religious festivals, they also provide valuable, often unique, and secured habitats for plants and animals, and green spaces for people. Church compounds in fact serve as in situ conservation sites for biological resources, mainly indigenous trees and shrubs. These, in turn, give prestige to the religious sites.

Over recent centuries the traditional conservation task of forest resources by the EOTC was negatively influenced by several prevalent political and social events (Wassie et al, 2005a and b). For example, large parts of forests got lost after major regime changes – and periods of low social control – in the early 1990s (Bekele, 2003). These events also have considerably reduced the number and extent of church forests, and probably resulted in drastic changes of the species composition and species richness. While most people seem to respect the forests and their protection during the more stable political periods, the ever-increasing demand for agricultural lands still results in large-scale transformations into agricultural lands. Despite the actual and potential significance of the church forests, only a few studies have analysed their biophysical and social dimensions (Bingelli et al, 2003; Aerts et al, 2006b; Wassie et al, 2005a and b; Wassie, 2007).

# VEGETATION STATUS OF CHURCH FORESTS

Recently, 28 church forests were inventoried in South Gondar, Northern Ethiopia (Wassie, 2007). These forests (Figure 6.2) were located between 1816m and 3111m asl and ranged in area from 1.6ha to 100ha. A total of 160 different indigenous woody species, representing 69 plant families, were recorded in the church forests studied. Surprisingly, only 8 exotic species were found, which occurred in 5 of the forests.

The total number of woody species in each of the 28 church forests ranged from 15 to 78 and the number of families ranged from 12 to 44 (Table 6.1). *Juniperus procera* L., *Olea europaea* L. and *Maytenus arbutifolia* (A. Rich.) Wilczek were the most important species across all the forests, while the families with the highest number of species were Fabaceae (17), Euphorbiaceae (10), Moraceae (10) and Rubiaceae (7). At higher altitude, forests were less diverse and dominance by a single to few species was relatively strong.

The minimum plant density, calculated from size of individuals >1cm diameter, was 619/ha and the maximum was 2421/ha, while mean number of seedlings ranged between 0 and 5263/ha. The minimum basal area, calculated from woody plants with DBH ≥5cm, was 4.8m²/ha and the maximum 111.5m²/ha (Table 6.1). All forests had a large number of dead stumps (cut trees bigger than 5cm diameter), on average 81 (range 12–149) per hectare. This clearly indicates that people are harvesting trees regularly.

Similarities among the studied forests generally was low (Wassie, 2007; Wassie et al, 2010): only 10 pairs among the 378 pairs of forests compared scored higher than 0.5 $S_j$ (Chao-Jaccard abundance-based similarity, see Chao et al, 2005), where $S_j$ ranges from 0 (no similarity) to 1 (complete similarity). The numbers of common species and families increased when the forests were stratified by altitude. Maximum similarity in species composition was found with minimum altitude difference between forests, and geographical distance had only a weak effect (Wassie, 2007). This is in contrast to other studies reporting that geographical proximity was a key factor for similarity in species composition (Santos et al, 2007). The generally low similarity in species compo-

**Table 6.1** *Characteristics of the 28 church forests in South Gondar, Ethiopia*

|  | Mean | SD | Minimum | Maximum |
|---|---|---|---|---|
| Year of establishment | 1448.7 | 301.1 | 360 | 1984 |
| Altitude | 2388.9 | 351.0 | 1616 | 3111 |
| Forest area (ha) | 18.5 | 20.9 | 3.0 | 100 |
| Species number | 37.9 | 14.8 | 15 | 78 |
| Family number | 26.4 | 7.0 | 12 | 44 |
| Basal area (m²/ha) | 43.0 | 25.3 | 4.8 | 115.5 |
| Density ≥5cm DBH | 579.5 | 291.9 | 267 | 1553 |
| Density ≥1cm DBH | 1253.6 | 463.4 | 619 | 2412 |
| Number of dead stumps per ha | 79.9 | 43.6 | 12.5 | 149 |

sition among church forests suggests that church forests have their own relatively unique species composition. This qualifies church forests for programmes aiming at both in situ and ex situ conservation of woody plant species.

Unpublished records of the South Gondar zonal Department of Agriculture (SGAZDA, unpublished) claim there are 125 woody species in the study area, but in 28 church forests covering only 500ha a total of 168 species were found (Wassie, 2007; Wassie et al, 2010). Both Chao's and Jackknife's Indices estimated higher species richness for all forests compared to the observed number of species, suggesting that a considerable number of species for these forests might have been missed (Wassie, 2007; Wassie et al, 2010). The fact that only 28 churches were surveyed out of 1404 available in the study area strongly suggests that it is most likely that more species can be found if more church forests are assessed. Obviously, the species diversity of the area has not been exhaustively identified and the potential of these church forests for species conservation may also have been overlooked.

The overall species richness of these remnant forests was higher than that of Wof-Washa Forest (51 woody species) (Teketay and Bekele, 1995), which is the biggest continuous forest in the central highlands of Ethiopia. This clearly reveals that the church forests are relatively rich in woody species composition, and demonstrates the vital role that they can play in the restoration of Ethiopia's forest ecosystems. However, most species, including *Ekebergia capensis* Sparrm., *Juniperus procera* L., *Podocarpus falcatus* (Thunb.) Mirb., and *Schefflera abyssinica* (Hochst. ex. A. Rich.) Harms, exhibited a relatively poor regeneration (Wassie, 2007). This calls for management interventions that promote successful regeneration in and around the remaining forests (Teketay, 1997; Bongers et al, 2006; Wassie et al, 2005a and b; Wassie et al, 2009a and b). For example, we expect that regeneration would benefit from protecting adult seed trees against harvest, reducing grazing intensity, protection against transformation into agricultural lands, and avoiding the construction of monuments and improper grave houses, which reduces available sites for regeneration (Wassie, 2007).

Other studies in northern Ethiopia show comparable results. In Tigray, at ca. 2000m elevation in an area of 13,000 hectares, only ten forest fragments were found, and these were very small (mean 6.56ha, range 0.40–20.95) and had on average only 9.9 woody species per 400m$^2$ plot and a total of 40 species (Aerts et al, 2006a). Woody species richness increased with fragment size, and forest composition was strongly determined by plot position in the landscape. In a study covering a large area of the Ethiopian highlands, 38 sites between 1400m and 3000m asl were investigated (Bingelli et al, 2003). Rapid biodiversity assessment (RBA) of the woody plant species was carried out and a total of 223 species were recorded. The sizes of the forests ranged from 0.5 to 50 hectares and an average forest had 41 species of woody plants (range 21–68). On average 26.0 (range 10–55) species, roughly two-thirds of the total, were used by the local communities. Out of the 223 recorded species, 42 species were consistently rare or uncommon, in other words usually had fewer than five individuals per RBA, and over three-quarters of these species were used by local people. Similarly to the church forests considered earlier, the majority of these sacred groves are under threat mainly because of the lack of natural regeneration, often due to overgrazing, but also to unsustainable tree harvesting (for timber and fuel in particular) (Bingelli et al, 2003). The very limited area and associated threats are indeed general for church forests: Aerts and co-workers (2008b) digitized 394 church forests in randomly selected areas (total area half a million hectares) in northern Ethiopia and found an average forest size of 2.5 ± 0.2ha, with a total forest cover of only 0.2 per cent of the area.

## THEOLOGICAL, RELIGIOUS AND SOCIAL PERSPECTIVES OF CHURCH FORESTS

The main theological and religious perspective of the EOTC in conserving forest resources stems from the Holy Scriptures of the Church (Anonymous, 1989). The EOTC perceives nature in a holistic manner, favouring the respect and veneration of nature and biodiversity conservation. The importance of trees and their holy services were documented in many religious books, locally referred to as 'Gedel', 'Tamre-Mariam' and 'Dersan' (Wassie et al, 2005a and b). The conservation and sustainable use of forest resources and other forest services and the development of these resources is in general propagated in the EOTC philosophy. Trees for example provide the people of the churches with wood and charcoal for church buildings, internal services, and making sacramental and sacred utensils; fruits, leaves and other parts as food; medicine and raw materials to make ink and dyes (which are, in turn, used to write religious scriptures); protection of the church building; shade, sweet and pleasant smells around churches; grace and esteem to churches; and privacy and tranquillity for hermits and monks (Wassie et al, 2005a and b).

The church forests are primarily reserved for the church services. However, there are some other forest benefits, namely collection of tree seeds, plant parts for medicinal purposes, wild edible fruits, hanging beehives and spiritual contemplation, which are allowed for the followers with the permission and recognition of the church administrators. Other potential benefits, such as collection of wood for fuel and construction, collection of fodder, wildings and honey, and non-tangible benefits such as non-spiritual recreation and enjoyment, are entirely forbidden (Wassie et al, 2005a and b).

The church people protect and conserve their forest resources through religious sanctions (known as *Gizet*) and/or civil law. However, religious commitment and respect for the EOTC are the primary reasons for individual community members to preserve the forests. Four churches were selected for a household socio-economic survey. From each selected church 5 per cent of the church community members were randomly chosen and interviewed. Accordingly in total 122 household heads in the four churches, living close to the church forests, were interviewed. A majority of the respondents (93 per cent) confirmed that the survival of church forests is due to their ownership and/or association with the EOTC. Most respondents (92.6 per cent) expressed willingness to plant seedlings around the church, with 94 per cent setting a precondition that the newly planted seedlings must be kept under the owner- ship of their church. Of the public instruments governing the local communities, more respondents respected religious rules/beliefs and sanctions (82 per cent) than civil legislation (<1 per cent) or a combination of both (17.2 per cent). In addition, 98 per cent said they trusted and respected the clergy/churches more than the Bureau of Agriculture and other government bodies that oversee public conservation activities and regulation (Wassie, 2002). These results confirm the strong role played by the church in conservation of the remaining forest patches.

Based on this strong commitment of people to the church and its rules for forest conservation, a wide variety of stakeholders, including administrative bodies, the Bureau of Agriculture, religious institutions, elders and the commu- nity at large, agreed that the conservation of these forests should largely be maintained under the traditional church forest policy. However, stakeholders also exhibited several divergent views, mainly on ownership and use rights, which may lead to mistrust and poor cooperation among themselves (Wassie, 2002). Participatory approaches are needed to develop conservation rules that are well accepted and respected by the local stakeholders.

## PROSPECTS AND CHALLENGES FOR CHURCH FORESTS

Ethiopian Church forests may best represent the old growth forest in the area, but only for the last centuries. A thousand years and more ago forests in the area had other species compositions (Darbyshire et al, 2003). Church forests may serve as species-diverse nuclei, from where management actions might effectively restore forests over larger, often degraded, areas (Bongers et al, 2006; Wassie, 2007; Aerts et al, 2006a). Active management actions are recom- mended, such as planting, seeding and soil scarification (Wassie et al, 2009a), reduction of seed predation (Wassie and Teketay, 2006; Wassie et al, 2009c), reduction of grazing (Wassie et al, 2009b), planting of nurse plants (Aerts et al, 2006b and 2008a), and connecting forest patches (Aerts et al, 2008a). These aspects are highly important in regional environmental development plans.

The role of church forests in forest restoration and conservation requires management aiming at long-term sustainability. However, many tree species populations are typically small and unstable, often lack regeneration (Wassie, 2007; Aerts et al, 2006a), and may face extinction in the future. From these observations we expect that church forests will deteriorate to poorer forest or even non-forest systems when no active management and protection actions are undertaken (Wassie, 2007; Aerts et al, 2006b; Bingelli et al, 2003, Hobbs and Suding, 2009). Large-scale restoration is unlikely to occur, however, because of current economic, social and political constraints.

Here we suggest the following:

1  Legal protection of church forest is highly needed and should be strengthened; church forest areas have to be gazetted after being clearly demarcated and marked in the field.
2  Training is needed on silvicultural and forest management techniques and species-based tree propagation.
3  Interventions on church forests should be designed and applied in line with community traditions.
4  Species-specific silvicultural interventions are needed to facilitate species regeneration, for instance opening of trampled soil and maybe also opening of dense thickets of shrubs and lianas, addition of seeds and seedlings from outside, liberation of specific tree seedlings and saplings in dense shrub or liana vegetation, and protection against grazing (Bongers et al, 2006; Wassie, 2007).
5  That church forests should be expanded in area, because their current small patch size might introduce extinctions of tree species in the longer run.

We conclude that the ecological and social status of church forests provides a strong opportunity to conserve natural forests and to restore degraded areas into more productive and diverse natural forests.

# REFERENCES

Aerts, R., van Overtveld, K., Haile, M., Hermy, M., Deckers, J. and Muys, B. (2006a) 'Species composition and diversity of small Afromontane forest fragments in northern Ethiopia', *Plant Ecology*, vol 187, pp127–142

Aerts, R., Maes, W., November, E., Negussie, A., Hermy, M. and Muys, B. (2006b) 'Restoring dry Afromontane forest using birds and nurse plant effects: Direct sowing of *Olea europaea* ssp. cuspidata seeds', *Forest Ecology and Management*, vol 230, pp23–31

Aerts, R., Lerouge, F., November, E., Lens, L., Hermy, M. and Muys, B. (2008a) 'Land rehabilitation and the conservation of birds in a degraded Afromontane landscape in northern Ethiopia', *Biodiversersity and Conservation*, vol 17, pp53–69

Aerts, R., Pankhurst, R., Van Overtveld, K., November, E., Hermy, M. and Muys, B. (2008b) 'Historical deforestation patterns and the conservation value of church forests in the northern Ethiopian highlands', International Conference, Mountain Forests in a Changing World. Advances in Research on Sustainable Management and the Role of Academic Education, UNI BOKU, Vienna, 2–4 April 2008, Book of Abstracts, p14

Anonymous (1989) *The Holy Bible (Containing the Old and New Testament), King James Version*, World Publishing, Iowa Falls, IA

Bekele, M. (2003) 'Forest property rights, the role of the state, and institutional exigency: The Ethiopian experience', PhD dissertation, Acta Universitatis Agriculturae Sueciae Agraria 409, Swedish University of Agricultural Sciences, Uppsala, Sweden

Bhagwat, S. A. and Rutte, C. (2006) 'Sacred groves: Potential for biodiversity management', *Frontiers in Ecology and Environment*, vol 4, pp519–524

Bingelli, P., Desalegn, D., Healey, J., Matt, P., John, S. and Tekelhaimanot, Z. (2003)
'Conservation of Ethiopian sacred groves', *European Tropical Forest Research
Network (ETFRN) Newsletter*, vol 38, pp37–38

Boerma, P. (2006) 'Assessing forest cover change in Eritrea – A historical perspective',
*Mountain Research and Development*, vol 26, no 1, pp41–47

Bongers, F., Wassie, A., Sterck, F. J., Bekele, T. and Teketay, D. (2006) 'Ecological
restoration and church forests in northern Ethiopia', *Journal of the Drylands*, vol 1,
pp35–45

Chao, A., Chazdon, R. L., Colwell, R. K., and Shen, T. J. (2005) 'A new statistical
approach for assessing similarity of species composition with incidence and
abundance data', *Ecology Letters*, vol 8, pp148–159

Darbyshire, I., Lamb, H. and Umer, M. (2003) 'Forest clearance and regrowth in
northern Ethiopia during the last 3000 years', *The Holocene*, vol 13, no 4,
pp537–546

Ewers, R. M. and Didham, R. K. (2006) 'Confounding factors in the detection of
species responses to habitat fragmentation', *Biological Reviews*, vol 81, pp117–142

Hill, J. L. and Curran, P. J. (2001) 'Species composition in fragmented forests:
Conservation implications of changing forest area', *Applied Geography*, vol 21,
pp157–174

Hobbs, R. J. and Suding, K. N. (eds) (2009) *New Models for Ecosystem Dynamics and
Restoration*, Society for Ecological Restoration International, Island Press,
Washington, DC

McCann, J. C. (1997) 'The plow and the forest: Narratives of deforestation in Ethiopia,
1840–1992', *Environmental History*, vol 2, pp138–159

Ross, K. A., Fox, B. J. and Marilyn, D. (2002) 'Changes to plant species richness in
forest fragments: Fragment age, disturbance and fire history may be as important as
area', *Journal of Biogeography*, vol 29, pp749–765

Santos, K., Kinoshita, L. S. and dos Santos, F. A. M. (2007) 'Tree species composition
and similarity in semi deciduous forest fragments of southeastern Brazil', *Biological
Conservation*, vol 135, pp268–277

Teketay, D. (1997) 'Seedling population and regeneration of woody species in dry
Afromontane forests of Ethiopia', *Forest Ecology and Management*, vol 98,
pp149–165

Teketay, D. and Bekele, T. (1995) 'Floristic composition of Wof-Washa natural forest,
central Ethiopia: Implications for the conservation of biodiversity', *Feddes
Repertorium*, vol 106, pp127–147

Turner, I. M. (1996) 'Species loss in fragments of tropical rain forest – a review of the
evidence', *Journal of Applied Ecology*, vol 33, pp200–209

Wassie, A. (2002) 'Opportunities, constraints and prospects of Ethiopian Orthodox
Tewahido Churches in conserving forest resources: The case of churches in South
Gondar, northern Ethiopia', MSc thesis, Swedish University of Agricultural
Sciences, Skinnskatterberg, Sweden

Wassie, A. (2007) 'Ethiopian church forests: Opportunities and challenges for restoration', PhD dissertation, Wageningen University, Wageningen, The Netherlands

Wassie, A. and Teketay, D. (2006) 'Soil seed banks in church forests of northern
Ethiopia: Implications for the conservation of woody plants', *Flora*, vol 201,
pp32–43

Wassie, A., Teketay, D. and Powell, N. (2005a) 'Church forests in North Gondar Administrative Zone, northern Ethiopia', *Forests, Trees and Livelihoods*, vol 15, pp349–374

Wassie, A., Teketay, D. and Powell, N. (2005b) 'Church forests provide clues to restoring ecosystems in the degraded highlands of northern Ethiopia', *Journal of Ecological Restoration*, vol 23, p2

Wassie, A., Sterck, F. J., Teketay, D. and Bongers, F. (2009a) 'Effects of livestock exclusion on tree regeneration in church forests of Ethiopia', *Forest Ecology and Management*, vol 257, pp765–772

Wassie, A., Sterck, F. J., Teketay, D. and Bongers, F. (2009b) 'Tree regeneration in church forests of Ethiopia: Effects of microsites and management, *Biotropica*, vol 41, pp110–119

Wassie, A., Bekele, T., Sterck, F. J., Teketay, D. and Bongers, F. (2009c) 'Post-dispersal seed predation and seed viability in forest soil: Implications for the regeneration of tree species in Ethiopian church forests', *African Journal of Ecology*, vol 48, pp461–471

Wassie, A., Sterck, F. J. and Bongers, F. (2010) 'Species and structural composition of church forests in a fragmented landscape of Northern Ethiopia', *Journal of Vegetation Science* (in press)

Zeleke, G. and Hurni, H. (2001) 'Implications of land use and land cover dynamics for mountain resource degradation in the northwestern Ethiopian highlands', *Mountain Research and Development*, vol 21, no 2, pp184–191

# 7

# Incense Woodlands in Ethiopia and Eritrea: Regeneration Problems and Restoration Possibilities

*Abrham Abiyu, Frans Bongers, Abeje Eshete,*
*Kindeya Gebrehiwot, Mengistie Kindu,*
*Mulugeta Lemenih, Yitebitu Moges,*
*Woldeselassie Ogbazghi and Frank J. Sterck*

## INTRODUCTION

Tropical dry forests (TDF) are forests in dryland areas and are distinguished by their marked dry season and low water availability. Their annual rainfall ranges from 500mm to 2000mm, they have a ratio of potential evapotranspiration to precipitation of between one and four, and experience typically three to eight months of drought (Murphy and Lugo, 1986). They account for 40–42 per cent of the tropical forests worldwide (Murphy and Lugo, 1986; Mayaux et al, 2005). These dry forests contain a wealth of unique biodiversity (Janzen, 1988; Gentry, 1995; Groombridge and Jenkins, 2002), but suffer severe threats from anthropogenic and economic pressures. These forests are also among the least protected ecosystems (Janzen, 1988; Bullock et al, 1995). It is estimated that close to one billion people worldwide directly depend upon dryland products for their livelihoods. Plant species of this ecosystem have slower growth rates and restricted reproductive episodes compared to other tropical forests, which may make them more susceptible to disturbance.

The largest proportion of dry forest is found in Africa, where it accounts for 70–80 per cent of the forested area (Murphy and Lugo, 1986). The *Acacia-Commiphora* woodlands, which form part of the dry tropical forests of Africa, occur over large parts of the Horn of Africa and the Sudano-Sahelian zone (White, 1983; Friis, 1986 and 1991). These forests are endowed with diverse species of *Acacia*, *Boswellia* and *Commiphora*, important tree species that are known to yield commercial gums and resins such as gum arabic, frankincense, myrrh and opopanax. These gums and resins have very diverse uses, principally in food, pharmaceuticals, cosmetics/perfumery, adhesives, fabrics, paper, lithographic ink and dye industries. Indeed wood and non-wood products from these species play a significant role in the national economy and local livelihood of many people in Africa.

*Boswellia papyrifera* (Del.) Hochst. is one of the several species of trees that provide commercial gum resin called frankincense or gum olibanum (Figure 7.1). It is a deciduous multipurpose tree species with a papyraceous bark that peels in flakes. It is one of the 17 genera described in the family *Burseraceae*, which is estimated to encompass about 500–600 species (Vollesen, 1989; Hedberg and Edwards, 1989). It is found in woodlands and wooded grasslands, on steep rocky slopes, on lava flows, and in sandy river valleys at an altitude between 950m and 1800m asl, with an annual mean temperature of 20–25°C, and an annual mean precipitation of less than 900mm (Von Breitenbach, 1963; Azene, 1993; Anonymous, 1997; Ogbazghi et al, 2006b).

*B. papyrifera* has a large ecological and economic significance (Lemenih and Teketay, 2003; Gebrehiwot et al, 2003). Ecologically the species is important since it grows in harsh environments where most tree species may not grow. A recent study in northern Ethiopia (Gebrehiwot et al, 2005) showed that *B. papyrifera* is highly suitable for future reforestation (restoration) efforts in moisture-deficit arid and semi-arid areas. In addition, the species is one of the most economically important plants in the drylands of eastern Africa because of its oleo-gum resin product frankincense (Chikamai, 2002; Lemenih and Teketay, 2003).

Currently, improper land use is leading to three major processes of overexploitation that threaten the existence of the woody vegetation:

1   clearing and conversion to unproductive agricultural land use;
2   overgrazing by livestock; and
3   improper tapping for the extraction of gum resin (Gebremedkin, 1997; Marshall, 1998; Rijkers et al, 2006).

The prospects of these woodlands are also threatened by increasing population and livestock resulting from national migration/settlement programmes, advancing desertification and global climatic changes (Tadesse et al, 2002; Lemenih et al, 2007). Recent studies have consistently reported concerns for the decline of the population of the species due to lack of regeneration (Ogbazghi, 2001;

**Figure 7.1** *The frankincense tree* Boswellia papyrifera *in Eritrea*

*Notes:* (a) Tall trees in full leaf in the middle of the wet season; (b) Scattered frankincense trees in a dry landscape on a stony slope; (c) Frankincense tree being tapped in the middle of the dry season when the trees do not bear any leaves.
*Source:* Woldeselassie Ogbazghi

Gebrehiwot, 2003) and that concerted efforts are needed to conserve and promote the economic and ecological significances of the *Boswellia* forest of the regions. The objective of this chapter is to describe the regeneration status of *Boswellia* in the dry forests of Ethiopia and Eritrea, analyse possible factors for the loss of regeneration, and suggest restoration activities.

## POPULATION STRUCTURES

To assess the current status of *B. papyrifera* populations, we have compiled data for a large number of sites in Eritrea and northern Ethiopia (Figure 7.2). These sites lie within the Sudanian and Sahelian zones (White, 1983; Ogbazghi, 2001) and represent the main *Boswellia papyrifera* growing region in the Horn of Africa. In all of these sites, inventories of the standing populations have been made, in most cases in a large number of plots. Tree diameters have been measured and densities calculated. In Table 7.1 a number of characteristics of the sites and of the *Boswellia* populations are provided.

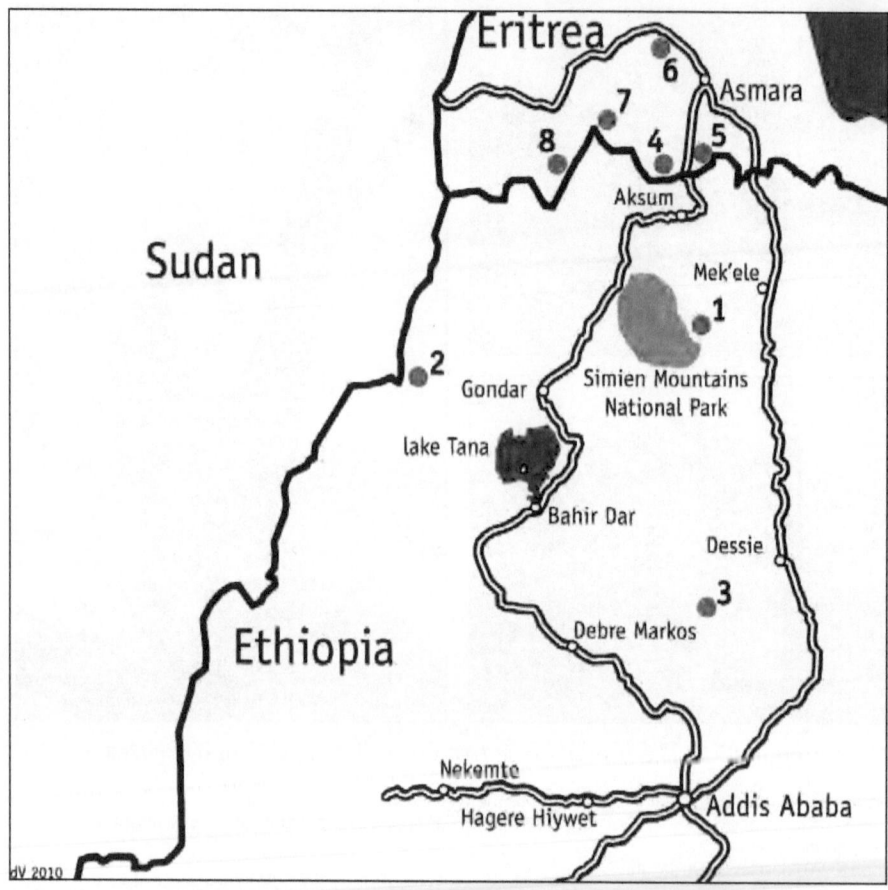

**Figure 7.2** *Map of Ethiopia and Eritrea with the location of the study sites*

*Notes:* Ethiopia: Abergelle (1), Metema (2), Wogdi (3); Eritrea: Adi-Ketina (4), Atawen (5), Ferhen (6), Molki (7), Shilalo (8).

The sites lie along an altitudinal gradient from 800m to 2000m above sea level, and in areas with average temperatures of between 14°C and 35°C. Most of the sites have tree densities of between 200 and 300 individuals per hectare (trees with DBH >10cm) with the exception of Ferhen, which had only 83 individuals per hectare.

To describe and analyse the *Boswellia* population status we examined stem-size distributions by fitting Weibull distribution curves (Bailey and Dell, 1973):

$$F(x) = 1 - e^{-(x/\beta)\alpha}$$

This distribution is characterized by the distribution function F(x), the number of trees at DBH class x, the scale parameter β, the slope parameter α and the DBH(x) in cm. This function is very successful in fitting stem-size distribution

**Table 7.1** *Overview of the* Boswellia papyrifera *sites covered in this chapter, with site-specific information*

| Study area | Adminis-trative zone | Altitude range (m) | Annual rainfall (mm) | T (°C) | Dominant soils | No of sample plots | Plot size (m) | No of trees per ha |
|---|---|---|---|---|---|---|---|---|
| Abergelle I | Central Tigray | 1400–1650 | 657 | 22.3 | Cambisol, Arenosols | 16 | 50 × 50 | 254 |
| Abergelle II | Central Tigray | 1400–1650 | 657 | 22.3 | Cambisol, Arenosols | 16 | 50 × 50 | 325 |
| Abergelle III | Central Tigray | 1400–1650 | 657 | 22.3 | Cambisol, Arenosols | 16 | 50 × 50 | 219 |
| Abergelle IV | Central Tigray | 1400–1650 | 657 | 22.3 | Cambisol, Arenosols | 16 | 50 × 50 | 109 |
| Metema I | North Gondar | 950–1100 | 660–1130 | 19.1–35.6 | Cambisol, Lithosol | 6 | 50 × 50 | 332 |
| Metema II | North Gondar | 950–1100 | 660–1130 | 19.1–35.6 | Cambisol, Lithosol | 6 | 50 × 50 | 334 |
| Metema III | North Gondar | 950–1100 | 660–1130 | 19.1–35.6 | Cambisol, Lithosol | 6 | 50 × 50 | 217 |
| Metema IV | North Gondar | 950–1100 | 660–1130 | 19.1–35.6 | Cambisol, Lithosol | 6 | 50 × 50 | 257 |
| Wogdi | South Wello | 950–1650 | 1200 | 14–25 | Vertisol, Andosol | 32 | 20 × 20 | 254 |
| Adi-Ketina | Debub | 1600–2000 | 500–700 | 15–21 | Cambisol, Lithosol | 34 | 20 × 20 | 253 |
| Atawen | Debub | 1600–1900 | 500–700 | 15–21 | Cambisol, Lithosol | 24 | 20 × 20 | 259 |
| Ferhen | Anseba | 1600–2000 | 200–500 | 15–21 | Cambisol, Lithosol | 26 | 20 × 20 | 83 |
| Molki | Gash-Barka | 1100–1600 | 500–800 | 21–28 | Cambisol, Lithosol | 31 | 20 × 20 | 296 |
| Shilalo | Gash-Barka | 800–1100 | 500–800 | 21–28 | Cambisol, Lithosol | 29 | 20 × 20 | 303 |

data (Vanclay, 1994; Alder, 1995) and is popular with modellers dealing with uneven-age stands (Kamziah et al, 2000; Zhang et al, 2001). Most forms of the distribution show either a simple decline or a unimodal form. Depending on the shape parameters, the distribution is skewed to the left, symmetrical or skewed to the right. The scale parameter ($\beta$) is approximately equal to the median DBH, while the shape parameter controls the skewness of the distribution. When the shape parameter becomes less than one the curve approaches an inverse J-shape distribution. Model parameters were determined by means of linear regression and maximum likelihood methods (Sheil and Salim, 2004). The disparities between the observed and the predicted distributions and between sites are explained by the responsible ecological factors (Lykke, 1998; Swaine, 1998). We excluded the first class (0–2cm DBH) from our modelling analysis because the class includes all the seedlings and germinants, the count of which is very dependent on year and season.

After fitting the function, all the sites studied showed a similar hump-shaped distribution (see Figures 7.3 and 7.4). The hump-shaped curves

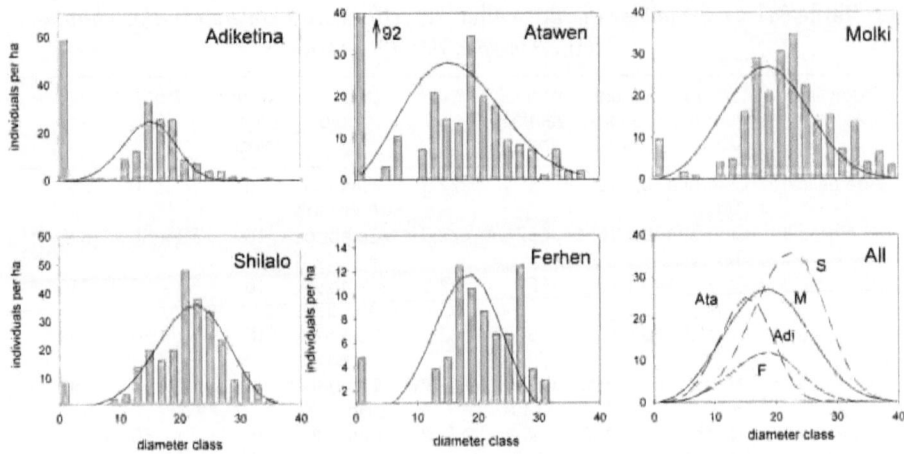

**Figure 7.3** *Frequency diameter distributions of* Boswellia *trees in five sites in Eritrea*

*Note:* Diameters are given in classes 2cm wide. The curves are Weibull models based on all frequencies except the class 0–2cm (see text for further explanation).

overestimated the frequency of the small individuals in the populations and underestimated the large individuals. In most populations the number of small individuals was extremely low, suggesting a lack of regeneration. Only Adi-Ketina and Atawen in Eritrea (Figure 7.3) and enclosed sites at Abergelle in Ethiopia (Figure 7.4) had a reasonable number of small individuals (germinants and small seedlings; see Ogbazghi, 2001; Gebrehiwot, 2003). At Metema, fenced plots within each of the four sites had germinants and small seedlings (data not shown; see Moges and Kindu, 2006), but at site level this was negligible (Figure 7.4b).

Although the form of most curves is rather similar, the absolute amounts and also the median sizes vary, and so does the skewness (Figures 7.3 and 7.4). For instance, in Eritrea (Figure 7.3), the highest tree diameter median value (23cm) was obtained from Shilalo and the lowest from Atawen (15cm). In terms of number of trees obtained from each site, the highest number of trees was obtained from Shilalo and the lowest from Ferhen. In Abergelle (Figure 7.4a) the highest number of trees per hectare was obtained from Site II and the lowest from Site IV, while the highest median diameter is located at Site IV and the lowest at Site II. At Metema (Figure 7.4b), the bulk of the population is found in the diameter range of 12–16cm, showing that it is composed of the smallest sized population from all the sites compared in the two countries. In Wogdi (Figure 7.4c) the median is 18–20cm. In all the cases the total number of trees forming the population decreased with the average size of the trees, as expected.

The modelled diameter distributions at Metema show a narrower range of sizes compared to the other areas, but the observed values are not very differ-

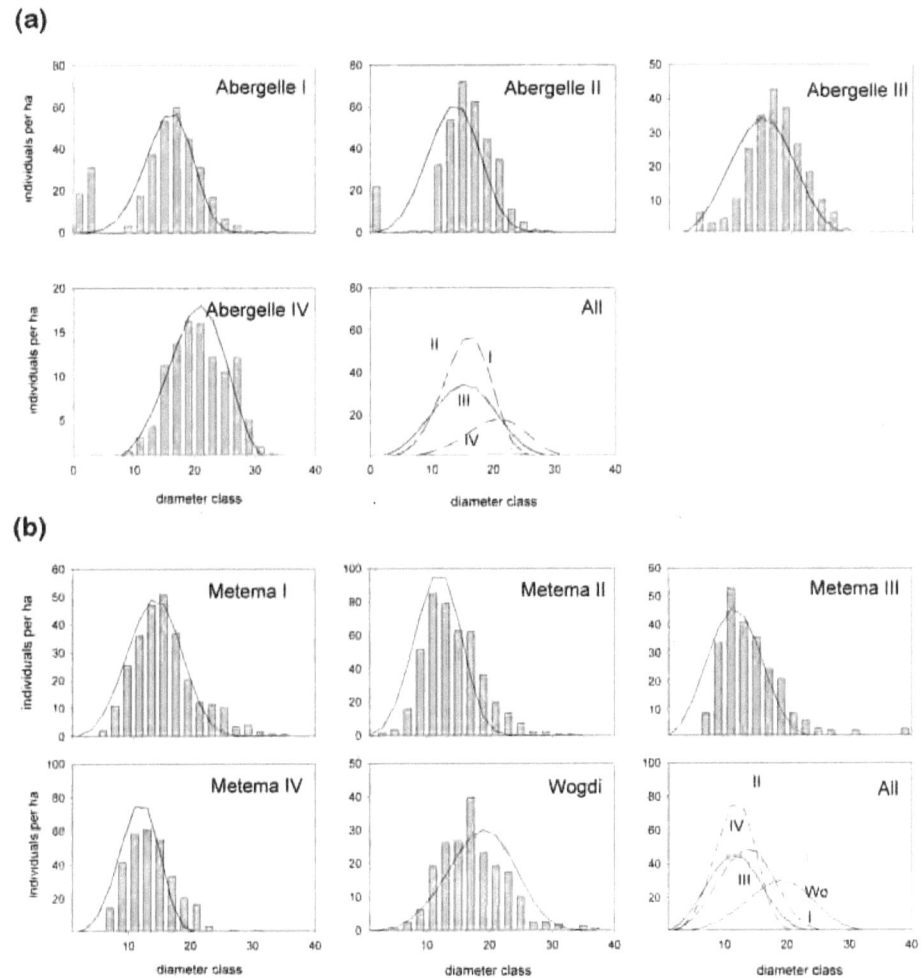

**Figure 7.4** *Frequency diameter distributions of* Boswellia *trees in Ethiopia*

*Notes:* (a) Four Abergelle sites; (b) four Metema sites and one Wogdi site. Diameters are given in classes 2cm wide. The curves are Weibull models based on all frequencies except the class 0–2cm (see text for further explanation).

ent. Probably the relatively large number in the 10–16cm classes largely determines the form of the curve. Possibly a regeneration wave occurred some time in the past. Unfortunately no data on size/age relationships of *Boswellia* trees are yet available. Anyhow, small individuals are strikingly lacking in all sites. To discuss the apparent lack of regeneration we further studied seedlings.

# SEEDLING REGENERATION

To analyse regeneration of *Boswellia* seedlings and possible reasons for their lack, we determined the density of seedlings under different treatments. We studied populations in enclosures (no grazing by animals) and open areas (free grazing), sometimes in combination with soil cultivation and sometimes without. In general the effect of enclosing an area is large and the density of seedlings in such plots is higher than in free-grazing areas (Ogbazghi, 2001; Gebrehiwot, 2003). At Abergelle, the two sites that were enclosed for eight years (between 1994 and 2002) had 2559 ± 359 and 2231 ± 245 seedlings (<30cm height) while the two open sites had only 66 ± 11 and 64±10 seedlings (each site based on 16 plots of 50 × 50m each) (Gebrehiwot, 2003). In the wet season of 2004 seedling densities in Abergelle were 8331/ha and 3325/ha in closed and in open-grazed *Boswellia* woodland (Negussie et al, 2008). At the start of the dry season, 28.2 and 22 per cent of these seedlings were still present. At the Metema sites, plots without fencing and with ground cultivation after seed dispersal had the lowest amounts of seedlings (7 seedlings per hectare), lower than in control plots without any treatment (21) and far lower than in fenced plots (32) (Moges and Kindu, 2006). In the latter case it is probable that the seedbed preparation that was made after seed dispersal might have destroyed the emerging seedlings. In general, we conclude that grazing and cultivation of crops both reduce seedling regeneration, and both factors may contribute to the lack of small individuals.

Of special interest is the fact that most seedlings were found within a distance of 2m from the nearest mature tree: in Abergelle 77 per cent (Gebrehiwot, 2003) and in Metema 78 per cent (Moges and Kindu, 2006) of the seedling population was within that distance. In addition to this, the seedling population showed a tendency to increase with slope. At Metema, slopes steeper than 7° have 38 seedlings/ha, while slopes of less than 7° had only 20 seedlings/ha (Moges and Kindu, 2006).

However, a low number of seedlings does not necessarily imply low regeneration. This is because the number of seedlings observed at any given time depends on mortality rate and the carrying capacity of the specific site (Condit et al, 1998). In most of the study sites, plant mortality is far higher for individuals in smaller size classes than in larger size classes. For instance in Abergelle, juvenile mortality was 36 per cent, while mortality in the saplings was 6.3 per cent (Gebrehiwot, 2003). In Eritrea, the mortality rate for smallest plants (DBH <0.5cm) was 67 per cent but was only 18 per cent for plants in the DBH 2–3cm range (Ogbazghi, 2001). Seedling mortality also increases with distance from the nearest mature tree. In Abergelle, mortality increased from 64 to 76 to 85 per cent as the distance increased from <2m to 2–5m and to 5–10m (Gebrehiwot, 2003).

# DISCUSSION

## Size distributions suggest lack of regeneration

The number of mature *Boswellia* trees per hectare varied from region to region and from site to site, but is generally between 200 and 300 individuals (except for Ferhen). This may be related to the size of the individual trees (more smaller trees can be packed on the same area), but also to the number of trees of other species in the same community. Ogbazghi and co-workers (2006a) show that at higher altitudes the trees are shorter with smaller crowns and, hence, densities could be higher, but they also show that a large range of woody species (total 58) are found co-occurring in these woodlands. Also Gebrehiwot (2003) found 18 families (42 species) co-occurring with *B. papyrifera*. The most abundant family was Fabacaeae and the most abundant non-*Boswellia* species were *Acacia etbaica* (17–44 trees/ha), *Lannea fruticos* (7–10 trees/ha), *Terminalia brownie* (4–16 trees/ha), *Combretum hartmannianu* (2 trees/ha) and *Ximenia americana* (2 trees/ha).

The most striking result from the stem diameter size distributions of *B. papyrifera* in all the study sites is the under-representation of individuals in the lower diameter classes and the over-representation of individuals in the higher diameter classes. More than half (ca. 65 per cent) of the total population of *B. papyrifera* in Eritrea and Metema is in the ranges of 16–24cm and 13–15cm DBH respectively. The extremely low density of individuals in the lower diameter classes suggests that recent regeneration is severely lacking and that the population is under serious threat in the long run.

Seedlings were mainly confined to areas very close to mature trees, both in enclosures and in open grazing areas, and seedling densities decreased with distance. This might be the result of higher seed levels close to the mother tree, and may also result from increasing mortality with distance from mature trees. Once a sapling becomes larger (for example 3cm collar diameter in Eritrea), plant mortality diminished to a negligible level. This is probably due to the increased endurance of the saplings against damaging agents such as fire and grazing.

Should the observed regeneration structures in *B. papyrifera* raise the alert level? It should indeed, if regeneration is lacking. However, it still is possible that the populations are in a steady-state condition, as occurs in many dry woodlands. In some *Acacia* woodlands, for instance, very few individuals can sufficiently replace the standing vegetation without negatively affecting the viability of the population (Wiegand et al, 1999). In many cases, however, population structures as shown for *Boswellia* indeed indicate lack of regeneration. Thus dynamic monitoring of the regeneration status of *Boswellia* is highly recommended.

# Possible reasons for lack of regeneration

A lack of regeneration can be caused by a number of factors. First we have to distinguish between regeneration via seeds or via root suckers. If via seeds, several steps can lead to reduced regeneration. Trees may not produce sufficient seeds, or seeds may not be viable due to poor seed quality (for example lack of embryo) or to high seed infestation by insects while on the mother tree. Once on the ground, seeds may not be able to further disperse because they are highly infested by insects and/or eaten by vertebrate herbivores and destroyed. Germination may be hampered by sub-optimal environmental conditions (for example too cold or too dry). Once germinated, the resulting seedlings may be trampled or eaten by grazers, or show very low growth rates due to low light availability, low water or low nutrient availability. Also seedlings may be susceptible to herbivorous insects or fungi. The alternative route of regeneration is via root suckers. In Box 7.1 and the following sections we treat most of these possibilities for the case of *Boswellia*.

# Regeneration via seeds or via root suckers

There is no agreement on the origin of the observed seedling and sapling populations in different areas. Bond and Midgley (2001) distinguished two modes of regeneration: recruitment by means of seeds and persistence by means of root sprouting. Root sprouting is thought to be the preferred mode of regeneration in frequently disturbed environments. For *B. papyrifera* both seedlings reproduced sexually from seed and asexually from root suckers are found (Ogbazghi, 2001; Eshete, 2002, Moges and Kindu, 2006). However, the main route of regeneration is variable. Some studies reported root suckers as the main route of regeneration (Moges and Kindu, 2006) while others found sexually produced seedlings to be the main route of regeneration (Ogbazghi, 2001; Eshete, 2002).

In view of the seed properties of this species, such as absence of a soil seed bank reserve (Eshete, 2002), absence of seed dormancy (Gebremedhin and Negash, 1999), the susceptibility of the seeds to insect attack, and the production of non-viable and embryo-lacking seed (Ogbazghi, 2001), regeneration from root suckers as a main route of regeneration is a sensible speculation. This is substantiated, though not strongly quantified, by several authors and personal communications. However, there is not enough information on either the origin of observed seedlings (either from seed or roots sucker) or the differential performance of these seedling groups. We hypothesize that seedlings from root sucker origin may perform better as we expect these seedlings to be continuously nursed by the mother tree for moisture and nutrients (Vieira and Scariot, 2006).

Up to now research focus on *B. papyrifera* has been on the importance of safe sites, seed and seedling banks, dispersal and germination. It has long been suggested that recruiters maximize their fitness by ensuring they are reproductively mature before the next disturbance (Bell and Pate, 1996; Bell, 2001). Conversely, resprouters maximize their fitness by allocating resources to struc-

---

**BOX 7.1 QUESTIONS FOR ANALYSING REASONS FOR LACK OF REGENERATION**

*Regeneration via seeds*

- Sufficient seeds?
- Seeds are viable?
- Seeds are dispersed (to 'safe sites')?
- Seeds are surviving (not destroyed by insects, fungi or vertebrates)?
- Seeds germinate?
- Seedlings are surviving (not trampled, eaten or attacked by insects or fungi)?
- Seedlings are growing (enough water, light and nutrients)?

*Regeneration via sprouts*

- Enough sprouts?
- Resource flow from mother tree to sprouts?
- When are sprouts independent from mother tree?
- Sprouts are surviving (not trampled, eaten or attacked by insects or fungi)?
- Sprouts are growing (enough water, light and nutrients)?

---

tures that increase their chance of surviving the next disturbance. Reviews have generally concluded that obligate seeders devote more resources to growth and reproduction, whereas resprouters devote more resources to below-ground storage structures to facilitate resprouting following disturbance (Bellingham and Sparrow, 2000; Bell, 2001; Bond and Midgley, 200a). Several studies have also found that resprouters devote proportionally larger amounts of root tissue to starch storage than do obligate seeders (Patc et al, 1990; Bell and Ojeda, 1999; Verdaguer and Ojeda, 2002). We suggest that this distinction between resprouters and reseeders in *Boswellia* deserves further attention.

# Are seeds a bottleneck?

Seedling abundance depends, among other factors, on the availability of viable seeds and on germination capacity. We know that a large proportion of the seed reaching the ground is destroyed by insects or infected by fungi. For instance, about 18–25 per cent of the bulk seeds of *B. papyrifera* collected in Ethiopia (Gebremedhin and Negash, 1999) and Eritrea (Ogbazghi, 2001; Rijkers et al, 2006) were found attacked by insects. This, however, is probably not inhibiting the population. In stands with healthy trees the available viable seeds were plentiful and as such not a bottleneck for regeneration (Ogbazghi, 2001). The challenge for natural regeneration is thus rather on providing better growing conditions for seedlings to grow into saplings and then into mature trees. Significant variation in the mortality rate of the seedlings in open grazing and closed sites demonstrates that the lack of growth into saplings is highly influenced by external factors such as grazing (cf. Negussie et al, 2008).

## Effects of tree tapping on reproduction

Germination tests on seeds collected from healthy-looking intensively tapped trees, have shown a very low germination percentage, as low as 14–16 per cent (Ogbazghi, 2001). Intensive tapping has led to the production of non-viable or embryo-lacking seeds, and hence to very low seedling recruitment. This is a common phenomenon in other tree species such as the rubber tree (Ros-Tonen et al, 1995).

Tapping creates a multifaceted undesirable negative impact on B. papyrifera trees. Small trees may not be able to recover their wounds after tapping. Under normal tapping regimes, larger trees are tapped at more tapping spots on the trees than smaller trees, generally this varies between 3 and up to 16 spots. Such intensive tapping may interfere with the carbon balance of the tree, and thus may reduce growth and/or reproduction. Rijkers and co-workers (2006) showed that intensive tapping leads to the production of non-viable seeds. In three sites in Eritrea, tapping strongly reduced the amount of inflorescences, fruits and seeds, and seed weight (Rijkers et al, 2006). This effect was tree size dependent. They also showed that reproduction was consistently higher in sites with a tapping rest. Particularly in large trees, a 14-year non-tapping period was more beneficial than a 4-year period.

Another type of damage is the infestation by insects. This is a general phenomenon and Eshete (2002) found that the proportion of trees damaged by 'an unidentified worm' was higher in tapped as compared to untapped stands in Gondor. These unidentified worms were later identified to be beetles of the genus Scarabaeidae and Cerambycidae (long horn beetles). In Tigray the main species is the long horn beetle Idactus spinipennis Gahan (Negussie, 2008), that had infested over 65 per cent and over 85 per cent of the adult Boswellia trees in two distant Boswellia areas. It is not yet clear if and how insect infestation interacts with tapping, but as tapping generally weakens the tree, the probability of insect infestation is expected to be higher.

Tapping during the rainy season leads to continuous flow of the resin. The resin so accumulated on the stems has been reported to become air-dried making the plants susceptible to fire (Eshete, 2002).

## Seed germination

Seeds from tapped trees have a low germination rate. In Eritrea these rates were ca. 14 per cent compared to rates above 80 per cent for seeds from untapped stands (Rijkers et al, 2006). Germination tests with combinations of temperature and water availability showed that germination was no problem: B. papyrifera has little or no dormancy (Gebremedhin and Negash, 1999; Ogbazghi, 2001).

## Grazing: Effect of enclosures on regeneration

Unregulated free-range grazing has a devastating effect on *Boswellia* regeneration. Through removal and trampling of the aerial parts of the plant and predation of fallen seed, grazing is a major factor hindering the natural regeneration of *B. papyrifera*. In all instances where the deleterious impact of grazing was studied, seedling establishment was better in enclosed or fenced experimental plots than in openly grazed sites. Seeds and seedlings of *B. papyrifera* are vulnerable to grazing because:

1  the seeds have epigeal growth;
2  the seedlings are succulent and palatable and are therefore preferred by livestock and wildlife for grazing and browsing; and
3  the seedlings grow too slowly to escape grazers rapidly.

In Eritrea, for instance, in three years seedlings attained a maximum height of only 15cm and basal diameter of 1.5cm (Ogbazghi, 2001). This elongated juvenile period increases the risk of being eaten, trampled or damaged. In Tigray, seedling densities were lower in open-grazed areas than in closed ones (Negussie et al, 2008).

Another factor in this respect is fire, which can kill most seedlings. However, no studies so far have been conducted to quantify the fire effects.

# RESTORATION POSSIBILITIES

Regeneration problems of *Boswellia* are evident and are clearly negatively impacting future populations of this important tree species, which represents a threat in both ecological and economic terms. In Box 7.2 and the following sections, we suggest a number of possible restoration activities.

## 1. Improve seed availability

Though the mechanism is little understood, undoubtedly intensive tapping negatively affects production of seeds (Rijkers et al, 2006). Less intense resin production approaches should be implemented by reducing the number of tapping points per tree and frequency of tapping per season, introducing resting periods or putting a minimum diameter size for tapping. We suggest that the number of tapping points per tree should not exceed four. The resting period after an intense tapping period should not be less than four years to allow for recuperation. If it is not possible to leave the trees untapped for the specified period of years, leaving untapped plus mother trees in a systematic way, so that regular seed dispersal will be possible, would be a good option. Yet the method, frequency and intensity of tapping (see Tadesse et al, 2004) and especially its impact on the physiology of the tree, needs further research.

---

**BOX 7.2 RESTORATION POSSIBILITIES FOR**
***BOSWELLIA PAPYRIFERA* IN ETHIOPIA AND ERITREA**

*1. Improve seed availability*
- Tap less intensely
- Introduce rest periods (intervals with and without tapping)
- Leave seed-producing trees, of a number of sizes, untapped

*2. Protect germinants, seedlings and sprouts, and saplings from grazing*
- Enforce enclosures (guards, fences)
- Determine period that the species needs to grow through the grazing-sensitive stage
- Determine growth and development aspects in the first years, and sensitivity to drought periods

*3. Increase availability by enrichment seeding and enrichment planting*
- Develop community programmes with enrichment actions where short- and long-term benefits are clear

*4. Protect germinants, seedlings and sprouts, and saplings from fire*
- Increase awareness of the damage of fire in these systems
- Stimulate and enforce community control of agricultural fire

*5. Change land and tree ownerships*
- Give farmers ownership over trees on their land

---

## 2. Protect germinants, seedlings, sprouts, and saplings from grazing

Grazing clearly has a detrimental effect on small *Boswellia* plants. The use of closed areas needs to be strengthened. Already promising results are being obtained from several areas (for Tigray, see Negussie et al, 2008; for Eritrea, see Ogbazghi, 2001). The next step should be to study the management of these areas in order to avoid conflict between the different land uses and benefit-sharing, and to maximize the economic value of enclosures (for example through enrichment planting).

Understanding the effects that environmental and natural changes have on the individual fitness of trees and on the overall population numbers in certain parts of a population life cycle is a key question. It is important to identify the most sensitive and vulnerable parts of a population life cycle. Management efforts should focus on these phases.

# 3. Increase availability by enrichment seeding and enrichment planting

Artificial regeneration by addition of seeds or by enrichment planting has to be seen as a serious possibility for restoring populations. However, it has received little attention so far. A number of nurseries in Ethiopia are attempting to raise seedlings of *B. papyrifera* for enrichment planting. However, success has been limited. For example, only 4.5 per cent (in 1999) and 8.7 per cent (in 2000) of nursery-raised seedlings planted in the field in northern Ethiopia have survived (Gebrehiwot et al, 2003). Similar low survival was also reported in Eritrea (Ogbazghi, 2001). The development during the seedling phase is quite slow, which makes the seedlings vulnerable to browsing and also to recurrent fires. The low survival rate can also be attributed to the lack of silvicultural knowledge of the species, including time of seed collection, nursery practices, choice of appropriate planting sites and post-planting care. Several companies are currently planting areas with *Boswellia*, but establishment and growth results are not yet available (Tousseyn, 2004; personal communication).

Recently, cuttings have been planted and the results are promising. Large cuttings perform better than small ones and the collection period also has a strong effect on rooting performance (Haile et al, submitted; Negussie, 2008). The latex of *Euphorbia abyssinica* was proved to stimulate rooting of such cuttings (Negussie et al, 2009).

# 4. Protect germinants, seedlings and sprouts, and saplings from fire

Deliberate or accidental fire has and will remain a major threat in the *B. papyrifera* woodlands. Deliberate fire should be avoided by local participation and benefit-sharing. Accidental fire can be avoided by using the appropriate tapping season; in other words avoiding tapping in the rainy season reduces fire incidence. Proper management of personnel (daily labourers) during cooking and other activities in the forest and proper training for farmers for the production of honey can also reduce the occurrence of fires that can result from human use of fire for different purposes.

# 5. Change land and tree ownerships

Forest ownership is a crucial issue and should be resolved for successful forest management. Despite the presence of different kinds of ownership, the general trend is decentralization. It is expected that 40 per cent of the world's forests will be managed or owned by communities and individuals by 2050 (FAO, 2003). Therefore it is important to recognize that without secure tenure rights, indigenous and other local groups will lack long-term financial incentives for converting their forest resources into economically productive assets for their own development. There is also widespread acknowledgement that govern-ments and public-forest management agencies have not been good stewards of public forests. Consequently many countries have taken steps towards develop-ing a more participatory and collaborative form of forest governance and management. The new governance structures recognize local communities as a primary stakeholder in forest conservation and offer them incentives, in the form of ownership/user rights, benefit-sharing mechanisms and so on, to participate in forest protection and management (see Amente et al, Chapter 12 of this volume; Schmitt, Chapter 16 of this volume). In this way less powerful groups, such as women and the poor, will be considered instead of marginalized when conflicts on forest resources need to be resolved.

The influx of hired daily labourers aggravates the negative impact of tapping. Many migrant workers lack the necessary awareness, and are not inter-ested in the long-term fate of the tree. Adequate promotion, training and salaries may increase willingness of local people to participate in *Boswellia* tapping. This creates a benefit-sharing mechanism for better forest manage-ment.

Lack of tree tenure is at the heart of the observed problems in the *Boswellia* growing areas (Ogbazghi, 2001). The tree-tenure problem should be resolved by means of ownership/user rights or benefit-sharing.

We conclude that very low levels of regeneration were seen almost all over the growing niche of *B. papyrifera*. Such a low regeneration rate is partic-ularly worrying given the failure to grow it as nursery-raised seedlings. The concerns put forward by different parties (for example TRAFFIC, see Marshall, 1998) are justified. There are a number of possible interpretations of the results of this study. It may be that regeneration is episodic, depen-dent, for example, on the availability of good climatic conditions. Climatic variability has been suggested as a major cause of episodic recruitment among woody plants in arid environments (see, for example, Jordan and Nobel, 1979 and 1981; Turner, 1990; Pierson and Turner, 1998; Bowers and Turner, 2002; Bowers et al, 2004). However, as we have shown in this chapter, human-controlled factors may be the more important impediments for proper *Boswellia* regeneration and thus for the long-term provision of its important non-timber product, frankincense.

# REFERENCES

Alder, D. (1995) 'Growth modelling for mixed tropical forests', Tropical Forestry Paper No. 30, Oxford Forestry Institute, Oxford, UK

Anonymous (1997) 'Technical manual for the development/propagation and production of incense resource base. Amhara National Regional State, Bahar Dar' (unpublished)

Azene, B-T., Birnie, A. and Tengnas, B. (1993) *Useful Trees and Shrubs for Ethiopia. Identification, Propagation and Management for Agricultural and Pastoral Communities*, Regional Soil Conservation Unit, SIDA, Nairobi

Bailey, R. L. and Dell, T. R. (1973) 'Quantifying diameter distributions with the Weibull function', *Forest Science*, vol 19, pp97–104

Bell, D. T. (2001) 'Ecological response syndromes in the flora of southwestern Western Australia: Fire resprouters versus reseeders', *Botanical Review*, vol 67, pp417–440

Bell, T. L. and Ojeda, F. (1999) 'Underground starch storage in *Erica* species of the Cape Floristic Region – Differences between seeders and resprouters', *New Phytologist*, vol 144, pp143–152

Bell, T. L. and Pate, J. S. (1996) 'Growth and fire response of selected Epacridaceae of south-western Australia', Australian Journal of Botany, vol 44, no 5, pp509–526

Bellingham, P. J. and Sparrow, A. D. (2000) 'Resprouting as a life history strategy in woody plant communities', *Oikos*, vol 89, pp409–416

Bond, W. J. and Midgley, J. J. (2001) 'Ecology of sprouting in woody plants: The persistence niche', *Trends in Ecology and Evolution*, vol 16, pp45–51

Bowers, J. E. and Turner, R. M. (2002) 'The influence of climatic variability on local population dynamics of *Cercidium microphyllum* (foothill paloverde)', *Oecologia*, vol 130, pp105–113

Bowers, J. E., Turner, R. M. and Burgess, T. L. (2004) 'Temporal and spatial patterns in emergence and early survival of perennial plants in the Sonoran Desert', *Plant Ecology*, vol 172, pp107–119

Bullock, S. H., Mooney, A. A. and Medina E. (eds) (1995) *Seasonally Dry Tropical Forests*, Cambridge University Press, Cambridge, UK

Chikamai, B. N. (ed) (2002) *Review and Synthesis on the State of Knowledge of Boswellia spp. and Commercialisation of Frankincense in the Drylands of Eastern Africa*, KEFRI, Nairobi

Condit, R., Sukumar, R., Hubbel, S. and Foster, R. B. (1998) 'Predicting population trends from size distributions: A direct test in a tropical tree community', *The American Naturalist*, vol 152, no 4, pp495–509

Eshete, A. (2002) 'Regeneration status, soil seed banks and socio-economic importance of *B. papyrifera* in two woredas of North Gondar Zone, northern Ethiopia', MSc thesis, Swedish University of Agricultural Sciences, Skinnskatteberg, Sweden

FAO (2003) State of the World's Forests, FAO, Rome

Fitwi, G. (2000) 'The status of gum arabic and resins in Ethiopia', report of the Meeting of the Network for Natural Gum and Resins in Africa (NGARA) 29–31 May, Nairobi, pp14–22

Friis, I. (1986) 'The forest vegetation of Ethiopia', *Acta Universitas Upsalienses, Symbolae Botanicae Upsaliensis*, vol 36, no 2, pp31–47

Friis, I. (1992) 'Forests and forest trees of northeast tropical Africa', HMSO Kew Bulletin Additional Series XV, London

Gebremedhin, T. (1997) '*Boswellia papyrifera* (Del.) Hochst. from Western Tigray. Opportunities, constraints and seed germination responses', MSc thesis, Swedish University of Agricultural Sciences, Skinnskatteberg, Sweden

Gebremedhin, T. and Negash, L. (1999) 'The effect of different pre-sowing seed treatments on the germination of *Boswellia papyrifera*, a key dry-land tree', *Ethiopian Journal of Natural Resources*, vol 1, no 1, pp37–55

Gebrehiwot, K. (2003) 'Ecology and management of Boswellia papyrifera (Del.) Hochst. dry forests in Tigray, Northern Ethiopia', PhD thesis, Georg-August-University of Göttingen, Germany

Gebrehiwot, K., Muys, B., Haile, M. and Mitloehner, R. (2003) 'Introducing *Boswellia papyrifera* (Del.) Hochst and its non-timber forest product, frankincense', *International Forestry Review*, vol 5, pp348–353

Gebrehiwot, K., Muys, B., Haile, M. and Mitloehner, R. (2005) 'The use of plant water relations to characterize tree species and sites in the drylands of northern Ethiopia', *Journal of Arid Environments*, vol 60, no 4, pp581–592

Gentry, A. H. (1995) 'Diversity and floristic composition of neotropical dry forest', in S. H. Bullock, A. A. Mooney and E. Medina (eds) (1995) *Seasonally Dry Tropical Forests*, Cambridge University Press, Cambridge, UK, pp146–194

Groombridge, B. and Jenkins, M. (2002) *World Atlas of Biodiversity*, prepared at UNEP World Conservation Monitoring Centre, California University Press, Berkeley, Los Angeles, CA

Haile, G., Gebrehiwot, K., Lemenih, M. and Bongers, F. (submitted) 'Time of collection and cutting sizes affect vegetative propagation of *Boswellia papyrifera* (Del.) Hochst through leafless branch cuttings', submitted to *Journal of Arid Environments*

Hedberg, I. and Edwards, S. (eds) (1989) *Flora of Ethiopia, Volume 3: Pittosporaceae to Araliaceae*, The National Herbarium, Addis Ababa University, Addis Ababa

Janzen, D. H. (1988) 'Tropical dry forests: The most endangered major tropical ecosystem', in E. O. Wilson and F. M. Peter (eds) *Biodiversity*, National Academy Press, Washington, DC, pp130–137

Jordan, P. W. and Nobel, P. S. (1979) 'Infrequent establishment of seedlings of *Agave deserti* (Agavaceae) in the northern Sonoran Desert', *American Journal of Botany*, vol 66, pp1079–1084

Jordan, P. W. and Nobel, P. S. (1981) 'Seedling establishment of *Ferocactus acanthodes* in relation to drought', *Ecology*, vol 62, pp901–906

Kamziah, A. K., Ahmad, M. I. and Ahmad Zuhaidi, Y. (2000) 'Modelling diameter distribution in even-aged and uneven-aged forest stands', *Journal of Tropical Forest Science*, vol 12, pp669–681

Lemenih, M. and Teketay, D. (2003) 'Frankincense and myrrh: Distribution, production, local uses, development potential, medicinal and industrial uses, and research needs in Ethiopia', *SINET: Ethiopian Journal of Science*, vol 26, no 1, pp63–72

Lemenih, M., Feleke, S. and Tadesse, W. (2007) 'Constraints to smallholders' production of frankincense in Metema district, northwestern Ethiopia', *Journal of Arid Environments*, vol 71, pp393–403

Lykke, A. M. (1998) 'Assessment of species composition change in savanna vegetation by means of woody plants' size class distributions and local information', *Biodiversity and Conservation*, vol 7, pp1261–1275

Marshall, N. (1998) *Searching for a Cure: Conservation of Medicinal Wildlife Resources in East and Southern Africa*, TRAFFIC International, Cambridge, UK

Mayaux, P., Holmgren, P., Achard, F., Eva, H., Stibig, H. J. and Branthomme, A. (2005) 'Tropical forest cover change in the 1990s and options for future monitoring', *Philosophical Transactions of the Royal Society B, Biological Sciences*, vol 360, pp373–384

Moges, Y. and Kindu, M. (2006) 'Effects of fencing and ground cultivation on natural regeneration of *Boswellia papyrifera* in Metema Wereda, Ethiopia', *Journal of the Drylands*, vol 1, no 1, pp45–51

Murphy, P. G. and Lugo, A. E. (1986) 'Ecology of tropical dry forest', *Annual Review of Ecology and Systematics*, vol 17, pp67–88

Negussie, A. (2008) 'The damage of long horn beetle (*Idactus spinipennis* Gahan) on dry deciduous *Boswellia* woodlands in central and western Tigray, northern Ethiopia', MSc thesis, Mekelle University, Mekelle, Tigray, Ethiopia

Negussie, A., Aerts, R., Gebrehiwot, K. and Muys, B. (2008) 'Seedling mortality causes recruitment limitation of *Boswellia papyrifera* in northern Ethiopia', *Journal of Arid*

*Environments*, vol 72, pp378–383

Negussie, A., Aerts, R., Gebrehiwot, K., Prinsen, E. and Muys, B. (2009) '*Euphorbia abyssinica* latex promotes rooting of *Boswellia* cuttings', *New Forests*, vol 37, pp35–42

Ogbazghi, W. (2001) 'The distribution and regeneration of *Boswellia papyrifera* (Del.) Hochst. in Eritrea', PhD dissertation, Wageningen University, Tropical Resource Management Papers No. 35, Wageningen University, Wageningen, The Netherlands

Ogbazghi, W., Bongers, F., Rijkers, T. and Wessel, M. (2006a) 'Population structure and morphology of the frankincense tree *Boswellia papyrifera* along an altitude gradient in Eritrea', *Journal of the Drylands*, vol 1, no 1, pp85–94

Ogbazghi, W., Rijkers, T., Wessel, M. and Bongers, F. (2006b) 'The distribution of the francincense tree *Boswellia papyrifera* in Eritrea: The role of environment and land use', *Journal of Biogeography*, vol 33, pp524–535

Pate, J. S., Froend, R. H., Bowen, B. J., Hansen, A. and Kuo, J. (1990) 'Seedling growth and storage characteristics of seeder and resprouter species of Mediterranean-type ecosystems of SW Australia', *Annals of Botany*, vol 65, pp585–601

Pierson, E. A. and Turner, R. M. (1998) 'An 85-year study of Saguaro (*Carnegiea gigantea*) demography', *Ecology*, vol 79, pp2676–2693

Rijkers, T., Ogbazghi, W., Wessel, M. and Bongers, F. (2006) 'The effect of tapping for frankincense on sexual reproduction in *Boswellia papyrifera*', *Journal of Applied Ecology*, vol 43, pp1188–1195

Ros-Tonen, M., Dijkman, W. and Lammerts van Bueren, E. (1995) *Commercial and Sustainable Extraction of Non-Timber Forest Products. Towards Policy and Management-Oriented Research Strategy*, The Tropenbos Foundation, Wageningen, The Netherlands

Sheil, D. and Salim, A. (2004) 'Forest tree persistence, elephants, and stem scars', *Biotropica*, vol 36, pp505–521

Swaine, A. M. (1998) 'Assessment of species composition change in savannah vegetation by means of woody plants' size class distribution and local information', *Biodiversity and Conservation*, vol 7, pp1261–1275

Tadesse, W., Teketay, D., Lemenih, M. and Fitwi, G. (2002) 'Country report for Ethiopia', in B. N. Chikamai (ed) *Review and Synthesis on the State of Knowledge of Boswellia Species and Commercialisation of Frankincense in the Drylands of Eastern Africa*, Food and Agriculture Organization of the United Nations, Rome, pp14–31

Tadesse, W., Feleke, S. and Eshete, T. (2004) 'Comparative study of traditional and new tapping method on frankincense yield of Boswellia papyrifera', Ethiopian Journal of Natural Resources, vol 6, pp287–299

Tousseyn, D. (2004) 'Gum and resin resources of Ethiopia. A prospect for a sustainable market development of gum resins from *Boswellia* and *Commiphora* species, propositions for an ecological revival', internship report, Wageningen University, Wageningen, The Netherlands

Turner, R. M. (1990) 'Long-term vegetation change at a fully protected Sonoran Desert site', *Ecology*, vol 71, pp464–477

Vanclay, J. K. (1994) *Modelling Forest Growth and Yield. Application to Mixed Tropical Forests*, CAB International, Wallingford, UK

Vollesen, K. (1989) 'Burseraceae', in I. Hedberg and S. Edwards (eds) *Flora of Ethiopia, Volume 3*, National Herbarium, Addis Ababa University, Addis Ababa, and Uppsala University, Uppsala, Sweden, pp442–478

Verdaguer, D. and Ojeda, F. (2002) 'Root starch storage and allocation patterns in seeder and resprouter seedlings of two cape *Erica* (Ericaceae) species', *American*

*Journal of Botany*, vol 89, pp1189–1196

Vieira, D. L. M. and Scariot, A. (2006) 'Principles of natural regeneration of tropical dry forests for restoration', *Restoration Ecology*, vol 14, no 1, pp11–20

Von Breitenbach, F. (1963) *The Indigenous Trees of Ethiopia*, Ethiopian Forestry Association, Addis Ababa

White, F. (1983) *The Vegetation of Africa: A Descriptive Memoir to Accompany the UNESCO/AETFAT/UNSO Vegetation Map of Africa*, UNESCO, Paris, France

Wiegand, K., Jeltsch, F. and Ward, D. (1999) 'Analysis of the population dynamics of *Acacia* trees in the Negev Desert, Israel, with a spatially-explicit computer simulation model', *Ecological Modelling*, vol 117, pp203–224

Zhang, L., Gove, J. H., Liu, C. and Leak, W. B. (2001) 'A finite mixture of two Weibull distributions for modeling the diameter distributions of rotated sigmoid, uneven-aged stands', *Canadian Journal of Forest Research*, vol 31, pp1654–1659

8

# Vegetation Variation in Forest–Woodland–Savanna Mosaics in Uganda and its Implication for Conservation

*Grace Nangendo, Frans Bongers, Hans ter Steege and Alfred De Gier*

## INTRODUCTION

Forest–woodland–savanna (FWS) mosaics are complex, highly varied and dynamic landscapes that cover extensive areas of the tropical world. The largest FWS formations in South America occur in Brazil and cover an area equivalent to Western Europe (Furley, 1999). Another big mosaic area stretches from Guyana into Brazil (Jansen-Jacobs and ter Steege, 2000). In Africa, FWS mosaics are prevalent in areas surrounding the Congo basin forests, including Uganda. These areas have been defined as transitional zones between the moist tropical forest and the drier savanna landscape typical of much of Africa. On the northern side, the transition occurs at about 8°N, with the exception of Togo and Benin and part of Côte d'Ivoire (Gautier and Spichiger, 2004). Many FWS mosaics occur in Uganda because of its location in a zone of overlap between the ecological communities characteristic of the dry east African savannas and the west and central African rainforests (Howard, 1991).

The FWS mosaics exhibit high beta-diversity due to their heterogeneity (Mayle et al, 2007). Their existence as a mixture of vegetation types enables them to hold a unique assemblage of flora (Ratter et al, 1997) and fauna (Skowno and Bond, 2003; Mayle et al, 2007; Coppedge et al, 2008) Also animals that require more than one vegetation type for their day-to-day survival (for example breeding in one and feeding in another) often find a home in mosaic landscapes.

Until recently, FWS mosaics were considered poor in terms of biodiversity (Furley, 1999). Most specialists considered them as either being mismanaged forest areas (Clayton, 1958) or an intermediate stage in a gradual forest degradation towards savanna. As a consequence, hardly any specific management plans exist for their conservation. Some studies indicate that recent conservation initiatives have neglected these high-biodiversity areas (Smith et al, 2001; Mayle et al, 2007). Approaches to prioritizing conservation efforts for the Earth's terrestrial biota also consider species richness of a specific habitat as a basis for deciding conservation areas (Oslon and Dinerstein, 1998). In Uganda, undisturbed forest areas were designated as nature reserves for the conservation of forest species, while national parks, where there is regular disturbance, were designated for the conservation of savanna species. This left out the areas where forest, woodland and savanna coexist, despite the fact that several forest reserves, classified as FWS mosaics, were identified as areas of high biodiversity value.

Although forest, woodland and savanna vegetations have been extensively studied, the areas where these three vegetation types co-occur have received little attention. This has resulted in an information gap, which is also reflected in biodiversity conservation planning where no specific plans exist for conservation of FWS mosaics. Assessing the species composition and the assemblage of the vegetation within an FWS mosaic and the factors that maintain such landscapes would provide key information for conserving such areas.

In FWS mosaics the woody plants' composition, canopy cover and vegetation structure are the main differentiating factors. Whereas forest and woodland vegetation have a near continuous woody plant cover, savannas have a discontinuous one. Forest, however, has a multilayered canopy and often hosts larger and taller woody plants than woodland.

In this chapter we present a synthesis of results obtained from a study of an FWS mosaic in Uganda. We analysed its vegetation cover, its woody plant composition variation and the changes therein. From the analysis of the environmental variables that were considered to have influence on the vegetation composition and distribution, we found that fire had a very high impact. We therefore assessed the impact of annual burning on woody plant composition of three vegetation types subject to a similar fire regime for 46 years. We then provide strategies that could guide the management of such landscapes.

# FWS MOSAIC OCCURRENCE AND CONSERVATION STATUS IN UGANDA

FWS mosaics are prevalent in many areas adjacent to closed forest. They are difficult to isolate and map because of the complex boundaries between them and the adjacent vegetation types, either closed forest or grassland areas. Where FWS mosaics occur and the climate and soil conditions would theoretically be suitable for forest, they are often considered relics of drier periods (Schwartz et al, 1996) and are maintained through purposeful use of fire (King et al, 1997).

More than half of Uganda is covered by FWS mosaics (Burgess et al, 2004), mostly because of its location in a zone of overlap between the ecological communities characteristic of the dry east African savannas and the central African rainforests. This ecoregion covers the whole area surrounding Lake Victoria, being bordered by the Albertine rift montane forests to the west, Mount Elgon to the northeast and stretching northwards to include the southern part of Murchison Falls National Park. The area is ranked among the richest in birds, mammals and butterfly populations (Burgess et al, 2004). There has, however, been high habitat loss and it is one of the few areas in Africa where the human population is expected to be highest by 2025 (between 200 to 1000 individuals per km$^2$). It is, therefore, grouped among the critically threatened areas and is of the highest conservation priority.

Budongo Forest Reserve, located in the northwestern part of Uganda, was selected for the study (Figure 8.1). It consists of two forest blocks, the Main Budongo (made up of Budongo, Siba, Busaju forests) and the Kaniyo-Pabidi

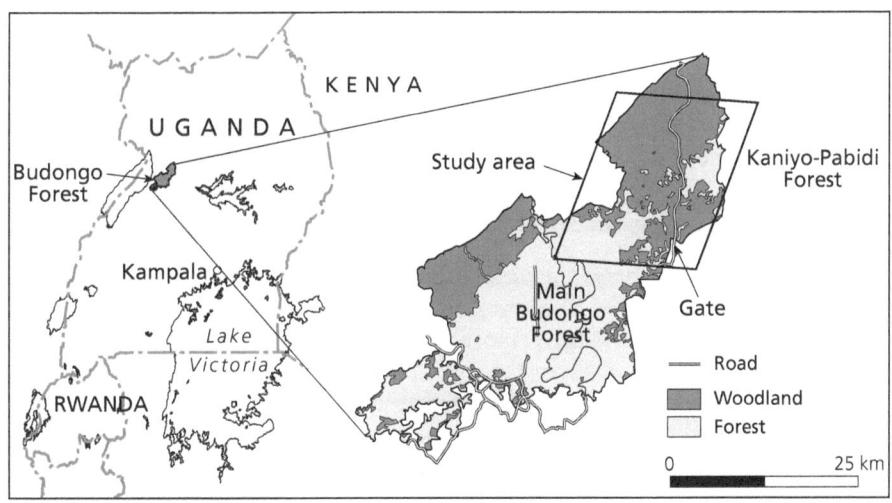

**Figure 8.1** *Study area map showing the location of the entrance gate to the conservation area since 1985 (gate location); before 1985, the gate was located at the escarpment, north of the forest reserve*

**Figure 8.2** *Some of the vegetation variation within the forest: (a) height and greenness differences between the woodland, foreground, and forest, background; (b) variation in vegetation cover in the woodland and savanna areas*

forest blocks. It is characterized by high temperatures (up to a monthly average of 32°C during December–February) and rainfall of 1400–1525mm annually on 100 to 150 rainy days. The forest is of exceptional importance in terms of biodiversity, ranking third in overall importance in the country (Forest Department Uganda, 1999). The reserve, which consists of 53.7 per cent forest and 46.3 per cent 'grassland', covers 82,530ha, making it Uganda's largest forest reserve. A woodland–savanna area, interspersed with forest patches, commonly referred to as Kaniyo Pabidi Woodland, separates the Main Budongo forest block from the Kaniyo-Pabidi forest block.

In the woodlands, fire has been prevalent for hundreds of years (Paterson, 1991). The causes of fire include land clearance, verge clearance, lightning strikes, hunting and refreshing grass for the ungulates (Wheater, 1971). Figure 8.2 shows some of the vegetation structure differences of the reserve.

## VARIATION IN VEGETATION COVER AND RELATED SPECIES-DISTRIBUTION DYNAMICS

We identified and mapped the vegetation cover classes that existed in the area using a combination of the Spectral Angle Mapper and the Expert system classification methods. Based on remotely sensed data, six major classes were recognized in the study area. In the field, however, data were collected for only five of these classes. The sixth class, burnt area, was no longer identifiable by the time the field data were collected and was therefore identified only on the image. Class identification was based on percentage canopy cover, main woody plant species making up the canopy, and type and thickness of main undergrowth (Table 8.1). The percentage canopy cover layer was constructed from data collected in the field on a total of 591 points (for details, see Nangendo et al, 2007).

Table 8.1 *Criteria used for identification of cover classes*

| Cover type | Canopy cover (%) | Canopy woody species | Main undergrowth |
|---|---|---|---|
| Forest | 81–100 | Typical forest species | Minimum to no undergrowth |
| Closed woodland | 51–80 | Woodland species interspersed with pioneer species | *Papea carpensi* dominant |
| Open woodland | 31–50 | *Terminalia*-dominated woodland | *Panicum maximum* (partial cover) |
| Very open woodland | 11–30 | Mixed woodland species | *Panicum maximum* (continuous cover) |
| Wooded grassland | 0–10 | Mixed woodland species but smaller, shorter (<5m) and scattered | *Setaria* species (continuous cover) |
| Burnt area | NA | NA | NA |

We also analysed the vegetation cover changes and the associated succession patterns of the woodland–savanna areas. Specifically, we evaluated vegetation changes over a 17-year period and used a chronosequence approach to analyse the related vegetation composition changes.

Landsat 5 TM (1985) and Landsat 7 ETM+ (2002) images were used. Field data was collected from 268 locations within the study area. At each plot, a geographical coordinate, percentage canopy cover and fire scar age (recent, old or none), species names, and diameter at breast height (DBH) for all woody plants with DBH ⩾2cm were collected. The vegetation cover change was evaluated using the Image Differencing Change Detection method. Before image differencing, a vegetation index (modified soil adjusted vegetation index (MSAVI)) was calculated for each image. The two MSAVI maps were then subtracted from each other. Results showed that between 1985 and 2002 a vegetation increase of 15.1 per cent and a decrease of 14.3 per cent occurred. The increase and decrease in vegetation cover occurred in distinctively separate areas. Whereas vegetation increase was mainly between the two forest blocks and on the western side of the Kaniyo-Pabidi forest, the decrease was mainly at the northwestern edge of the Main Budongo forest.

Axis 1 of a detrended correspondence analysis (DCA), determined using abundance counts per plot of trees ⩾10cm DBH, was related to the change values (1985–2002), fire scar age estimates, stem density and percentage canopy cover. A significant negative correlation was obtained with all variables except stem density, which was not significant. A correlation of −0.59 was obtained between DCA axis 1 and change. Sites that decreased in MSAVI were generally at earlier stages of succession, while sites that increased in MSAVI were generally at advanced stages of succession.

A second DCA, where all species that had individuals in the three classes (DBH <2cm, 2cm⩾ DBH <10cm and DBH ⩾10cm) were included, showed two main types of establishment patterns. Some species, for example *Pterygota mildbraedii* and *Khaya anthotheca*, established and grew to maturity at the late succession stage, while and others, for example *Grewia mollis*, *Terminalia velutina* and *Lonchocarpus laxiflorus*, established seedlings early in the succession development but followed one of several growth patterns. Some species, such as *Grewia mollis*, *Annona senegalensis* and *Lonchocarpus laxiflorus*, established early and remained prevalent only in the early succession stages. Other species, for example *Maesopsis eminii* and *Terminalia velutina*, established early but their adults occurred at later stages of succession. Species observed in the early succession stage, apart from *Maesopsis eminii*, have fire-resistant adaptations that allow them to survive the fire disturbance. For *Maesopsis eminii*, whose seeds are dispersed mainly by birds, particularly hornbills (Cordeiro et al, 2004), there is regular reseeding of the disturbed sites.

Factors that may have contributed to vegetation increase, in addition to regeneration of seeds of the already existing trees, include the existing trees acting as perch sites for birds that bring in seed of early invaders (Guevara et al, 1986; Archer et al, 1988; McClanahan and Wolfe, 1993) and the release of 'Gullivers', a word coined by Bond and Van Wilgen (1996) for individuals held in the grass layer by heavy grazing or regular fires. The floristic transition is gradual as species composition changes to suit the changing environment. Similar trends were also observed in the lower part of Budongo forest (Sheil et al, 2000) and in Kibale forest (Lwanga, 2003).

Increase in vegetation cover could be due to the access limitations enforced by the Uganda Wildlife Authority and National Forest Authority (NFA) that make it increasingly difficult for local people, who live to the south of the woodland–savanna area, to enter and set fires. Furthermore, in 1994 the Kaniyo-Pabidi ecotourism site was established on the western edge of the Kaniyo-Pabidi forest to attract tourists to visit resident chimpanzee populations. This further discouraged local people from accessing surrounding woodland–savanna areas and possibly provides part of the explanation why there was dramatic vegetation increase north of the ecotourism site.

For the late succession species, it is suspected that they established over a prolonged period after the canopy closed and conditions became suitable. If fire does not occur in these areas for the next 10 to 20 years, the established species will grow to overtake the early succession species. Hudak et al (2004) observed that reversion of such landscapes to the previous vegetation states is often almost impossible.

With frequent disturbance, only fire-resilient species may survive. Complete elimination of fire will provide opportunity for the forest species to get established and probably take over the area. In between these extremes, many vegetation types can exist and even be stable, although succession may still occur along varied pathways (Bongers and Blokland, 2004; Favier et al, 2004a). Maintenance of extreme protection restrictions risks losing many of the species typical for woodland–savanna areas, but reinstitution of intensive fire regimes, on the other hand, would halt any successional development present in these areas. Therefore, a balance between the two scenarios may have to be identified and implemented.

# COMPOSITIONAL VARIATION OF WOODY PLANTS AND THE ROLE OF FIRE

We analysed the existing vegetation types and characterized them in terms of woody plant composition. We assumed that all vegetation types present in the study area, and the species they supported, were an integral part of a compositional/successional gradient that stretched across the FWS mosaic. We quantified this gradient using DCA and assessed the variation of species composition along the gradient. Furthermore, we assessed whether satellite image classification obtained in the previous section could be used to adequately map the vegetation structure and woody plant composition of the area. We collected the following data from 591 plots: species names, DBH for all woody plants ⩾10cm DBH, percentage canopy cover and a fire indicator value.

A total of 26,076 individuals from 121 species, 89 genera and 38 families were found. The most species-rich family was Moraceae, with 11 per cent of all species found (13), followed by Euphorbiaceae and Mimosaceae with 8 per cent each (10). The most species-rich genus was *Ficus*, with 5 per cent of all species (6), followed by *Acacia*, *Albizia*, *Celtis* and *Combretum* with three per cent each (4). The most abundant genus, in terms of total individuals encountered, was *Combretum*, with close to 16 per cent of all individuals, followed by *Terminalia* (14 per cent) and *Grewia* (13 per cent). Further analysis of the species composition of the FWS mosaic in relation to the vegetation cover classes obtained in the previous section revealed that the vegetation cover classes were compositionally separable and each class had indicator species that could be used to identify it (see Nangendo et al, 2006, for details).

The DCA ordered the species mainly along the first axis (Figure 8.3 and Table 8.2). The effect of the second axis was only evident close to zero along axis 1, the forest side, where there appeared to be two groups of species. Fire best explained the gradient in species composition ($r^2 = 0.324$), followed by slope and distance from gate. All site variables individually had a positive correlation with the DCA, with the exception of distance from gate, which had a negative correlation. Plots of different fire classes also differed significantly in their species composition and indicator species were identified for each fire class.

Areas that had recent fires, and are probably most frequently burnt, had species that characteristically displayed fire-resistant traits, for example thick bark, peeling off of the old bark and good sprouting ability after a fire (Saha and Howe, 2003; Zida et al, 2007). In the old-fire class, seed dispersal, a factor not explored in this study, may have an important role. A number of the species that occurred in the old-fire class were most abundant in the no-fire class. Their seeds were probably dispersed into the old-fire class areas, by wind for example, and, when conditions became favourable, they got established.

To identify how regular fire would influence woody species composition, another study was carried out on three vegetation types to identify the effect of 46 years of annual burning. After 46 years, all large trees had died off and the

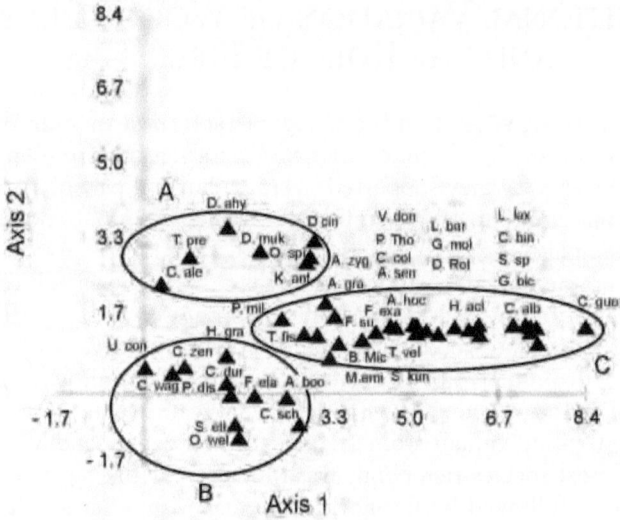

**Figure 8.3** *DCA graph showing species composition variation along the first two ordination axes*

*Note:* The full names of each of the species are listed in Table 8.3. A, B and C represent groupings of species that could be identified in the graph.

**Table 8.2** *Full names of the species in Figure 8.3*

| Graph name | Full name | Graph name | Full name |
|---|---|---|---|
| A. hoc | Acacia hockii | G. bic | Grewia bicolor |
| A. gra | Albizia grandibracteata | G. mol | Grewia mollis |
| A. zyg | Albizia zygia | H. gra | Holoptelea grandis |
| A. boo | Alstonia boonei | H. aci | Hymenocardia acida |
| A. sen | Annona senegalensis | K. ant | Khaya anthotheca |
| B. mic | Bridelia micrantha | L. bar | Lannea barteri |
| C. sch | Caloncoba schweinfurthii | L. lax | Lonchocarpus laxiflorus |
| C. alb | Carpololobia alba | M. emi | Maesopsis eminii |
| C. dur | Celtis durandii | M. dis | Margaritaria discoidea |
| C. wag | Celtis waghtii | O. wel | Olea welwitschii |
| C. zen | Celtis zenkeri | O. spi | Oncoba spinosa |
| C. bin | Combretum binderanum | P. dis | Phyllanthus discoideus |
| C. col | Combretum collinum | P. tho | Piliostgma thonningii |
| C. gue | Combretum guenzii | P. mil | Pterygota mildbreadii |
| C. mol | Combretum molle | S. ell | Sapium ellipticum |
| C. ale | Cynometra alexandri | S. spp | Securidaka spp. |
| D. cin | Dichrostachys cinerea | S. kun | Stereospermum kunthianum |
| D. aby | Diospyros abyssinica | T. fis | Tapura fischeri |
| D. muk | Dombeya mukole | T. vel | Terminalia velutina |
| D. rot | Dombeya rotundifolia | T. pre | Trichillia preiureana |
| F. ela | Funtumia elastica | U. con | Uvariopsis congensis |
| F. exa | Ficus exasperata | V. don | Vitex doniana |
| F. sur | Ficus sur | | |

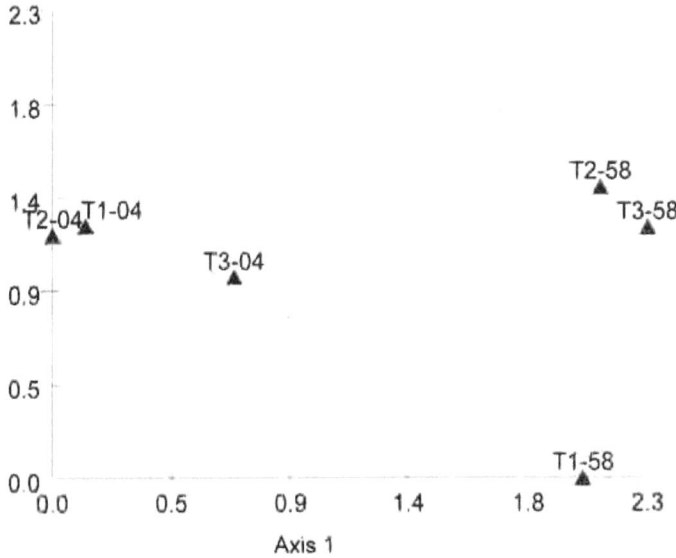

**Figure 8.4** *DCA graph indicating the relationship between the transects in terms of their species composition*

*Notes:* T1-58, T2-58 and T3-58 are the three transects in 1958 and T1-04, T2-04 and T3-04 are the respective transects in 2004.

three areas now shared many species. A comparison of the 1958 and 2004 woody plant composition of the respective vegetation types using the Sørensen similarity measure resulted in low similarities. The highest compositional similarity value was r = 0.62. A DCA run on the same data showed that the 2004 transects were all far from the 1958 transects (Figure 8.4) (Nangendo et al, 2005). The shift in location of all the 2004 transects, away from the 1958 transects, and their being compositionally close in 2004 shows that the vegetation is adjusting to suit the existing fire regime. Similar results were observed elsewhere (van Wilgen et al, 1998). If the current fire regime is maintained into the future, a homogeneous vegetation class that is compositionally different from any of the 1958 vegetation types may emerge.

Whereas high numbers of herbivores, especially elephants, have been associated with destruction of woody vegetation (Laws, 1970; Lewis, 1986), their elimination often leads to increase in vegetation cover but with lower diversity of both grasses and woody species (Smart et al, 1985). Increase in grass cover in turn leads to more intense fires (Huston, 1994). The hotter fires in turn favoured more fire-adapted woody species. To counterbalance this unifying effect, and since one of the aims of the park is to create a diverse habitat (Uganda Wildlife Authority, 2001), fire regimes need to be varied over the area. For future management support, permanent plots for vegetation monitoring in the park need to be (re-)established and maintained. Thresholds of vegetation change need to be set which, if reached, should trigger an assessment of the causes of the change and possibly to corrective actions. Additional data sets from other parts of the conservation area ought to be analysed and interpreted to substantiate the observed trend.

Despite the variation in rainfall over Budongo Forest Reserve, with the northern part receiving less rain than the south, the north still receives over 1200mm a year (Forest Department Uganda, 1997), which is sufficient for forest maintenance. Elephants that previously restricted forest expansion (Laws et al, 1975) no longer exist. The species turnover could possibly be explained by an additive effect of the environmental variables considered in this study, the historical impact of elephants and probably others that were not considered in this study, for example seed dispersal mechanisms, which have been shown to favour establishment of species with higher dispersal ability in the post-disturbance period (Ohsawa et al, 2002). However, just like in other studies where FWS occur (Hovestadt et al, 1999), fire plays a major role in controlling the species distribution pattern but it does not explain all the variation.

The Simpson index of the vegetation classes did not differ with the exception of wooded grassland, which was significantly higher than the others (Table 8.3). The forest class had the highest Fisher's alpha, followed by the closed woodland. The wooded grassland had the lowest Fisher's alpha. The basal area decreased from the forest (highest) to the wooded grassland (lowest). The stem density for the forest, closed woodland and open woodland were very close and differences were not significant. Whereas many individual trees occurred in each cover class, they varied in size, with the forest having larger diameter trees than the other cover classes. This study has thus shown that although the values for the forest, especially basal area, were often much higher, a major gradient stretching from the forest to the wooded grassland exists and species composition and forest structure vary along this gradient.

Eggeling (1947) and Sheil (1999) identified successional stages within the forest, with ironwood (*Cynometra alexandri*) at the climax end of the spectrum and the colonizing (woodland) forest as the starting point. In their study, the lowest diversity occurred in the ironwood forest. In our study the highest diversity was within the forest area and it gradually reduced until the wooded grassland. The colonizing forest, identified by Eggeling as the starting point of the succession, occurs somewhere towards the middle of the current gradient.

Although the forest side of the gradient had more species, and ought to be preserved (Sheil and Burslem, 2003), other species occur away from the forest environment. Hence, if the whole succession gradient occurs in an area, there would be more species (Connell, 1978; Huston, 1994; Massada et al, 2009) than if one or a few stages of the succession gradient were conserved. The maintenance of the high diversity of Budongo may be more attributed to the existence of all stages of the succession gradient, as was also observed elsewhere by Shea et al (2004), than to acquisition of more forest species from elsewhere, which often takes a long time (Chapman et al, 1997). For purposes of conserving woody plants in a dynamic landscape, it is important that each vegetation

**Table 8.3** *Comparison of cover class mean values for the Simpson index, Fisher's alpha, basal area and stem density*

|  | FO | CW | OW | VOW | WG |
|---|---|---|---|---|---|
| Simpson index | 0.28[a] | 0.30[a] | 0.31[a] | 0.30[a] | 0.49[b] |
| Fisher's alpha | 5.62[a] | 4.42[b] | 3.15[c] | 3.40[b,c] | 2.22[c] |
| Basal area | 28.43[a] | 14.87[b] | 12.25[b,c] | 9.43[c] | 2.62[d] |
| Stem density | 517.62[a] | 527.02[a] | 528.67[a] | 417.19[b] | 125.64[c] |

*Notes:* The class numbers consistently represent FO = forest, CW = closed woodland, OW = open woodland, VOW = very open woodland and WG = wooded grassland. Values of a variable, for example Simpson index, with a same letter (in superscript) are not significantly different from each other (ANOVA: $P = 0.05$).

class represented is included and maintained within the conservation area.

In areas where fire may be applied, the vegetation type and its development stage may affect the potential for ignition and spread of the fire (Everett et al, 2000; Burrows, 2008). Although no evidence exists of fires having destroyed tropical rainforests in Uganda, it has been observed elsewhere that tropical forests can burn (Laurence, 2003). It is, therefore, important that fire be used cautiously. Learning to use burning methods that have been used in the past (Goma et al, 2001; Massada et al, 2009) will be a prerequisite for improving current fire management.

In this respect, conservationists need to put more attention on the current vegetation management practices of the local people surrounding conservation areas (Leone and Lovreglio, 2004) since, in addition to possessing valuable fire management knowledge, they have been noted to also use fire destructively (Condit et al, 1998). A well-balanced management, including a controlled fire management system that will prevent forest from colonizing the whole area, or vice versa, yet allowing the existence of varying disturbance regimes, is a prerequisite for species and species diversity maintenance.

# CONSERVATION IMPLICATIONS

This study has shown that FWS mosaics have high vegetation variation. The vegetation cover classes identified were compositionally separable. Each vegetation cover class had indicator species that could typify it. These are therefore unique vegetation cover classes that need to be treated as important entities of the FWS mosaic when planning for conservation – a portion of each of the vegetation cover classes needs to be conserved to ensure a comprehensive species diversity coverage.

Knowledge of the compositional dynamics is central to understanding the ecological processes (Sheil et al, 2000). The species occurred as a continuum along a compositional gradient. Some of the forest species, for example *Uvariopsis congensis* and the *Celtis* species, had a restricted distribution and occurred at one end of the compositional gradient. Many savanna species, for example *Combretum collinum* and *Lonchocarpus laxiflorus*, occurred at the other end of the gradient. Yet others, for example *Funtumia elastica*, occurred over a wide range but had a peak abundance in a specific location along the gradient (Nangendo et al, 2006). To conserve the variation in species observed in such FWS mosaics, a representative portion of each part of the gradient has to be maintained.

A large number of the species identified in this study were wide ranging, in other words they occurred in more than one vegetation cover class. This means the species composition of a particular vegetation cover class is not confined to that vegetation cover class and that the species diversity found in a specific vegetation cover class is dependent on other vegetation cover classes for its maintenance. Maintenance of diversity of such a vegetation cover class thus requires conservation of the adjacent vegetation cover classes as well. Additionally, wide species distribution also indicates that identification of a vegetation cover class should not be based on species occurrence alone but also on their proportional abundance.

The forest edge provides a habitat for species that do not occur in extremely closed forest environments and for species whose juveniles occur along the forest edge but whose big trees occur inside the forest only (Favier et al, 2004b). Laws (1975) observed, during the period of high disturbance of the woodland areas in the Budongo forest, that *Cynometra alexandri*, a climax species (Eggeling, 1947), occurred at the forest edge. This indicates that, at the time, there was limited regeneration of forest edge species, or that the edge changed during the lifetime of this species. Also the species mixture (forest and savanna species) encountered in the closed woodland during the current study did not exist in the late 1960s. To maintain a varied range of species that occur along the forest edge, its disturbance levels need to vary, thus providing for forest progression in some areas while keeping a relatively open canopy environment in others.

The relatively high correlation coefficient ($r = -0.59$) observed between change in vegetation cover (1985–2002) and the gradient in species composition in the woodland areas suggests that the spatial gradient in species composition also reflects the temporal changes in vegetation taking place in the area. This spatial gradient in species composition was also highly correlated with the fire regime, and we thus expect that temporal changes in vegetation are the result of changes in fire management in the area. To maintain the current species composition, the forces that drive change need to be controlled.

Whereas some species only occurred in the early part of the species succession gradient (Nangendo, 2005), others occur in the late succession part. For another group of species, the different size classes (DBH $\geqslant$10cm, 2cm $\geqslant$ DBH <10cm, and DBH <2cm and $\geqslant$50cm in height) occupied different positions along the succession gradient. Whereas for the early and late succession species the specific environment in which they occurred needs to be conserved, for the ones whose diameter classes were more spread along the successional gradient, their continuity depends on conserving all succession stages that accommodate these diameter classes.

A combination of local people's burning practices, where small areas are burnt in several sites over time, and the NFA and the UWA restrictions have complemented each other to create a complex mosaic in this area. In north-central Oklahoma in the US, areas with a mosaic vegetation consisting of various recovery stages after disturbance held higher numbers of bird species than areas with more regular, uniform burning (Coppedge et al, 2008). This type of disturbance-related variation has facilitated the establishment and maintenance of a variety of species. On the other hand, it is evident from this study and work done elsewhere in Uganda (Lwanga, 2003) that in the absence of fire the forest takes over the woodland and savanna areas. Eggeling (1947) also argued that the woodland and savanna areas provided areas for forest expansion and that this maintained the succession of the forest communities. To maintain a mosaic landscape, a balance between forest establishment and woodland and savanna maintenance is critically needed. This requires a well-defined and well-implemented balance between fire establishment and fire restriction, including a need to control the extent of areas burnt in any single burning event.

Fire is vital for the control of bush encroachment. For the Budongo forest, as for many other forests in Uganda, early burning is recommended (Forest Department Uganda, 1997) but it is hardly practised by the field staff, mostly due to inadequate staffing at the forest stations (Forest Department Uganda, 1999), lack of explicit fire management plans and lack of training in fire management. Few funds are allocated for these, indeed expensive, activities (Skowno and Bond, 2003). Policymakers, both at ministry and institutional level (for example the NFA), need to recognize the value of FWS mosaics. This will then allow for planning for such landscapes, including allocation of funds and personnel, including training.

According to the Forestry Nature Conservation Master Plan (1999), the part of the Budongo forest where this study was carried out includes a recreation area (the Kaniyo-Pabidi forest), a nature reserve (north of the Kaniyo-Pabidi forest) and a buffer zone (south and west of the Kaniyo-Pabidi forest). In this study we observed much vegetation variation, with high conservation value in the nature reserve and the buffer zone. If the 'hands off' type of management currently practised in much of this area is continued, much of the vegetation variation will get lost through bush encroachment. For maintenance of the conservation values (for which these areas were set aside in the first place), active management of the area, including fire use, has to be revived.

The use of fire is beneficial to certain vegetation cover classes but may also destroy the to-be-conserved vegetation. At the same time, no single fire regime is optimal for the conservation of the full range of species along the gradient (Burrows, 2008). Whereas a vegetation cover class like the *Terminalia* woodland requires fire, which prevents establishment of forest species (Favier et al, 2004a), annual burning carried out for 46 years drastically reduced the abundance of *Terminalia velutina*, which was previously the most abundant species in that area (Nangendo et al, 2005). The areas that currently host abundant *Terminalia velutina* (the closed woodland and the open woodland)

(Nangendo et al, 2006) require less frequent fires: 3–5 year intervals and preferably early burning (note that the burning intervals suggested are estimates based on field observation and discussions with field staff). These fire regimes should be refined based on experimental studies. Seedlings and saplings of fire-intolerant species will be removed by fire (Zida et al, 2007; Levick et al, 2009.

The forest does not require fire disturbance; in fact it would not tolerate it. The very open woodland (2–3 year interval) and wooded grassland (1–2 year interval, preferably late dry season) do need fire. Grass, in addition to recovering fast after a fire, grows tall and thick and would not easily burn with early season fires. For the conservation of a landscape that varies in vegetation cover, varying fire regimes are needed. These findings corroborate earlier experimental work carried out in Uganda (Spence and Angus, 1971) and in north-central Oklahoma (Coppedge et al, 2008).

For landscapes where the management goal is to replace the FWS mosaic with a closed forest and other factors, for example nutrient supply and soil hydrology, are conducive to tree growth, exclusion of fire is recommended. At sites with a depleted seed bank, purposeful reseeding using the desired species is required.

## CONCLUSIONS

- The FWS mosaic has distinct vegetation cover classes which are sequentially organized (from forest to wooded grassland) in terms of species composition.
- To conserve the biodiversity in the FWS mosaic, a part of each vegetation cover class needs to be conserved. Preferably, an area including all cover classes should be selected, since the cover classes complement each other in terms of species composition.
- Vegetation classes differing in species composition converge towards species better adapted to the existing fire regime, especially when subjected to a repeated (annual) fire regime for a long time (over 40 years).
- Fire is essential for conserving FWS mosaics, and a well-balanced fire management system is needed to control forest expansion yet allowing the existence of varying fire disturbance regimes.

## REFERENCES

Archer, S., Scifres, C., Bassham, C. R. and Maggio, R. (1988) 'Autogenic succession in a subtropical savanna: Conversion of grassland to thorn woodland', *Ecological Monographs*, vol 58, pp111–127

Bond, W. J. and van Wilgen, B. W. (1996) *Fire and Plants*, Chapman and Hall, London

Bongers, F. and Blokland, E. (2004) 'Forêts secondaires: Stades de succession écologique et multiples chemins', in *FAO/UICN Atelier Regional sur la Gestion de Forets Tropicales Secondaires en Afrique Francophone: Realite et Perspectives*, FAO, Rome, pp19–33

Burgess, N., Hales, J. D. A., Underwood, E., Dinerstein, E., Oslon, D., Ituoa, I., Schipper, J., Ricketts, T. and Newman, K. (2004) *Terrestrial Ecoregions of Africa and Madagascar*, Island Press, Washington, DC

Burrows, N. D. (2008) 'Linking fire ecology and fire management in south-west Australian forest landscapes', *Forest Ecology and Management*, vol 255, pp2394–2406

Chapman, C. A., Chapman, L. J., Wrangham, R., Isabirye-Basuta, G. and Ben-David, K. (1997) 'Spatial and temporal variability in the structure of a tropical forest', *African Journal of Ecology*, vol 35, pp287–302

Clayton, W. D. (1958) 'Secondary vegetation and the transition to savanna near Ibadan, Nigeria', *Journal of Ecology*, vol 46, pp217–238

Condit, R., Sukumar, R., Hubbell, S. P. and Foster, R. B. (1998) 'Predicting population trends from size distributions: A direct test in a tropical tree community', *The American Naturalist*, vol 152, pp495–509

Connell, J. H. (1978) 'Diversity in tropical rain forests and coral reefs', *Science*, vol 199, pp1302–1310

Coppedge, B. R., Fuhlendorf, S. D., Harrell, W. C. and Engle, D. M. (2008) 'Avian community response to vegetation and structural features in grasslands managed with fire and grazing', *Biological Conservation*, vol 141, pp1167–1444

Cordeiro, N. J., Patrick, D. A. G., Munisi, B. and Gupta, V. (2004) 'Role of dispersal in the invasion of an exotic tree in an East African submontane forest', *Journal of Tropical Ecology*, vol 20, pp449–457

Eggeling, W. J. (1947) 'Observations on the ecology of the Budongo rain forest, Uganda', *Journal of Ecology*, vol 34, pp20–87

Everett, R. L., Schellhaas, R., Keenum, D., Spurbeck, D. and Ohlson, P. (2000) 'Fire history in the ponderosa pine/Douglas-fir forests on the east slope of the Washington Cascades', *Forest Ecology and Management*, vol 2000, pp207–225

Favier, C., Chave, J., Fabing, A., Schwartz, D. and Dubois, M. A. (2004a) 'Modelling forest–savanna mosaic dynamics in man-influenced environments: Effects of fire, climate and soil heterogeneity', *Ecological Modelling*, vol 171, pp85–102

Favier, C., de Namur, C. and Dubois, M. A. (2004b) 'Forest progression modes in littoral Congo, Central Africa', *Journal of Biogeography*, vol 31, pp1445–1461

Forest Department Uganda (1997) 'Forest management plan for Budongo Forest Reserve for 1997 to 2007', Forest Department, Kampala

Forest Department Uganda (1999) *Forestry Nature Conservation Master Plan*, Ministry of Water, Lands and Environment, Kampala

Furley, P. A. (1999) 'The nature and diversity of neotropical savanna vegetation with particular reference to the Brazilian cerrados', *Global Ecology and Biogeography*, vol 8, pp223–241

Gautier, L. and Spichiger, R. (2004) 'The forest–savanna transition in West Africa', in L. Poorter, F. Bongers, F. N. Kouamé and W. D. Hawthorne (eds) *Biodiversity of West African Forests: An Ecological Atlas of Woody Plant Species*, CAB International, Wallingford, UK, pp33–40

Goma, H. C., Rahim, K., Nangendo, G., Riley, J. and Stein, A. (2001) 'Participatory studies for agro-ecosystem evaluation', *Agriculture, Ecosystems and Environment*, vol 87, pp179–190

Guevara, S., Purata, S. E. and Maarel, E. V. D. (1986) 'The role of remnant forest trees in tropical secondary succession', *Vegetatio*, vol 66, pp77–84

Hovestadt, T., Yao, P. and Linsenmair, E. K. (1999) 'Seed dispersal mechanisms and the vegetation of forest islands in a West African forest savanna mosaic (Comoé National

Park, Ivory Coast)', *Plant Ecology*, vol 144, pp1–25

Howard, P. C. (1991) *Nature Conservation in Uganda Tropical Forest Reserves*, IUCN, Gland, Switzerland

Hudak, A. K., Fairbanks, D. H. K. and Brockett, B. H. (2004) 'Trends in fire patterns in a southern African savanna under alternative land use practices', *Agriculture, Ecosystems and Environment*, vol 101, pp307–325

Huston, M. A. (1994) *Biological Diversity. The Coexistence of Species on Changing Landscapes*, Cambridge University Press, London

Jansen-Jacobs, M. and ter Steege, H. (2000) 'Southwest Guyana: A complex mosaic of savannahs and forests', in H. ter Steege (ed) *Plant Diversity in Guyana: With Recommendations for a National Protected Area Strategy*, Backhuys publishers, Leiden, The Netherlands, vol 18, pp147–158

King, J., Moutsinga, J. B. and Doufoulon, G. (1997) 'Conversion of anthropogenic savanna to production forest through fire protection of the forest–savanna edge in Gabon, Central Africa', *Forest Ecology and Management*, vol 94, pp233–247

Laurence, W. F. (2003) 'Slow burn: The insidious effects of surface fires on tropical forests', *Trends in Ecology and Evolution*, vol 18, pp209–212

Laws, R. M. (1970) 'Elephants as agents of habitat and landscape change in East africa', *Oikos*, vol 21, pp1–15

Laws, R. M., Parker, I. S. C. and Johnstone, R. C. B. (1975) *Elephants and Their Habitants. The Ecology of Elephants in North Bunyoro, Uganda*, Oxford University Press, London

Leone, V. and Lovreglio, R. (2004) 'Conservation of Mediterranean pine woodlands: Scenarios and legislative tools', *Plant Ecology*, vol 171, pp221–235

Levick, S. R., Asner, G. P., Kennedy-Bowdoin, T. and Knapp, D. E. (2009) 'The relative influence of fire and herbivory on savanna three-dimensional vegetation structure', *Biological Conservation*, vol 142, pp1693–1700

Lewis, D. M. (1986) 'Disturbance effects on elephant feeding: Evidence of compression in Luangwa Valley, Zambia', *African Journal of Ecology*, vol 24, pp227–241

Lwanga, J. S. (2003) 'Forest succession in Kibala National Park, Uganda: Implications for forest restoration and management', *African Journal of Ecology*, vol 41, pp9–22

Massada, A. B., Carmel, Y., Koniak, G. and Noy-Meir, I. (2009) 'The effects of disturbance based management on the dynamics of Mediterranean vegetation: A hierarchical and spatially explicit modelling approach', *Ecological Modelling*, vol 220, pp2349–2602

Mayle, F. E., Langstroth, R. P., Fisher, R. A. and Meir, P. (2007) 'Long-term forest–savannah dynamics in the Bolivian Amazon: Implications for conservation', *Philosophical Transactions of The Royal Society Biological Sciences*, vol 362, pp291–307

McClanahan, T. R. and Wolfe, R. W. (1993) 'Accelerating forest succession in a fragmented landscape: The role of birds and perches', *Conservation Biology*, vol 7, pp279–288

Nangendo, G. (2005) 'Changing forest–woodland–savanna mosaics in Uganda, with implications for conservation', PhD dissertation, Wageningen University, Wageningen, The Netherlands

Nangendo, G., Stein, A., ter Steege, H. and Bongers, F. (2005) 'Changes in woody plant composition of three vegetation types exposed to a similar fire regime for over 46 years', *Forest Ecology and Management*, vol 217, pp351–364

Nangendo, G., ter Steege, H. and Bongers, F. (2006) 'Composition of woody species in a dynamic forest–woodland–savannah mosaic in Uganda: Implications for conservation and management', *Biodiversity and Conservation*, vol 15, pp1467–1495

Nangendo, G., Skidmore, A. K. and van Oosten, H. (2007) 'Mapping East African tropical forests and woodlands: A comparison of classifiers', *ISPRS Journal of Photogrammetry and Remote Sensing*, vol 61, pp393–404

Ohsawa, K., Kawasaki, K., Takasu, F. and Shigesada, N. (2002) 'Recurrent habitat distribution and species diversity in a multiple-competitive species system', *Journal of Theoretical Biology*, vol 216, pp123–138

Oslon, D. M. and Dinerstein, E. (1998) 'The Global 200: A representation approach to conserving the Earth's most biologically valuable ecoregions', *Conservation Biology*, vol 12, pp502–515

Paterson, D. J. (1991) 'The ecology and history of Uganda's Budongo forest', *Forest and Conservation History*, vol 35, pp179–186

Ratter, J. A., Ribeiro J. F. and Bridgewater, S. (1997) 'The Brazilian cerrado vegetation and threats to its biodiversity', *Annals of Botany*, vol 80, pp223–230

Saha, S. and Howe, H. F. (2003) 'Species composition and fire in a dry deciduous forest', *Ecology*, vol 84, pp3118–3123

Schwartz, D., de Foresta, H., Mariotti, A., Balesdent, J., Massimba, J. P. and Girardin, C. (1996) 'Present dynamics of the savanna–forest boundary in the Congolese Mayombe: A pedological botanical and isotopic ($^{13}$C and $^{14}$C) study', *Oecologia*, vol 106, pp516–524

Shea, K., Roxburgh, S. H. and Rauschert, E. S. J. (2004) 'Moving form pattern to process: Coexistence mechanisms under intermediate disturbance regimes', *Ecology Letters*, vol 7, pp491–508

Sheil, D. (1999) 'Developing tests of successional hypotheses with size-structured populations, and an assessment using long-term data from a Ugandan rain forest', *Plant Ecology*, vol 140, pp117–127

Sheil, D., and Burslem, F. R. P. D. (2003) 'Disturbing hypothesis in tropical forests', *Trends in Ecology and Evolution*, vol 18, pp18–26

Sheil, D., Jennings, S. and Savill, P. (2000) 'Long-term permanent plot observations of vegetation dynamics in Budongo, a Ugandan rain forest', *Journal of Tropical Ecology*, vol 16, pp765–800

Skowno, A. L. and Bond, W. J. (2003) 'Bird community composition in an actively managed savanna reserve, importance of vegetation structure and vegetation composition', *Biodiversity and Conservation*, vol 12, pp2279–2294

Smart, N. O. E., Hatton, J. C. and Spence, D. H. N. (1985) 'The effect of long-term exclusion of large herbivores on vegetation in Murchison Falls National Park, Uganda', *Biological Conservation*, vol 33, pp229–245

Smith, T. B., Kark, S., Schneider, C. J., Wayne, R. K. and Moritz, C. (2001) 'Biodiversity hotspots and beyond: The need for preserving environmental gradients', *Trends in Ecology and Evolution*, vol 16, p431

Spence, D. H. N. and Angus, A. (1971) 'African grassland management – Burning and grazing in Murchison Falls National Park, Uganda', in E. Duffey and A. S. Watt (eds) *The Scientific Management of Animal and Plant Communities for Conservation*, Blackwell, London

Uganda Wildlife Authority (2001) *Murchison Falls National Park, Bugungu Wildlife Reserve, Karuma Wildlife Reserve: General Management Plan 2001–2011*, Printers Den, Kampala

van Wilgen, B. W., Biggs, H. C. and Potgieter, A. L. F. (1998) 'Fire management and research in the Kruger National Park, with suggestion on the detection of thresholds of potential concern', *Koedoe*, vol 41, pp69–87

Wheater, R. J. (1971) 'Problems of controlling fires in Uganda national parks', in *Proceedings of the 11th Annual Tall Timbers Fire Ecology and Erosion Control Conference*, Tallahassee, Florida, pp259–275

Zida, D., Sawadogo, L., Tigabu, M., Tiveau, D. and Odén, P. C. (2007) 'Dynamics of sapling population in savanna woodlands of Burkina Faso subject to grazing, early fire and selective tree cutting for a decade', *Forest Ecology and Management*, vol 243, pp102–115

9

# The Role of Plantation Forests in Fostering Ecological Restoration: Experiences from East Africa

## Mulugeta Lemenih and Frans Bongers

## INTRODUCTION

Eastern Africa harbours 3 of the 34 global biodiversity hotspots: the Eastern Afromontane, the Horn of Africa and the Coastal Forests of Eastern Africa hotspots. Most importantly, east Africa is a world-renowned region for its high concentrations of diverse wildlife resources. Unfortunately, the biological richness of the region is degrading at an alarming rate (Nair, 2006). This highlights the urgency for large-scale conservation efforts to avert irreversible degradation. The challenge, however, is how these rich bio-resources of national, regional and international heritage can best be conserved. The fact that the biodiversity resources are largely degraded implies that restoration is needed; conservation alone is not enough. Degradation does not only imply the disappearing vegetation but also the deterioration of biophysical conditions of a site that may become barriers for self-recovery. The abiotic and biotic ecological conditions that characterize degraded lands include:

- low stock and quality of seeds and other propagules of native species in the soil seedbank;
- primacy of seeds of aggressive (competitive) grasses in the soil seedbank;
- low numbers of animal-dispersed seed as a result of both low visitation by

171

disperser animal community and increased isolation of degraded sites from remnant forest patches;
- high incidence of herbivory; and
- unfavourable ecological conditions such as poor soil quality, high desiccation and strong direct irradiance (Ashton et al, 2001; Lamb et al, 2005; Cummings et al, 2005; Lemenih and Teketay, 2005; McNamara et al, 2006; Wassie et al, 2009a and b).

As degraded lands expand, options for livelihoods shrink and poverty prevails, leading to further natural resources degradation and conflicts over uses of resources. It is in this context that restoring degraded tropical lands is a necessity so as to enhance sustainable rural livelihoods, economic development, conservation of biodiversity and resolving conflicts over resources (Brown and Lugo, 1994; Lemenih, 2004).

Success in restoring degraded lands requires the accurate identification and management of the biotic and abiotic barriers that restrict the rapid restoration of a site (Whisenant, 1999; Cummings et al, 2005; Chazdon, 2008). Restoration can be achieved relatively fast and with less effort or costs when strategies that foster the natural breakdown of restoration barriers are identified and employed to ameliorate their conditions, rather than directly managing the barriers (Dobson et al, 1997; Lamb et al, 2005; Lugo, 2007; Cummings and Reid, 2008). One such fostering technique is the establishment of plantation forests, which are now widely recognized as an important tool to foster ecological restoration on degraded tropical lands (Lugo, 1997; Parrotta et al, 1997; Lindenmayer, 2002; Lamb et al, 2005; Montagnini and Jordan, 2005; McNamara et al, 2006; Chazdon, 2008). This chapter summarizes the available experiences and provides ample evidence for the potentials of plantation forests as foster ecosystems. By plantation forests we refer to forest types in which the standing trees are artificially established by planting or deliberate seeding. The species used could be native or exotic, and planted either as monoculture or mixed species. The stands could be established on lands where forests historically existed but are degraded (reforestation) or on historically non-forested lands such as grasslands (afforestation). Fostering, as defined in this chapter, refers to the use of tree plantations to catalyse the natural (re)generation and/or (re)introduction[1] of native flora, vertebrate and invertebrate fauna and to improve soil quality (Harrington, 1999; Otsamo, 1998 and 2000; Ashton et al, 2001; Yirdaw, 2002; Lamb, 2003; Lemenih, 2004).

Driven mainly by the decline in the areas of natural forests and their timber and non-timber products' supply, plantation forests are increasing worldwide (Evans and Turnbull, 2004; FAO, 2005), including in Africa (Chamshama and Nwonwu, 2004). This also dictates a global interest to understand the range of possible environmental effects, particularly the impact on biodiversity, of such developments (Christian et al, 1998; Kelty, 2006; Gómez-Aparicio et al, 2009). It is quite obvious that plantations replacing intact natural ecosystems will depauperate the biodiversity richness of such ecosystems (Elliot, 2003; Kelty, 2006) and affect the indigenous people that depend on their bio-richness

(UNFF, 2003; Scherr et al, 2004; Wiersum, Chapter 15 of this volume). On the other hand, plantation forests established on degraded lands have a 'catalytic effect' in fostering natural succession and restoring ecosystem functions (Parrotta et al, 1997; Cusack and Montagnini, 2001; Tyynelä, 2001; Yirdaw, 2002; Lemenih and Teketay, 2005; Butler et al, 2008) and realize diverse benefits ranging from the contributions to sustainable forest management to restoring soil quality, improving the living conditions of rural communities, and poverty reduction to minimizing communities' environmental and economic risks (UNFF, 2003; Lamb et al, 2005).

Quite a large number of studies have documented the fact that tropical plantation forests established after severe site degradation foster the recolonization of diverse native flora and fauna significantly better than sites left unplanted. Ample studies are now available from diverse climatic regions (moist and dry tropical) and geographical areas such as Africa (Chapman and Chapman, 1996; Geldenhuys, 1997; Bone et al, 1997; Moges, 1998; Woldu, 1999; Bernhard-Reversat, 2001; Huttel and Loumeto, 2001; Tyynelä, 2001; Senbeta et al, 2002a and b; Yirdaw, 2002; Duncan and Chapman, 2001 and 2003; Lemenih et al, 2004a; Lemenih and Teketay, 2004a and 2005); Asia (Fang and Peng, 1997; Lamb 1998; Otsamo, 1998 and 2000; Chen et al, 2003; Igarashi and Kiyono, 2008); the Caribbean (Lugo, 1992; Parrotta, 1992 and 1993; Harrington and Ewel, 1997; Inman et al, 2007); Australia (Dobson et al, 1997; Keenan et al, 1997; Harrington, 1999; Lindenmayer, 2002; Zobrist, 2005) and Latin America (Guariguata et al, 1995; Parrotta et al, 1997; Powers et al, 1997; Montagnini and Jordan, 2005; Corley et al, 2006; Zurita et al, 2006; Paritsis and Aizen, 2008). In these studies exotic, native, monoculture and mixed species established under both wet and dry climatic conditions have been covered, including the most controversial genus – *Eucalyptus* (Calder et al, 1992; Bone et al, 1997; Bernhard-Reversat, 2001; Senbeta and Teketay, 2001; Senbeta et al, 2002a and b; Yirdaw, 2002; Lemenih et al, 2004a; Lemenih and Teketay, 2005). The studies have also explored diverse aspects such as floristic diversity, faunal diversity (small and big including soil macro-, meso- and micro-organisms) and soil properties that provide the opportunity to assess the impacts of plantation developments from a broader ecosystem perspective, that of ecosystem restoration. The contributions of plantation forests to biodiversity restoration was also a subject addressed and discussed during the World Bank Biodiversity Rehabilitation project symposium held in Washington, DC, in 1995 (Parrotta, 1995b; Parrotta et al, 1997; Geldenhuys, 1997; Harrington and Ewel, 1997). This wealth of information is here synthesized to first provide evidence on the potential of plantation forests in re-establishing a sufficiently rich diversity of flora and fauna and improving soil conditions and, second, show that plantation establishment and management techniques and networks that promote sufficient diversity need to be communicated to promote successes in practical applications. Although this chapter focuses on experiences from east Africa, it also covers a broad range of studies from elsewhere to capture the major findings of recent biodiversity and soil property studies in plantation forests, and thus provides a more comprehensive review.

# PLANTATION FORESTS OF EAST AFRICA

Although east Africa comprises about 11 countries, we focus here on only 4: Ethiopia, Uganda, Kenya and Tanzania, primarily because of their high plantation forest resources (FAO, 2005) and the high number of studies available. The history of plantation forestry in east African states goes back to the late 19th and early 20th centuries (Evans, 1992; Chamshama and Nwonwu, 2004). In Ethiopia the first recorded plantation establishment was in 1894–1895, when introduced *Eucalyptus* species were planted around Addis Ababa (Pohjonen and Pukkala, 1988 and 1990; Bekele, 1992). The main expansion of plantation forestry in east Africa occurred from 1940 to 1980 simultaneous with its increase in the tropics and subtropics (Chamshama and Nwonwu, 2004). Today, the plantation forests of Ethiopia, Kenya, Uganda and Tanzania altogether cover 2.9 million hectares of land (Table 9.1). Although east Africa is ecologically well suited for plantation forest development, the current plantation area is very low compared to the potential (Chamshama and Nwonwu, 2004; Anonymous, 2006). Furthermore, against the trend of the rapid global expansion over the last two decades, plantation forest development has almost stagnated in most of east African (FAO, 2005). The significant tree planting efforts in Ethiopia, for instance, were those in the 1960s and 1970s (Pohjonen and Pukkala, 1988 and 1990; Teklu, 2003), and in Kenya and Uganda too, significant areas were established during the 1970s and 1980s (Kenya MENR, 1994; Chamshama and Nwonwu, 2004).

A characteristic feature of plantation forests development in east Africa is the dominance by a few exotic species that were selected for their fast growth, high adaptability and production of woods of good quality (Evans, 1992; Chamshama and Nwonwu, 2004). *Eucalyptus* is the most widely planted genus across the region, followed by *Cupressus*, *Pinus*, *Acacia* and *Tectona*, whose shares vary from country to country (Chamshama and Nwonwu, 2004). In Ethiopia 59 per cent of the plantation stands comprise *Eucalyptus* spp., followed by *Cupressus lusitanica* (26 per cent) (Teklu, 2003), and in Kenya *Cupressus lusitanica* occupies 46 per cent of the total area of planted forests (Kenya MENR, 1994; Anonymous, 2003).

**Table 9.1** *Extent and trends of plantation forests development in east Africa*

| Country | Land area (000 ha) | High forests (000 ha) | Plantation forests (000 ha) | | | Annual change rate (ha/yr) | |
|---------|------|------|------|------|------|------|------|
| | | | 1990 | 2000 | 2005 | 1990–2000 | 2000–2005 |
| Ethiopia | 110,430 | 13,000 | 509 | 509 | 509 | 0 | 0 |
| Kenya | 56,915 | 1245 | 238 | 212 | 202 | −2600 | −2000 |
| Uganda | 19,964 | 928 | 35 | 35 | 36 | 0 | 1000 |
| Tanzania | 88,359 | 2437 | 135.2 | 135.2 | 135.2 | 0 | 0 |
| Total | 275,668 | 17,610 | 2907.2 | 2891.2 | 2887.2 | | |

*Source:* FAO (2005)

The heavy reliance on fuelwood by the vast majority of the population in the region, increasing commercial fuel prices, the decline in natural forest cover, raising global and domestic demands for timber, C-trade opportunity and other current national and international issues are likely to inspire governments in the region to reinitiate plantation forest development (Chamshama and Nwonwu, 2004, Anonymous, 2006; Lemenih, 2008). Renewed interest in plantation forestry is already visible in Uganda, Tanzania and Kenya (Chamshama and Nwonwu, 2004). Particularly, Uganda is actively seeking the involvement of the private sector in plantation development, including establishing a loan scheme by the National Forestry Authority for tree farmers and offering leases on its own reserve land to encourage private plantation developments. Many of the countries have also recently revised their forest policies to create a conducive environment for the sector's development (Nair, 2006; see also Chapters 2–4 of this volume).

# PLANTATION FORESTS AS A TOOL FOR ECOSYSTEM RESTORATIONS AND THE MECHANISMS BEHIND

## Ecosystem restoration strategies

There are wide ranges of technical management options to recover degraded lands and their vegetations (Ashton et al, 2001; Perrow and Anthony, 2002; Lamb et al, 2005); these are generally grouped into 'passive' and 'active', based on the intensity of human involvement (Allen et al, 1995, Laycock, 1995; McInvar and Starr, 2001). A passive strategy underlines a process of ecosystem restoration principally through natural causes with limited human support. It relies most on the natural self-regenerating potential of the degraded sites (Lemenih and Teketay, 2004b; Lemenih, 2006). Restoration under passive approaches is slow and less effective for heavily degraded ecosystems as these have a poor self-regenerating potential (Parrotta, 1992; Kuusipalo et al, 1995; Laycock, 1995; Dobson et al, 1997; Lamb, 1998; Harrington, 1999; Montagnini, 2001; Lemenih and Teketay, 2004a and b; McNamara et al, 2006). Such restrictive characteristics for self-restoration include low soil fertility, shallow soil, low soil moisture and few seeds in the soil seedbank (Lamb, 1998; Ashton et al, 2001; Lemenih and Teketay, 2004a, 2005 and 2006; Zamora and Montagnini, 2007). Furthermore, forest successional processes on heavily degraded lands are often hindered because of the unappealing environmental conditions to dispersers, high herbivory levels and seed predation, low dispersal as a result of increasing isolation from seed sources, competition from aggressive weeds and grasses, desiccating winds and strong irradiances (Harrington, 1999; Cubina and Aide, 2001; Lemenih and Teketay, 2004a and b and 2006; Cummings et al, 2005; McNamara et al, 2006). For instance, in their review of soil seedbank studies in Ethiopia, Lemenih and Teketay (2004a) concluded that (i) scarcity or complete absence of a viable soil seedbank for woody species in environments affected by humans such as abandoned farmlands, (ii) poor seed rain/dispersal, and (iii) site-impoverishment due to soil degradations would be severe limitations to ecological restoration. These limitations also intensify with the length and intensity of disturbances. Lemenih and Teketay (2006) showed that soil seedbanks not only have very low stocks of viable seeds of woody species but also that this declined from 5.7 per cent after 7 years to none after 53 years of continuous cultivation following conversion of natural forest to farmlands in Ethiopia. This could be one of the biotic hindrances for the rapid self-recovery of such farmlands upon abandonment (Lemenih et al, 2008). The same was found in Costa Rica (Aide et al, 1995), where pastureland was rapidly dominated by grass, fern and herbaceous species, inhibiting the re-establishment of secondary forest species.

Seed dispersal of most tropical woody species are limited to very short distances from forest edges, and even when a rare dispersal event occurs, most species fail to accumulate viable seeds because of short-term seed viability and high seed predation. Aide and Cavelier (1994), for instance, captured seeds of 30 species and 300 seeds/m$^2$ in plots located in the interior of a forest, while none were found in the plots established on degraded grassland located just 20m away from the forest edge. Cubina and Aide (2001) showed that out of 35 species that produced fruits in the nearby forest, only 14 species were captured in seed rain. Only 0.3 per cent of the seeds belonging to just 3 species were able to disperse to 4m distance from the forest edge, a trend that was also reflected in the soil seedbank composition.

Another common constraint to the self-recovery of heavily degraded tropical lands is the prevalence of invasive and highly competitive grasses. These grasses not only thwart emergence and survival of tree seedlings from competition but also provide excellent habitat for seed and seedling predators such as rodents and advance soil quality impoverishment (Nepstad et al, 1991; Jim, 2001; Cummings et al, 2005). Moreover, the re-establishment of most tropical tree species, particularly tree species of the later seral types, requires some degree of shade (Chapman et al, 1999; McNamara et al, 2006; Wassie et al, 2009a), and the harsh climatic conditions of open degraded lands are unfavourable to the survival and germination of their seeds (Teketay, 1996; Wassie et al, 2009a) and cause very high seedling mortality even when planted (Chapman and Chapman, 2004; Wassie et al, 2009b). These abiotic and biotic limitations discussed above generally retard the rapid self-restoration of heavily degraded sites (Whisenant, 1999; Cummings et al, 2005), and most of them can persist over a long timescale (Aide and Cavelier, 1994; Dobson et al, 1997; Parrotta et al, 1997), which means that leaving such ecosystems for self-recovery will either take an unnecessarily long time (Dobson et al, 1997; Parrotta et al, 1997) or even an undesirable course (Gibson and Brown, 1992; Sarmiento, 1997), while these barriers can effectively be removed to enhance the restoration process with more 'active' involvement of humans (Dobson et al, 1997; Montagnini, 2001; Lamb et al, 2005).

'Active' restoration refers to a case where human support to assist ecosystem recovery is high. The support can cover a wide range of activities such as (re)seeding of propagules, planting of either single or mixed species as nurse/facilitator crops or framework species, redressing of the soil with mulches and other organic substrates, constructing soil and water conservation structures, (re)introduction of disperser animals, inoculation of beneficial micro-organisms such as mycorrhizae, or combinations of some or all of the above. Planting is the most intentional human action towards ecosystem recovery (Lamb et al, 2005), and is one of the most frequently employed active restoration technique (Lamb et al, 2005; Ruiz-Jaen and Aide, 2005).

## The mechanisms behind plantation forests as fostering ecosystems

The role of plantation forests in restoration ecology and biodiversity conservation can be direct and indirect. Directly, plantation forests bolster biodiversity restoration on degraded lands by providing facilitational (playing a nursing role) and/or modificational (remedying detrimental biotic and abiotic constraints) services. These services include:

1   the provision of conducive microclimatic conditions at forest floor for regenerates;
2   provision of habitats, moving corridors and perches for dispersers;
3   improvement of soil quality through increased litter fall, improved soil organic matter content, nutrient recycling, soil stability, improved soil moisture, protection against erosion, and assisting mineral weathering or pedogenesis; and
4   suppression of aggressive and competitive weeds (Parrotta et al, 1997; Dozier et al, 1998; Lemenih, 2004; Cummings et al, 2005; Lamb et al, 2005).

Indirectly, plantation forests assist biodiversity restoration and conservation by playing a role of 'substitution'. They compensate for wood and non-wood products that are otherwise obtained from natural forests (Sedjo and Botkin, 1997; Montagnini, 2001; Carnus et al, 2003). That means plantations play a 'buffering role' by providing the possibility for taking hands off the natural forests and/or regenerating stocks, enabling the setting aside of these natural forests as reserves (Sedjo and Botkin, 1997; UNFF, 2003; Victor, 2003), a function that is increasingly acknowledged today (Kelty, 2006). Without dedicated plantations for wood and non-wood products supply, biodiversity-rich remnant natural forests and woodlands or biodiversity stocks in restoration landscapes will undoubtedly continue to degrade. The further degradation will certainly be the worst barrier to future restoration managements as it expands fragmentation or isolation of degraded lands and further restricts seeds and other propagules' dispersal (Lemenih and Teketay, 2004a and b).

## Plantation forests as corridors, stepping stones and habitats for dispersers

To initiate flora restoration on degraded lands, seeds and other propagules of the target species should either be available in the soil seedbanks or arrive at a site in enough quantity and quality through dispersal. This, however, is one of the major limitations for rapid natural succession on degraded sites.

Many plant communities in the tropics depend on animal dispersal, mainly forest-dwelling animals, as their most important form of seed dispersal (Howe and Smallwood, 1982). On the other hand, the nature of habitat, whether wooded or non-wooded, is an important factor affecting the distance over which animals disperse seeds (Weins, 1992). In other words, decisions of foraging animals on where to feed, what to feed on and for how long to feed are determined based on the nature of the habitat. The common hypothesis is that bare lands are avoided by disperser communities of forest habitat types (MacArthur, 1972), and those forest animals crossing into open areas may venture only a short distance from the forest edge and for a short time (van Ruremonde and Kalkhoven, 1991; Duncan and Chapman, 2001; Zamora and Montagnini, 2007; Butler et al, 2008). For instance, van Ruremonde and Kalkhoven (1991) showed that animal-dispersed seeds to wooded areas are inversely correlated to the degree of isolation of the source and recipient sites. This isolation can effectively be bridged by establishing either strips of forest (corridors) or a series of small patches (stepping stones) between forest relicts or between a forest relict and a degraded site intended to be restored. Corridors are habitats that permit the movement of organisms between ecological isolates (Newmark, 1993), while stepping stones are a series of small patches connecting otherwise isolated patches (Baum et al, 2004).

Plantations can provide effective corridor or stepping stone services as well as habitat for dispersers to catalyse enhanced seed dispersal for a number of reasons. First, they provide better cover for dispersers to safely move through and thus improve contiguity between natural forest relicts and/or between natural forest patches and degraded lands. Second, they provide safer dwelling habitat and perch sites for dispersers than open degraded sites. Third, they provide food/forage for dispersal animal agents (Montagnini, 2001; Castellón and Sieving, 2006; Zamora and Montagnini, 2007; Butler et al, 2008). And fourth, they also provide protection and safety for the dispersed seeds that otherwise would be blown away by wind, dry out or even be liable to high predation if left in the open sites (Wunderle, 1997). For instance, Zamora and Montagnini (2007) showed that plantations of *Balizia elegans*, *Dipteryx panamensis* and *Jacaranda copaia* received 2263, 2091 and 5522 seeds per m$^2$ (collected in 6 months) respectively, compared to non-planted abandoned pasture control that received only 593 seeds, and the seeds received in the plantations plots were from many more species than the control unplanted plots. In a review of Latin American studies, Montagnini (2001) showed that while open pastures had the highest proportion of wind-dispersed seeds, in plantations dispersal was predominantly from birds and bats, demonstrating that plantations attract disperser avifauna. In doing so, plantations act both as keystone (supporting) habitat or species and as corridors to facilitate native biodiversity replenishment.

## Plantation forests suppress competitive grasses and improve microclimatic conditions of degraded sites

As they grow tall and begin to close canopy, planted forests shade out the ground and thus suppress grass species common to open lands. This in turn increases the chances for seedlings of native shade-loving species to successfully emerge, survive and grow (Guariguata et al, 1995; Otsamo, 2000; Cusack and Montagnini, 2001). Plantations also accumulate quite large amounts of litter, which diminishes grass cover while encouraging woody species growth (Montagnini, 2001). Reduction of grass vegetation minimizes root competition for emerging woody seedlings. This makes conditions for seedling survival and growth more favourable under the canopies of plantation trees. For instance, Cusack and Montagnini (2001) reported that plots with high light incidence generally had understoreys dominated by ferns and grasses and few or no woody species, while those with low light incidence due to canopy shade hosted more woody species. Plantation canopy shading also moderates microsite conditions of the forest floor such as soil moisture, soil and air temperature, and air humidity to offer a more favourable growth environment for woody species seedlings (Yirdaw, 2002; Lemenih et al, 2004a). The shade also protects young seedlings from direct sun radiation and desiccating winds. These microclimatic elements are essential environmental factors for the regeneration and early growth of climax tropical woody species (Keenan et al, 1997).

## Plantation forests as improvers of degraded soils

Low soil quality common to degraded lands is another important barrier to forest succession as it hinders establishment, growth and survival of tree seedlings (Aide and Cavelier, 1994; Lamb, 1998). Besides their contribution to biodiversity restoration, forest plantations improve soil fertility and reduce erosion on degraded sites, making them conducive for recolonization of native species (Montagnini, 2000; Lemenih et al, 2004b and c; Glenday, 2006). Generally, the soil effects of planted forests are inconsistent, but depends on the pre-plantation site soil status, the intrinsic characteristics of the planted species, the mode of establishment (monoculture or mixed species), the age of the plantation and the ecosystem used for comparison (Binkley and Resh, 1999; Binkley et al, 2003; Lemenih et al, 2004b and c). When planted on degraded lands, the effects of plantation forests on soil quality are always beneficial, whereas when planted on relatively fertile soils the effects could be negative for some time (Post and Kwon, 2000; Zerfu, 2002; Lemenih et al, 2004b and c). In the highlands of Ethiopia, within 15 years of plantation establishment, a stand of C. *lusitanica* had resulted in improved soil conditions in terms of bulk density, soil C, total N and other properties compared to soils subject to different farming intensities (mechanized or traditional low-intensity farming) (Lemenih et al, 2004c). Similarly, *C. lusitanica* and *E. saligna* plantations established on abandoned farmland improved soil conditions compared to soil subject to continuous farming (Lemenih et al, 2004b). This is consistent with other studies that showed increments in soil C following conversion of former agricultural lands to plantations (Fisher, 1995; Smith et al, 1997; Polglase et al, 2000; Post and Kwon, 2000; Silver et al, 2000; Garten, 2002; Glenday, 2006). A study in east Africa that assessed carbon densities in various types of land cover, including plantations, showed that the area-weighted mean carbon density for indigenous forest was 330 ± 65Mg C/ha, which was greater, but not significantly so, than that of hardwood plantations with 280 ± 77Mg C/ha (Glenday, 2006). Glenday (2006) indicates that the difference in carbon density between indigenous forest and hardwood plantations is due to age-related effects, with the carbon density of old hardwood plantations of 370 ± 90Mg C/ha being very similar to the 360 ± 63Mg C/ha of old indigenous forest, while the young hardwood plantations had even greater mean carbon density (240 ± 71Mg C/ha) than young secondary indigenous forest (170 ± 78Mg C/ha). Interestingly, some studies confirm that stand harvesting and logging even when frequently (every six or seven years) applied is not preventing build-up of total organic matter, although it occurs at a lower rate (see, for example, Loumeto and Bernhard-Reversat, 2001).

Time since reforestation/afforestation and tree species involved both significantly affect the extent and rate of soil improvements resulting from plantation forests. A review of several studies revealed that soil carbon accumulation in the early stages of plantation development (<10 years) could be either low or negative as there is relatively little input of carbon from the plantation biomass, a trend that gradually changes to continuously improved soil carbon accumulation (Trouve et al, 1994; Post and Kwon, 2000; Jaiyeoba, 2001; Paul et al, 2002). For instance, in Congo and Nigerian savanna plantations, soil carbon content increased with stand age (Mboukou-Kimbatsa and Bernhard-Reversat, 2001; Jaiyeoba, 2001). In fact, soil impacts of plantation forests vary based on the planted species because tree species differ in their intrinsic characteristics and mechanisms by which they affect soil. Such variations include the rate of nutrient inputs and outputs, canopy density, which moderates temperature and humidity at soil surface, water use traits, ability of N-fixation, and litter quality (Binkley and Giardina, 1998; Jaiyeoba, 2001; Lemenih et al, 2004b). For instance, Lemenih et al (2004b) showed that *C. lusitanica* accrued nearly five times greater organic C than *E. saligna* (156 vs. 37g C/m$^2$/yr), and also differed in N accrual.

In general, trees always affect soil development, but not with the same rate and magnitude. Pedogenesis – the soil formation process – as a natural process is heavily shaped by vegetation (Jenny, 1941; Brady and Weil, 1999), and whether planted or naturally regenerated, the roles of trees/vegetation in soil formation and the mechanisms behind it do not fundamentally differ (Fisher, 1995; Binkley and Giardina, 1998). The mechanisms by which trees/vegetation contribute to soil formation processes is both direct and indirect. Some of the direct mechanisms include:

- organic matter addition;
- enhanced mineral weathering;
- nutrient pumping and recycling;
- symbiotic N-fixation;
- interception of particles and dusts (such as aerosols) from the air; and
- improved soil structure through root action (Binkley and Giardina, 1998, Kelly et al, 1998; Raulund-Rasmusson et al, 1998; Olsson, 2001; Augusto et al, 2002).

Trees/vegetation also affect soil conditions indirectly through:

- changes in microclimatic conditions in the understorey;
- decreasing soil erosion;
- fostering the recolonization of diverse flora and fauna that add to litter mass, litter diversity and decomposability; and
- increased population and diversity of decomposer organisms (Fisher, 1995; Dobson et al, 1997; Mboukou-Kimbatsa and Bernhard-Reversat, 2001).

The most noticeable soil improvement effect of plantation forests established on degraded lands is through soil organic matter (SOM) addition (Fisher, 1995, Lemenih et al, 2004b and c). Addition of SOM, in turn, gives life to soils (Katyal et al, 2001) by substantially improving physical, chemical and biological soil properties that render environments conducive for plant establishment and growth (Dobson et al, 1997). Added SOM improves soil structure, decreases soil bulk density, binds soil particles together to form stable aggregates that are resistant to erosion, allow water infiltration and thereby reduce runoff, improves water-holding capacity, prevents the effect of soil dispersion by raindrops, and reduces surface crusting and hard setting (see, for example, Paul and Clark, 1996; Fernandes et al, 1997). Decomposition and mineralization of SOM is the major source of soil N in the tropics, particularly on degraded lands (Dobson et al, 1997), and contributes significantly to the increased availability of other essential soil plant nutrients such as calcium (Ca), magnesium (Mg) and potassium (K), and the quantities of nutrients required in small amounts like zinc (Zn), iron (Fe), copper (Cu) and manganese (Mn) are also increased. Normally, 95 per cent of N and S reside in organic matter, and large proportions of Zn and Cu in soils occur in organic form (Stevenson, 1982; Stangel, 1991). Thus SOM addition represents a considerable improvement of soil fertility (Stangel, 1991). By doing so, it enhances plant succession and ecosystem development (Dobson et al, 1997). Plantations established on degraded lands also contribute to soil fertility by enhancing large macrofauna and mesofauna populations (Mboukou-Kimbatsa and Bernhard-Reversat, 2001). Such plantation effects increase with stand age as older plantations produce more litter and organic matter (Mboukou-Kimbatsa and Bernhard-Reversat, 2001).

Soil erosion is one of the severe soil degraders, and is the most common phenomenon on degraded sites (Aide and Cavelier, 1994; Brady and Weil, 1999). Erosion leads to reduction of soil depths, mainly the top layer, the layer where most soil organic matter accumulates, which further decreases soil nutrient availability, soil moisture-holding capacity and rooting depth (Aide and Cavelier, 1994; Omiti et al, 1999). The major underlying factor for soil erosion is removal of the protective vegetation cover (Brady and Weil, 1999; Omiti et al, 1999), and plantation forests can effectively reverse soil erosion by providing the necessary cover (Helles and Linddal, 1996; UNFF, 2003). The canopy soil cover from standing tree crops and their litter fall conserve soil, protect gully formation, prevent landslides, reduce direct splash and runoff, facilitate infiltration and raindrop interception, reduce soil sediment transport, and maintain desirable physical soil properties (McCulloch and Robinson, 1993; UNFF, 2003), all reducing soil erosion. Field investigations have demonstrated that afforestation leads to a reduction in runoff by as much as 20 per cent, mainly due to interception of rainfall by forest canopies (Bosch and Hewlett, 1982; McCulloch and Robinson, 1993). Reforestation/afforestation sites, particularly when they have good undergrowth, exhibit reduced water erosion by:

- reducing peak runoff and sediment loss;
- lessening risk of flooding;
- improving water infiltration;
- reducing impacts from raindrop splashes;
- intercepting rain;
- physically stabilizing soil by their roots;
- favouring soil development processes by improving SOM; and
- improving soil structure and soil water-holding capacity (Miller and Jastrow, 1992).

Miller and Jastrow (1992) have also shown that roots and mycorrhizal hyphae are involved in the creation of water-stable soil aggregates, which are important preconditions to soil stabilization and resistance to erosion. This means that plantation roots and their associated micro-organisms are important for restoration of vegetation and re-establishment of normal soil processes needed for the formation of soil structure and redevelopment of nutrient cycles (Atkinson, 2000)

# CASE STUDIES FROM EAST AFRICA

## Restoration of floristic diversity

A number of studies from east Africa, particularly from Ethiopia, Uganda and Tanzania, have reported the biodiversity-fostering potentials of forest plantations. These studies confirmed that most of the recolonizing species are woody and timber species of early and late succession types (Chapman and Chapman, 1996; Senbeta et al, 2002a; Yirdaw, 2002). From Uganda, Fimbel and Fimbel (1996) reported that the species richness of regenerates that recolonized the plantation forests of *P. caribaea* and *C. lusitanica* approaches those found in the surrounding natural forest, and nearly 60 per cent of these regenerating species were trees. In terms of density, the stem number in the sapling class under the pine stand was 2168 stems/ha, which was 70 per cent as numerous as those in the adjacent natural forest (3077 stems/ha), while the C. *lusitanica* stand contained 1111 stems/ha, 36 per cent of the level in nearby natural forest. Other *P. caribaea* and *P. patula* plantation stands in Uganda had 47 regenerating indigenous tree species, 60 per cent of the 78 species in nearby intact forest (Chapman and Chapman, 1996).

In most Ethiopian plantation studies, which covered dry and humid Afromontane regions (Table 9.2 and Figure 9.1), species richness of the regenerates, even in the plantations of *Eucalyptus* spp., was over 75 per cent of that recorded in the nearby natural forests (Moges, 1998; Woldu, 1999; Senbeta and Teketay, 2001; Senbeta et al, 2002a and b; Yirdaw, 2002; Lemenih et al, 2004a; Lemenih and Teketay, 2005; Alem, 2006). The study of Woldu (1999) in Entoto, near Addis Ababa, shows that regeneration of *Juniperus procera* under *Eucalyptus globulus* is 1750 stems/ha, 44 per cent of which is seedling, 13 per cent sapling and 43 per cent pole size. Alem (2006) recorded 90 per cent of the woody plant species, including *C. arabica*, found in the natural forest to also be recolonizing the stands of *E. grandis* plantation in a tropical humid forest region in southwestern Ethiopia. *C. arabica* did not differ significantly in density (number of coffee plant stems/ha) between natural forest (1101 stems/ha) and plantations (801 stems/ha). Similarly, Yirdaw (2002) found in the dry Afromontane forest of Menagesha that 83 per cent of the woody species recorded in the adjacent natural forest were represented in the understorey of eucalyptus plantations. In some plantations the woody regenerates are even more diverse and dense than in adjacent natural forest (Senbeta and Teketay, 2001). Senbeta et al (2002a) recorded an understorey woody plant density of 18,650 individuals/ha in the 27-year-old coppice stand of *E. saligna*, compared to 9658 individuals/ha in adjacent natural forest (Table 9.2). Lemenih et al (2004a) showed significantly higher regenerates species richness in natural forest (10.2) than in plantations of *C. africana* (8.7), *E. saligna* (8.6) and *P. patula* (7.5), but not densities (6370 individuals/ha in natural forest against 6280 in *E. saligna* and 5320 in *C. africana* plantations). Lemenih and Teketay (2005) showed that nearly 78 per cent of woody species recorded in nearby natural forest were found in the seedling and sapling population of the understorey of plantations, but soil seedbanks did not differ. This implies that seed dispersal facilitation due to the establishment of the plantations was more important than soil seedbanks in the recolonizing process.

Similar studies from other African countries provide similar evidences. Plantation forests of different species (eucalypt, acacia and pine) in Congo, for instance, have their understoreys invaded by many native woody species, forming dense thickets in stands older than 10 years (Loumeto and Huttle, 1997; Bernhard-Reversat, 2001). In South Africa *Eucalyptus* and *Pinus* plantation stands had 170 species in their understoreys, including all major growth forms found in the surrounding forest, with trees representing over 30 per cent of the regenerates (Geldenhuys, 1997). These results present strong evidence

**Figure 9.1** *Naturally regenerating native flora under plantation forests in east Africa*

*Notes: Juniperus procera* regeneration under *E. globulus* coppice stand at Entoto near Addis Ababa, Ethiopia.
*Source:* F. Bongers

of the potential of plantation forests to act as foster ecosystems to induce a rapid succession of native biodiversity on degraded lands in the region.

## Restoration of faunal diversity

Plantation ecosystems in east Africa also host a high diversity of vertebrates (birds and mammals) and invertebrates (insects and soil macro-, meso- and micro-organisms), a phenomenon that naturally follows the recolonization of native plant species and safe habitat (Chapman and Chapman, 1996; Hinde et al, 2001; Zanne et al, 2001; Jenkins et al, 2003). Jenkins et al (2003) showed that all four trophic herbivore groups (grazers, bulk feeders, gleaners and rooters) were abundant in plantation stands in Tanzania, although their distribution and abundance differed between plantations of different age. Older plantations had more bush pig, bushbuck and duikers, whilst younger plantations had more open-habitat specialists such as warthog and waterbuck. Those plantation stands close to flood plain contained significantly more large grazers and bulk feeders such as zebra, buffalo and waterbuck. Hinde et al (2001) did not find significant overall differences in the abundance of large mammal signs between miombo forest (mean 42.3 ± 8.7), evergreen forest (25.4 ± 3.1) and teak plantation (25.6 ± 0.55) in Tanzania. Animal species of different feeding habits tend to occupy different niches: grazers preferred the most open and grass-rich miombo; gleaners and rooters were most abundant in the teak plantations (Figure 9.2a).

In Uganda, pine plantations did not differ from adjacent natural forest in duiker or bush pig tracks, but bushbucks were more frequent in plantation than

**Table 9.2** *Density and number of regenerating species under plantation forests established on degraded natural forests sites in the highlands of Ethiopia*

| Locality | Plantation Species | Species richness of regenerates | Density of regenerate/ha | Reference |
|---|---|---|---|---|
| Entoto (near Addis Ababa) | Eucalyptus globulus | – | 1750 | Woldu (1999)* |
| Wondo Genet | Natural forest | 48 | 7275 | Yirdaw (2001) |
| | Pinus patula | 30 (62.5%) | 4500 | |
| | Cupressus lusitanica | 34 (70.8%) | 8225** | |
| | Juniperus procera | 32 (66.7%) | 5475 | |
| | Grevillea robusta | 29 (60.4%) | 5500 | |
| Munessa-Shashamane | Natural forest | 27 | 9658 | Senbeta et al (2002a) |
| | C. lusitanica (9 yrs) | 30 (111.1%)** | 7325 | |
| | C. lusitanica (17 yrs) | 22 (81.5%) | 7375 | |
| | C. lusitanica (25 yrs) | 16 (59.2%) | 5950 | |
| | E. globulus (13 yrs) | 16 (59.2%) | 6550 | |
| | E. globulus (16 yrs) | 13 (48.1%) | 2300 | |
| | E. globulus (22 yrs) | 17 (63%) | 13,400** | |
| | E. saligna (11 yrs) | 18 (66.7%) | 3575 | |
| | E. saligna (22 yrs) | 23 (85.2%) | 10,100** | |
| | E. saligna (27 yrs) | 25 (92.6%) | 18,650** | |
| | P. patula ( 10 yrs) | 18 (66.7%) | 2325 | |
| | P. patula (21 yrs) | 16 (59.2%) | 3750 | |
| | P. patula (28 yrs) | 15 (55.6%) | 2525 | |
| Menagesha-Suba | Natural forest | 26 | 11,680 | Senbeta and Teketay |
| | C. lusitanica (14 yrs) | 18 (69.2%) | 5770 | (2001) |
| | E. globulus (17yrs) | 27(103.8%)** | 7730 | |
| | C. lusitanica (24 yrs) | 11 (42.3%) | 1630 | |
| | P. radiata (24 yrs) | 15 (57.7%) | 3130 | |
| | P. patula (24 yrs) | 17 (65.4%) | 3940 | |
| | J. procera (42 yrs) | 27 (103.8%)** | 18,270** | |
| Munessa-Shashamane | Natural forest | 10.2 (1.75) | 6370 | Lemenih et al (2004a) |
| | C. Africana (28 yrs) | 8.7 (1.87) | 5320 | |
| | E. saligna (31yrs) | 8.6 (2.95) | 6280 | |
| | P. patula (31 yrs) | 7.5 (2.12) | 3680 | |
| | C. lusitanica (31 yrs) | 6.6 (2.22) | 3320 | |

*Notes:* * Woldu studies specifically the regeneration of *Juniperus procera*, a native species, under *E. globulus*; ** Plantations with higher density or species richness than the natural forest.

in natural forest (Figure 9.2b) (Zanne et al, 2001). Interestingly, this study found that plantations located nearby natural forest and those located in an isolated landscape are both capable of attracting wildlife, although the diversity and abundance for the isolated stand is considerably lower. Another study in Uganda reported frequent use of plantation forests by many animals including the red tail monkey (*Cercopithecus ascanius*), chimpanzee (*Pan troglodytes*), duiker (*Cephalophus harveyi* and C. *moniticola*), bushbuck (*Tragelaphus scriptus*), bush pig (*Potomochoerus porous*), civet (*Viverra civetta* and *Nandinia*

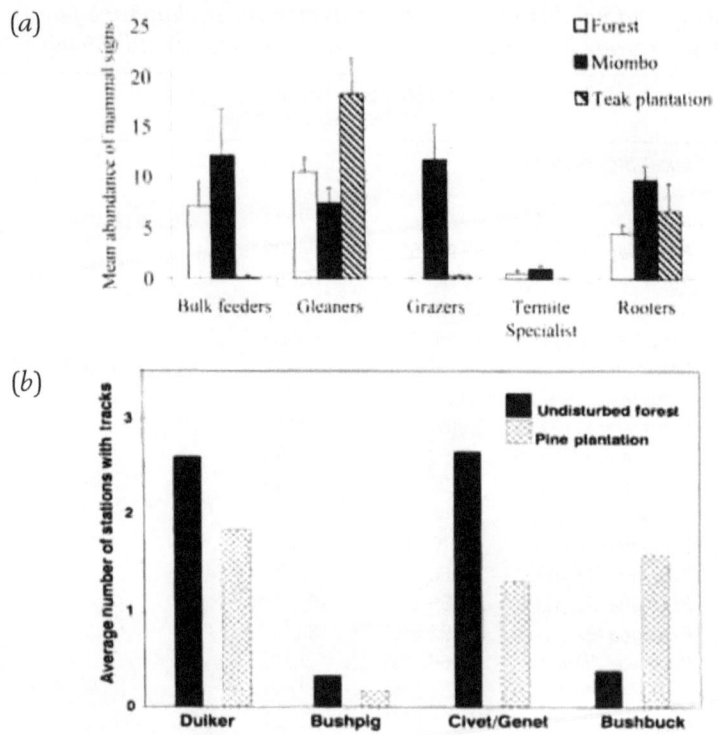

**Figure 9.2** *Abundance of mammal by (a) feeding habit and (b) species assessed in plantations and adjacent ecosystems in Tanzania and Uganda*

*Sources:* Hinde et al (2001) and Zanne et al (2001)

*binotata*) and genet (*Genetta* sp.) (Chapman and Chapman, 1996). These results demonstrate that plantations replacing miombo woodland can provide a possibly suitable refuge for wildlife, particularly in the agriculturally fragmented landscapes common to east Africa (Zanne et al, 2001).

Similar results have been found elsewhere. In Congo, for instance, abundant vertebrate fauna (birds and mammals) and large populations of insects and soil macro-, meso- and micro-organisms have been found to colonize plantation forest ecosystems (Dowsett-Lemaire and Dowsett, 1991; Brosset, 2001; Mboukou-Kimbatsa and Bernhard-Reversat, 2001). Brosset (2001) recorded as many as 53 per cent of bird species occurring in a nearby natural ecosystem to also occur in plantation stands of eucalypt, pine and acacia species, while the frequency of earthworms, termites, myriapods and cockroaches in these plantation forests reached a level close to that of the adjacent natural forest (Mboukou-Kimbatsa and Bernhard-Reversat, 2001). Densities of the soil macro-fauna increased with plantation age, and the taxa frequency pattern under acacia plantation resembled more that of the natural forest than those under pine and eucalyptus stands (Mboukou-Kimbatsa and Bernhard-Reversat, 2001).

Birds were also studied in plantations and adjacent ecosystems. Johnson and Freedman (2002) reported that bird occurrence in plantations (abundance and diversity) was similar to natural stands, but community composition was highly dissimilar, depending on the age and canopy density of the plantations. Younger plantations had birds typical of open and early successional habitats, while in older and taller plantations, birds typical of mature forest began to invade the habitat, resulting in a mixed species composition. In Hong Kong, secondary forest had a similar numbers of bird species (44 spp.) to plantations (46 spp.), but birds in plantations were typical of non-forest habitats while the secondary forest had more forest-associated species (Kwok and Corlett, 2000). White-footed mice were more found in Australian plantations of pine than in natural stands (Mengak et al, 1989). These results from different sites suggest that plantation forests can be used as a habitat or refuge by diverse species of wild animals, including birds and insects.

Since habitat preference varies from species to species, it might be necessary to carefully study original fauna species in the area and provide key (critical) niche requirements for these species through proper design and stand management of plantation forests at any given site (Inman et al, 2007), such as proper selection of plantation tree species and diversity of recolonizing species. For instance, Brosset (2001) indicated that trees in Congo like acacia, with abundant herbivorous and detritivorous insects, attract a greater number of birds than other plantation tree species. Similarly the abundance and diversity of undergrowth attracts diverse birds (Brosset, 2001), mammals (Zanne et al, 2001) and macro and micro soil organisms (Loubana and Reversat, 2001). Free-living nematode populations in plantations were significantly correlated (r = 0.733 for 12 plots) with the percentage of recolonizing plant species of the habitat prior to plantation establishment (Loubana and Reversat, 2001). Similarly, total frequency, as well as the frequency of earthworms, termites, myriapods and cockroaches in eucalypt plantations, increased with the percentage of forest plant species in the understorey (Mboukou-Kimbatsa and Bernhard-Reversat, 2001).

In general, the studies reviewed above prompt the claim that plantation developments can augment existing strategies of wildlife conservation in the region or may serve as park buffer zones (Chapman and Chapman, 1996; Fimbel and Fimbel, 1996).

## PLANTATION MANAGEMENT AND BIODIVERSITY RESTORATION

Plantation forests established on degraded tropical lands are able to catalyse the re-assemblage of diverse flora and fauna, and thus can be used as a tool for biodiversity restoration, including species whose population in the nearby natural ecosystems are severely depauperated or endangered (Tyynelä, 2001; Cummings and Reid, 2008). While plantations in general support a lower flora and fauna diversity than intact natural forests, there is little doubt regarding their fostering role in the rapid assemblage of a diverse flora and fauna on degraded lands. Nor is there any doubt regarding the biological richness they hold compared to most other land uses such as pasture, unplanted degraded lands and crop fields (Parrotta et al, 1997; Duncan and Chapman, 2003; Sayer et al, 2004; Montagnini and Jordan, 2005). The flora diversity assembled in plantation forest ecosystems can be managed and redirected to establish secondary forest regrowth (Lamb, 1998; Ashton et al, 2001; Sturm and Apel, 2006): plantations can be considered a 'transitional ecosystem' between degraded land and secondary forest regrowth (Figure 9.3). A range of tailored plantation management practices (see earlier sections) is available that can be applied to significantly enhance the efficacy of plantation forests to foster biodiversity and improve soil conditions either to re-establish secondary forests (Parrotta et al, 1997; Lamb et al, 2005; Montagnini and Jordan, 2005; Erskine et al, 2006; Sturm and Apel, 2006) or to serve as an in situ biodiversity conservation strategy without incurring appreciable loss of plantation productivity (Tyynelä, 2001; Sayer et al, 2004; Cummings and Reid, 2008).

The available evidence for plantation forests supporting biodiversity originate mainly from plantations established for timber production, rather than from plantations deliberately designed and managed to rigorously test their efficacy for ecological (biodiversity) restoration, or from stands established to optimize mutual ecological and socio-economic objectives (Parrotta et al, 1997; Lugo, 1997; Lamb, 1998; Sayer et al, 2004; Sturm and Apel, 2006). In fact, the biodiversity interest in plantation forests is a development since the late 1980s following the popular belief that biodiversity conservation has to be integrated into managed landscapes (Brown and Lugo, 1994; Lamb, 1998; Harvey and Haber, 1999). In the past, tree plantations were mostly established for the single purpose of wood production, and species selection and stand management practices were all geared towards optimizing wood quality and quantity. Indeed the native forest vegetation under plantation canopies was considered as weeds and constantly removed. Given that the vast majority of tropical plantations, including those in Africa, are (i) fast-growing monoculture of very few genera, *Eucalyptus* being the dominant genus (Keenan et al, 1999; FAO, 2001; Carnus et al, 2003; Evans and Turnbull, 2004; Sayer et al, 2004; Chamshama and Nwonwu, 2004), (ii) established predominantly on degraded marginal lands (Hunter et al, 1998; Keenan et al, 1999; Lee et al, 2005), and (iii) subjected to intensive silvicultural management practices that maximize timber yield (Keenan et al, 1999; Bernhard-Reversat, 2001; FAO, 2001; Lamb et al, 2005), all of which significantly affect the biogeochemistry of the plantation forest ecosystems, the low level of flora and fauna diversity in plantation forests compared to adjacent natural or secondary forest regrowth should not be surprising. A simple plantation forest established on marginal land and designated solely for wood production cannot be expected to host the same level of biodiversity as natural forests (Keenan et al, 1999), at least not in the short time horizon of 20 to 30 years which marks the upper ranges of stand ages covered by the majority of the studies. Some studies assessed plantation forests

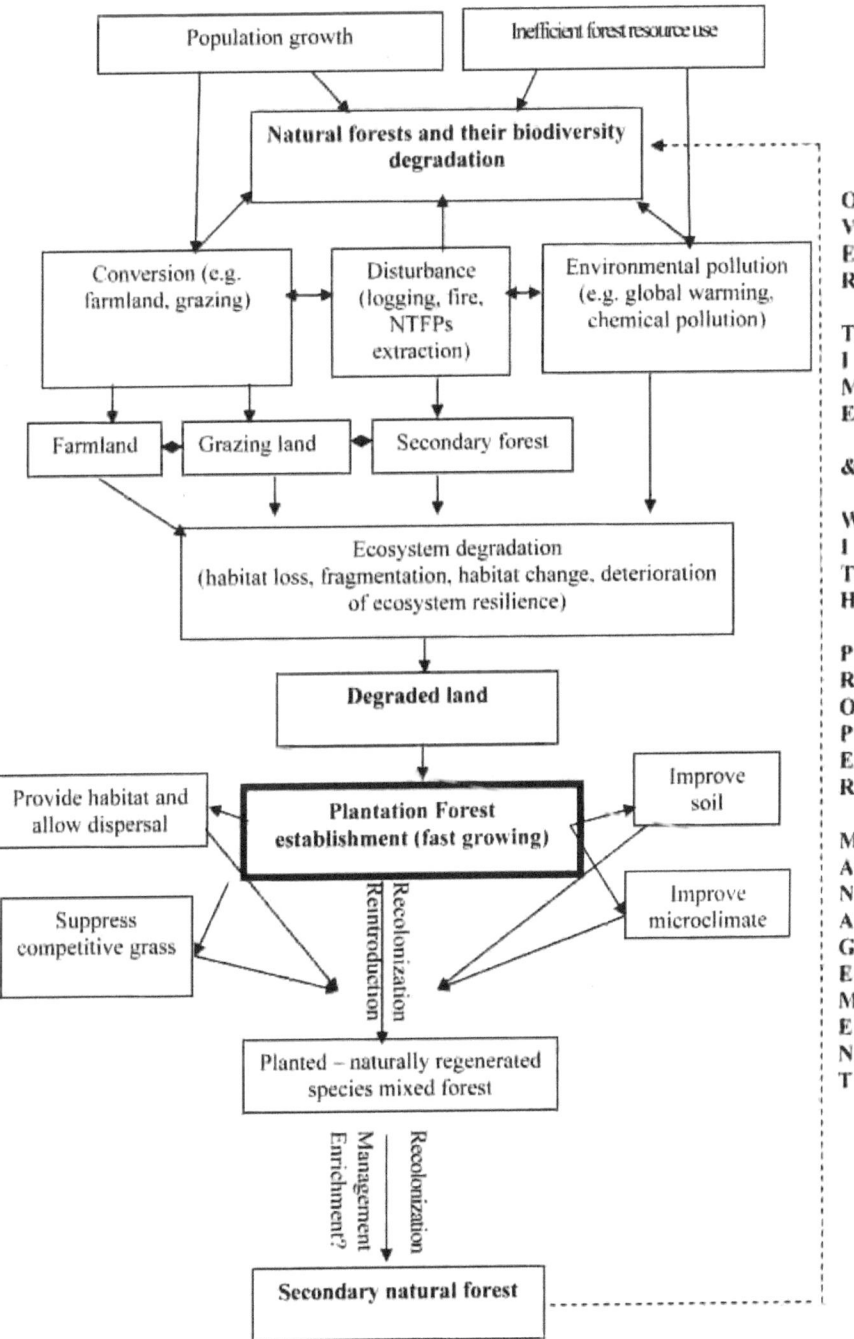

**Figure 9.3** *Conceptual framework demonstrating plantation forests as intermediary between degraded land and secondary natural forest*

less than 10 years old and some even as young as 4 of 5 (Parrotta, 1992; Cusack and Montagnini, 2001; Shono et al, 2006), although age-related diversity increment is an obvious reality (Fang and Peng, 1997; Keenan et al, 1997; Huttel and Loumeto, 2001; Lindenmayer and Hobbs, 2004).

In general, few studies have used reference ecosystems comparable to the plantation forest ecosystems with respect to age, prior site conditions, ecosystem management, level of seed or propagule availability, and other factors that affect flora and fauna assembly (Lee et al, 2005), and this is probably due to the fact that there are few such opportunities for an 'ideal' comparative study. In many cases the high diversity and density of flora and fauna populations in natural or secondary forests reflects the better overall site conditions of such environments in contrast to the severely degraded sites used for most (re)forestation programmes, making comparative analyses of the two unreasonable (Corlett, 1999; Lee et al, 2005). The management of plantation forests for wood production should differ from where the aim is to produce a wider range of conservation benefits and maintenance of more ecosystem functions (Lamb, 1998; Keenen et al, 1999; Sayer et al, 2004; Lamb et al, 2005). The few dedicated case studies that deliberately established and managed plantations on degraded lands for ecological restoration (Lamb et al, 2005; Sturm and Apel, 2006) showed strong evidence for the potentials of plantation forests to act as a transitional ecosystem between degraded lands and secondary forest regrowth. Sturm and Apel (2006) showed that a stand with *Pinus massoniana*, an exotic species established on degraded land, quickly developed towards mixed pine and indigenous broadleaved stands.

The studies discussed provide strong evidence and hope for the potentials of plantation forests as a tool for ecological restoration (Parrotta et al, 1997; Duncan and Chapman, 2003; Zobrist, 2005; Shono et al, 2006; Inman et al, 2007), particularly when complemented with proper design and adaptive management operations (Ashton et al, 2001; Sayer et al, 2004; Lamb et al, 2005; Sturn and Apel, 2006; Cummings and Reid, 2008; Igarashi and Kiyono, 2008). Plantation managements aiming to promote biodiversity restoration and other ecological and socio-economic values need to consider the following important attributes:

- understorey light availability;
- structural diversity of plantation stands at both stand and landscape levels;
- distance between the plantation stands and potential seed or propagule sources such as remnant natural forest patches;
- adaptive silvicultural treatments such as thinning, pruning, weeding and harvesting;
- the size and shape of the plantation forest stands;
- rotation age;
- when it is necessary to complement with enrichment plantings; and
- the management of biotic interactions (Diaz et al, 1998; Jenkins et al, 2003; Lemenih et al, 2004a; Lamb et al, 2005; Cummings and Reid, 2008).

The relationships between these factors and the level of recolonization, as well as various management prescriptions to manipulate plantation stands to promote a better fostering environment, are important. These management prescriptions can also benefit old plantation stands that have been established by traditional design to help them play increasing roles in biodiversity restoration (Hartley, 2002; Sayer et al, 2004).

## Structural diversity of plantation stands

Plantation forests that are diverse in both structure and composition, and at stand and landscape levels, provide a broad suite of habitats to host diverse flora and fauna (Chandrasekar-Rao and Sunquist, 1996; Hartley, 2002; Johnson and Freedman, 2002; Lindenmayer and Hobbs, 2004; Zobrist, 2005; Erskine et al, 2006), while also providing diverse products and services of socio-economic importance (Keenen et al, 1999). Structural diversification can be achieved by either mixing species, age and/or by creating mosaics of land use and land cover at landscape scale (Johnson and Freedman, 2002; Jenkins et al, 2003; Lindenmayer and Hobbs, 2004; Carnus et al, 2006). Such a diverse plantation closely resembles a natural ecosystem ('close to nature' plantation; Nichols and Carpenter, 2006). A mix of old and young stands in a plantation landscape provided habitat for both big and small mammals, the older and taller stands being preferred by larger mammals, while the young, short and open stands were used by mammals of open savanna habitat (Johnson and Freedman, 2002; Jenkins et al, 2003). Similarly, some studies recorded higher floristic diversity under mixed species plantations than in a monoculture (Guariguata et al, 1995; Carnevale and Montagnini, 2002; Carnus et al, 2003).

Landscape diversification is shown to foster flora and fauna diversity, while also preventing the development of pest insect outbreaks (George and Zack, 2001; Carnus et al, 2006). Better landscape-level design to achieve major biodiversity gains may include the retention or re-establishment of small areas of natural vegetation, riparian strips and preservation of natural forest corridors that can be linked to large-scale reforestation schemes (Lamb, 1998). Understanding its values, some countries such as Indonesia (Sayer et al, 2004) have constituted laws that govern such landscape mosaics in industrial plantation development schemes: establishment of industrial tree plantations demands one-third of the land to be retained under natural forest reserve.

Despite its apparent advantages, the practice of mixed species plantation is extremely limited. Monocultures are often preferred because of simplicity and low cost in establishment; simplicity in silvicultural management; and uniform products and thus good market sale. Unless the market for the products from the species in the mix is similar and attractive, the overall value of the mixed species plantation will be lower than that of monocultures (Lamb et al, 2005). Like in the cases of monocultures, mixed species plantations also differ significantly in their efficacy of fostering biodiversity depending on the species involved in the mixture (Lamb et al, 2005; Carnus et al, 2006), and recently the better efficacy of mixed species plantations over monocultures has been challenged (Powers et al, 1997; Parrotta, 1999; Sturm and Apel, 2006; Butler et al, 2008). These studies found greater plant diversity in the understorey of monocultures than in mixed species plantations: starting simple (in other words with single species) and complementing this with tailored management makes it possible to attain a diverse secondary forest re-establishment (Sturm and Apel, 2006). Indeed, mixing on the basis of a simple taxonomic difference alone is an insufficient prescription to foster better ecosystem restoration (Lamb, 1998; Lamb et al, 2005): complementarity (spatially and temporally), adaptation to site conditions and competitive balance (stable mix) is needed (Carnus et al, 2006). However, limited knowledge is available on these aspects (Lamb et al, 2005). The high establishment and maintenance costs of mixed species plantations (Erskine et al, 2006) also make them less attractive to plantation managers. Indeed, the establishment of plantation forests that are diverse either at stand or landscape level to promote heterogeneity needs to weigh the trade-offs between economic viability, level and speed of ecosystem functions recovery, and social desirability.

## Canopy structure and light admittance

Among the factors influencing biodiversity, light admittance through the plantation canopy may be the most important (Harrington and Ewel, 1997; Otsamo, 1998; Carnevale and Montagnini, 2002; Lemenih et al, 2004a; Inman et al, 2007). The level of light admittance through the canopy modifies air and soil temperature, soil moisture, and amount, duration and quality of photosynthetically active radiation (PAR) (Otsamo, 2000; Yirdaw and Luukannen, 2003; Lemenih et al, 2004a). Plantation stands with a dense canopy display low (cool) air and soil temperatures and PAR, which suppresses seedling growth following emergence. Stands with relatively open canopies and higher light admittance (Figure 9.4) bolster higher regeneration density as well as better regeneration growth than stands with dense canopies and thus darker forest floor (Lemenih et al, 2004a): C. *africana*, a deciduous broadleaf species with 54 per cent canopy closure fostered 25 per cent more species and almost 50 per cent more density of native species than C. *lusitanica* with 94 per cent canopy closure. In Costa Rica, the degree of plantation canopy shading was inversely related to the total number of regenerating native plants ($r = -0.72$) and the number of species ($r = -0.74$) (Carnevale and Montagnini, 2002). Similar results were found in Malawi (Bone et al, 1997). Inman et al (2007) showed that gap plots in plantations host more than twice the alpha and gamma diversity than understorey plots.

Canopy gaps are fundamental not only to initiate regeneration but also to advance the regenerates to grow (Otsamo, 1998; Lemenih et al, 2004a; Inman et al, 2007). The high light conditions (low canopy closure) are needed for regenerates to be able to reach maturity (Lemenih et al, 2004a; Inman et al, 2007). Seedlings planted under plantations with artificially controlled levels of canopy opening showed a fairly linear growth for up to 40–50 per cent of above-

**Figure 9.4** *Small open areas (gaps) in plantation forest provide extra opportunities for regeneration: Here* Podocarpus falcatus, Croton mycrostachys *and other native tree and shrub species regenerate in a gap in a* Pinus patula *stand at Wondo Genet, Ethiopia*

*Source:* F. Bongers

canopy photosynthetic photon flux density (PPFD), and tended to decrease as intensity level surpassed 50 per cent (Otsamo, 1998). Seedling planted in centres of plantation natural gaps showed greater growth than seedlings away from the gap centre (Otsamo, 2000).

Canopy opening also stimulates fauna diversity by influencing habitat components such as ground cover and environmental parameters (for example temperature) that in turn affect availability of diverse forage (food) and preferred perch in the regenerating understorey (Turner et al, 2002; Mengak and Guynn, 2003; Zobrist, 2005; Inman et al, 2007; Paritsis and Aizen, 2008). This effect depends on both the plantation species and on the inherent habitat preferences of the animals (Mengak et al, 1989; Mengak and Guynn, 2003; Jenkins et al, 2003).

Excessive canopy opening should be avoided as this will be unfavourable particularly to tree species of later seral types or as it encourages more grass dominance in the plantation understoreys (Kuusipalo et al, 1995; Powers et al, 1997; Cusack and Montagnini, 2001; Butler et al, 2008). For instance, Cusack and Montagnini (2001) reported that plots with the highest incident light generally had understoreys dominated by ferns and grasses, with few or no woody species. Similarly, having too dense canopies with low incident light results in lower or no woody regeneration (Lemenih et al, 2004a), due to low germination or to high seedling mortality due to carbon deficiency (Inman et al, 2007). Although the optimum canopy opening that fosters a high level of diversity could vary depending on the nature of vegetation to be restored and environmental conditions of the restoration site, we suggest that generally 25–45 per cent canopy opening should be maintained – the low openness for light-intolerant species and the high openness for less light-sensitive species.

The required canopy openness can be achieved by various techniques such as selecting species with appropriate crown openness (Lemenih et al, 2004a), planting at a wider spacing (Zobrist, 2005; Carnus et al, 2006), and/or starting dense but manipulating the canopy during the plantation lifetime using various silvicultural management practices such as thinning, selective girdling and pruning (Ashton et al, 1998; Otsamo, 2000; Zobrist, 2005; Inman et al, 2007). Species differ in their leaf size, orientation, density and persistence, branching habit, and branch sizes and persistence, and thus canopy shape, canopy depth and canopy width, which results in different canopy architecture and thus different performance with respect to biodiversity restoration (Lemenih and Teketay, 2004a and 2005). Artificial gap creation in plantations stimulates regeneration (Inman et al, 2007) and growth (Ashton et al, 1998; Otsamo, 2000).

## Plantation age (length of rotation)

With age of plantation forest, fauna and flora recolonization tends to diversify and show a successional transformation from species of open-site conditions to species characteristic of more closed forest. The understorey of young plantations generally shows low diversity, density and dominance of open-site species. With age this gradually changes towards the structure and composition of the adjacent natural forests (Fang and Peng, 1997; Bossuyt and Hermy, 2001; Huttel and Loumeto, 2001; Lindenmayer and Hobbs, 2004). In Congo, Huttel and Loumeto (2001) showed that the proportion of forest species increased from zero to nearly 80 per cent during the first 20 years after eucalypt planting, while the percentage of ruderal and savanna species decreased. They also found that basal area of undergrowth increased with age, from zero to 10 per cent of

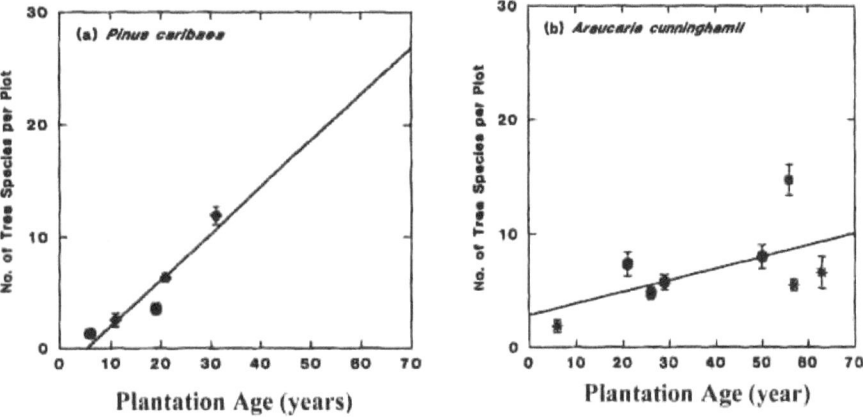

**Figure 9.5** *Relationship between plantation age and tree species richness under two plantation tree species*

*Source:* Keenan et al (1997)

total stand basal area over these 20 years of stand development. Senbeta et al (2002a) recorded six times more regenerates under an *E. saligna* stand 27 years old compared to a stand 11 years old. Senbeta and Teketay (2001) found a higher density of woody regenerates in a 42-year-old *J. procera* stand (18,270 plants per hectare), compared to a closeby 24-year-old *C. lusitanica* stand (1630 per hectare). Keenan et al (1997) showed that older plantations (>50 years) hold appreciably more regenerating species (350–1000 per cent) than younger plantations (Figure 9.5), and some of the regenerates in older plantations have grown high to form part of the plantation canopy. The same trend was observed for fauna diversity. Bird species diversity increased with age of *P. radiata* plantations and hoop pine plantations, mainly because older stands are structurally more diverse (Fang and Peng, 1997; Lindenmayer and Hobbs, 2004). Bird species of younger plantations are typical of open, early successional, upland habitat types, but with plantation age and higher vegetation, birds typical of conifer forest invade the habitat, resulting in a mixed species composition (Johnson and Freedman, 2002). In Tanzania, Jenkins et al (2003) showed that older plantations were predominantly used by mammals of dense forests such as bush pig and bushbuck, while younger plantations were mainly used by open-habitat specialists such as warthog.

The observed differences in biota with plantation age may imply that succession in the understorey of plantation forest is directional just like most natural successions, but it is relatively fast. Increased recruitment of woody species with age can be attributed to the increased chances and magnitudes of seed arrival and germination (Fang and Peng, 1997; Senbeta et al, 2002a), leading to accumulation of regenerates over time, and to the changing soil and microclimate conditions with age as a result of gap creation due to thinning and natural mortality (Oberhauser, 1997; Lindenmayer and Hobbs, 2004). Given that most flora regeneration depends on seed dispersal from outside the plantation forests, regeneration underneath plantation forests should be evaluated in terms of the probability of species pools that could actually reach the plantation stands as a result of dispersal over time. This in term depends on the degree of isolation of the plantation forests with respect to the natural source, mode of dispersal of the species and presence of dispersers. Time is, therefore, an important factor affecting the level of biodiversity in plantation forests, and this is particularly so on heavily degraded lands characterized by poor self-regenerating potential.

## Origin of the planted tree species (exotic vs. indigenous)

Recolonization biodiversity under canopies of different plantation species varies widely, even among those closely established (Harrington, 1999; Lemenih et al, 2004a). Although plantation species differ in many characteristics affecting their fostering possibilities, their origin (exotic versus indigenous) is a continuous matter of concern and debate (Keenan et al, 1999; Montagnini, 2001; Hartley, 2002). General performance differences between exotics and indigenous species in restoration remain contentious (Allen et al, 1995; Lee et al, 2005; Lemenih, 2006). Some studies show that exotics are either better or similar to native plantation species (Senbeta et al, 2002a; Yirdaw, 2002; Lemenih et al, 2004a), while others claim that exotics are inferior (Allen et al, 1995; Hartley, 2002). By far the most tropical plantations consist of exotic species, however (Keenen et al, 1999; Montagnini, 2001), particularly species that have done well in Africa (Chamshama and Nwonwu, 2004). Over 80 per cent (in Africa nearly 100 per cent, see Table 9.2) of current studies assessing biodiversity restoration in plantation forests have considered exotic monoculture plantations. And being exotic has not made plantations biodiversity deserts. Some studies have reported even a better performance of exotics over indigenous with respect to fostering biodiversity recolonization. In Ethiopia, greater abundance of naturally regenerated woody species was found under the exotic species C. *lustanica* than under the native J. *procera* (Yirdaw, 2001). No significant differences were found under indigenous C. *africana* compared to E. *saligna* and Pinus patula (Lemenih et al, 2004a). In Argentina species richness and composition of birds and understorey plants were similar between pine (exotic with 58 species of birds) and araucaria (native with 54 bird species) plantations (Zurita et al, 2006).

Studies so far do not present a clear superiority of natives over exotics concerning recolonization (Lugo et al, 1993; Haggar et al, 1997; Lemenih et al, 2004a). Claims are mostly based on theoretical arguments rather than empirical evidence (Montagnini, 2001). The choice for species should be based on species traits (McNamara et al, 2006). Preferred species should be light-demanding, have easy establishment on degraded lands, and have fast establishment and rapid early growth to produce rapid canopy closure that suppress competitive grass and allow quick introduction of native woody species

(Guariguata et al, 1995; Kamo et al, 2002). Exotic species may have more of these characteristics than native species (Lamb, 1998; Lamb et al, 2005). The criteria of fast growth, be it exotic or indigenous, is essential as it results in a quick financial return to meet the socio-economic needs of the rural community affected by restoration as well as to support the cost of restoration projects (McNamara et al, 2006). Unfortunately, only few, if any, known indigenous tropical species in many countries exist that can easily establish themselves and perform well in wood productivity on the hostile environments of degraded lands (Lamb et al, 2005). By starting with exotics and gradually redirecting the course of the ecosystem development, an ultimate ecosystem restoration or stands rich with productive native species can successfully be established. Alternatively a start with exotics of high economic value can be followed by gradual underplanting and/or tending of indigenous regenerates (Lamb et al, 2005; McNamara et al, 2006).

The environmental costs or undesirability of selective importation of some exotic species of high socio-economic significance can be challenged given today's fast and highly spontaneous intercontinental species invasibility. Several studies have shown that many plant species are crossing their political and geographical boundaries unintentionally (Lonsdale, 1999), and invading multiple biome regions, temperate and tropical, island and mainland, agriculture and urban territories, deserts and savannas (Lonsdale, 1999). In the light of changing global environments, no one can be sure of the boons and bans of the continuous exchange of germplasm across geographical regions, and exotics may be considered in the context of increasing gains of genetic pool, which in turn will increase the resilience of ecosystems to unforeseen environmental changes.

## Thinning and harvesting

Thinning and harvesting of plantation forests result in changes in the spatial and temporal environmental pattern and thus may have positive benefits for flora and fauna recolonization (Duncan and Chapman, 2003; Lindenmayer and Hobbs, 2004; Cummings et al, 2007; Cummings and Reid, 2008). Regenerates of *Afrocarpus falcatus* and *J. procera* in *E. saligna* and *E. globulus* stands in Ethiopia progressively increased in size in the understorey after successive clear felling and coppice resprouting (Senbeta et al, 2002b; Lemenih, 2006), until they suppress the plantation crop and take over. Huttel and Loumeto (2001) also showed that disturbed commercial eucalypt plantations hold greater species richness than undisturbed experimental plantations (38.6 versus 25.6 species respectively). In Malawi the diversity and vigour of regenerates increased in subsequent eucalyptus coppice stands (Bone et al, 1997), and in Uganda second rotation stands held more diversity and density than first rotation crops (Duncan and Chapman, 2003).

Thinned and harvested plots had five times more species richness and better growth of woody regenerates than untreated plots (Cummings et al, 2007; Cummings and Reid, 2008), mainly because of diversification of forest structure. Many studies show positive effects of thinning, logging or strip clearing in increasing native tree seedling growth and species diversity in plantations (Rubio et al, 1999; Otsamo, 2000; Inman et al, 2007). Harvesting releases the canopy shade and increases the availability of environmental resources such as light and soil nutrients to the regenerates (Bone et al, 1997; Senbeta et al, 2002b).

In Congo the density of termites decreased following harvest but quickly recovered as the stand cover recovered through coppice resprouting (Mboukou-Kimbatsa and Bernhard-Reversat, 2001), suggesting that the effect of harvesting and logging is short-lived. Harvesting, apart from physical impacts of harvesting vehicles, temporarily alters the microclimatic conditions of a site. This eventually promotes succession. Leaving large gaps as created by clear felling for a relatively long period of time may result in inconvenient environmental conditions such as soil moisture deficiency and high surface temperatures that could suppress many plant and fauna functional groups. Cummings and Reid (2008) showed that species richness, density and cover of regenerating native woody shrubs and trees in the understorey declined with time as a result of increased canopy opening by thinning or harvesting. Excessive canopy opening will lead to increase of grass cover with subsequent decline of tree species recruitment.

Anthropogenic silvicultural disturbances such as harvesting and post-harvest practices (felling, skidding and replanting) can also have a detrimental effect on animal populations due to the temporary destruction of their habitat, their forage and perch, and due to intense human interference during harvest. Adaptive silvicultural management practices that consider the environmental conditions of the site, plantation crop characteristics and nature of the recolonizing biological resources (flora and fauna), in order not to halt or reverse the restoration process, are recommended. Particularly in plantation establishment aimed at provision of wood products on a short rotation basis, quickly resprouting coppicing provides the necessary microclimatic environment (Rubio et al, 1999; Lemenih, 2006).

## Distance to remnant forest patches and natural habitats

For plant regeneration to occur in plantation forests, seeds and other propagules should either be available in the soil seedbank or dispersed by some means to the site. Similarly, for wildlife to enter plantation forests, they should translocate themselves from where they exist to the plantation ecosystem. It is this rationale that demands appropriate geographic synchrony between natural habitats and plantation stands. Many isolated plantation stands exhibited reduced recruitment or low seed rain, which may underline the effects of distance to seed sources and the lack of dispersal (Tucker and Murphy, 1997; Zamora and Montagnini, 2007). For instance, Huttel and Loumeto (2001) found 50 forest species in a stand only 10m away from the forest edge while a stand 50–100m away had only 32 forest species.

Large-seeded species adapted to animal and/or bird dispersal mechanisms tend to be more limited in abundance than small-seeded species as isolation between plantation sites and natural forest increases (Tucker and Murphy, 1997; Hardwick et al, 1997). Most soil seedbank studies indicated only rare occurrence of viable seeds of woody species (Teketay and Granstrom, 1995; Senbeta and Teketay, 2001; Lemenih and Teketay, 2005), and the recolonization of native woody species in plantation forests are from seed rains dispersed by birds or wild animals (Wunderle, 1997; Castellón and Sieving, 2006).

Plantations in close proximity to other important habitats are more frequently used by wildlife than those far away (Zanne et al, 2001). Particularly in exotic plantation stands, preserving some natural forest patches or at least some native trees in or adjacent to the plantation stands would make a big difference (Senbeta et al, 2002b). Establishing plantation forests nearer remnant natural forests – at a distance that is assumed to be easily transversed by disseminators – would enhance the rate, diversity and density of flora and fauna recolonization, and reduce the effort needed for enrichment planting (Lemenih, 2006).

## Size, shape and juxtaposition of plantations sites

Size, shape and juxtaposition of plantation forests (habitat patches) affect the recolonization, distribution and abundance of wildlife and birds (Faaborg et al, 1995; Lindenmayer et al, 1999). Some bird and wildlife species do not occupy a habitat when it is too small. This is extremely important from a restoration perspective, as it sets a lower limit on the size of patches that a species will occupy. Also patch context, in other words the type of habitat that surrounds a patch, may also have a strong influence on habitat suitability (George and Zack, 2001). For instance, birds nesting in patches surrounded by human-altered habitats (urban, suburban or agricultural) suffer higher nest predation than those nesting in patches adjacent to natural habitats (old fields or grassland) (Donovan et al, 1997). The size requirement approximation needs to also consider several behavioural requirements of wildlife such as breeding and other social considerations (many wild animals have social groups) (George and Zack, 2001). For many species the plantation forests must be at a sufficiently large scale: larger plantations provide a much safer habitat and allow disseminators to penetrate deep into the plantation. They should also be diverse in age or structure to provide heterogeneity of habitat that is attractive for various groups of wildlife (Jenkins et al, 2003). Larger plantations have a higher chance of covering topographic variability. In Spain, plantation size was the main factor explaining 67–75 per cent of the variation in bird species richness (Diaz et al, 1998): plantations of less than 25ha maintained less than 50 per cent of the regional forest bird pool, while stands of over 25ha comprised 68–86 per cent of the birds' pool.

# Complementing through enrichment planting and reintroduction

Large numbers of native flora and fauna species can be assembled in plantation forests, particularly using deliberate management measures that facilitate recruitment. A complete assemblage of all species of native forest ecosystem is rarely possible in plantations (Hutte and Loumeto, 2001; Lemenih and Teketay, 2005; Zamora and Montagnini, 2007). Some species may be absent in the recolonizing plant communities as a result of, for instance, endangerment of their population, of a decline in the population of their seed dispersers (Vieira and Scariot, 1996), or due to their restricted ecological requirements (Teketay, 2006). The absence of keystone species and their related disperser animals may negatively affect the success of restoration projects (Gibbs et al, 2008). Such important species may be introduced through enrichment plantings and reintroduction. For instance, Inman et al (2007) demonstrated that enrichment plantings of forages for Puerto Rican parrots in plantation forests boosted the density of the parrot population as well as that of many other frugivorous bird species. Similarly, Gibbs et al (2008) showed that reintroduced tortoises benefited from enrichment-planted cacti. Monitoring of the restoration ecosystem is constantly needed (Lemenih and Teketay, 2005; Lee et al, 2005; Inman et al, 2007).

# Managing the biotic interaction between fauna and flora

The two main factors attracting wildlife to plantation forests are forage availability in the plantation understorey and protection from predators (Tutin et al, 1997; Jenkins et al, 2003). Biotic interactions can be complementary or conflicting. Complementarity is achieved when the fauna component feed on elements of the ecosystem that otherwise may hamper the recolonization and growth of desirable woody species (de la Cretaz and Kelty, 2002), while conflict will arise when the fauna components feed on the flora and suppress their survival, growth and density (Cummings et al, 2005). Successful survival and growth of regenerating seedlings of desirable species are severely retarded as a result of browsing by wildlife species that co-inhabit the restoration forest (Cummings et al, 2005), a result that conforms with earlier studies (Whisenant, 1999; Sweeney et al, 2002). Herbivores can complement growth and survival of desirable woody species components by grazing on and weeding out competitive grasses, ferns and herbs (de la Cretaz and Kelty, 2002). Careful assessment of the interactions between these biotic components and adaptive management of the interaction are required in order to achieve an overall successful restoration. Controlled hunting, for instance, may be needed to reduce the community of browsers (Horsley et al, 2003).

## SOCIO-ECONOMIC CONSIDERATIONS OF PLANTING FOR BIODIVERSITY RESTORATION

Restoration efforts or projects hold socio-economic costs and benefits, which could facilitate or impede successes (McNamara et al, 2006). These can be analysed as an end and as a process, and at local project site level and beyond. Restoration projects encompassing ecosystem gains side by side with improved community wellbeing will enjoy wide political and community support and participation. Such projects are likely to be highly successful. Restoration efforts should thus be advanced as an economic sector, and target the provision of tangible direct and indirect economic and social benefits, especially in poor economies (such as the east African states). Economic gains and social benefits from restoration projects must outweigh the costs involved, and these benefits should be experienced at various spatial scales. Restoration benefits should not only be limited to the end product – a self-sustaining ecosystem – which has significant value by itself, but should also provide goods and services that contribute to local livelihoods and the regional and national economy, and that contribute to community capacity-building and development of institutional frameworks that are directed at restoration activities. Only under such conditions will restoration projects receive political and societal interest and broad participation.

Economic costs of restoration may comprise, above all, expenditures in technical works and the opportunity costs to the community due to land given up for restoration. The term 'degraded land' is subjective (Hunter et al, 1998) and is not equal to 'unusable land'. Hundreds of thousands of poor rural households in developing countries (Lamb et al, 2005), particularly those in east Africa (Lemenih et al, 2008), depend on lands that have largely lost their productivity capacity for subsistence. Restoration interventions must not target only the biophysical recovery (Hobbs, 2004), but also the recovery of products and services, and whether these benefits are socially desirable and economically attractive. Restoration projects should harmonize the recovery of biodiversity and ecosystem functions with a broader vision of rural development and the contribution to sustainable livelihoods (Hunter et al, 1998; Lemenih, 2004; Sayer et al, 2004). Options, advantages and disadvantages of a restored area for local people and for the region as a whole should be carefully analysed in close communication with stakeholders. Restoration efforts need to be a multipurpose venture: restoration or conservation of biodiversity and other ecosystem components and functions, provision of reliable food resources for the community, energy sources such as biomass, fresh water, products of traditional use values like herbal medicines, fodder, recreation, ecotourism and employment opportunities. Together these partial benefits may increase the sustainability and total benefits of a restoration project.

Social commitments and political will for restoration among peoples and nations struggling to overcome chronic poverty are achievable if and only if direct economic and other incentives (including any environmental and social services resulting from the restoration programmes) are provided to the local communities, otherwise their involvement is less likely to be sustained (Sayer et al, 2004). Many conservation and/or restoration attempts (for example nature reserves) failed in many parts of the developing world partly because of overlooking social and economic constraints (Montagnini, 2001). It will very often be useful from an early stage to adopt a participatory approach in which all stakeholders are identified and the socio-economic consequences and benefits well analysed, known and considered. In cases of multiple stakeholder groups with diverse and often conflicting values and opinions, the successful implementation of restoration projects will be a complex and controversial social issue, and this requires a careful identification and understanding of potential socio-economic issues. The realization of restoration projects is heavily dependent on the cooperation between land-users and those agents implementing the restoration projects, while fully convinced participation of all concerned stakeholders will result in restoration practices grounded in stewardship principles. The ability of restoration to establish relationships between people and the natural environment and to serve as a vehicle for the development of a resource stewardship and conservation ethic rests on the process of facilitation and type of participatory system employed.

Moreover, wood and other forest product scarcities at a national scale is becoming a serious problem in some of the countries in east Africa such as Ethiopia, and it has been the primary incentive in forest clearance (Lemenih, 2004; Mekonnen et al, 2007; Teketay et al, Chapter 2 of this volume). For instance, Ethiopia is experiencing an acute shortage of forest products supply, particularly that of fuelwood, where a modest estimate indicates over 33 million cubic metres per annum of deficit (FAO, 2001). This presses for restoration projects to provide sustainable wood products supply, while achieving the target of ecosystem recovery (Lemenih, 2004), and this is the advantage clearly seen in restoration programme based on fast-growing plantation forestry (Bone et al, 1997; Lamb, 1998; Montagnini, 2001). For instance, Bone et al (1997) reported that 33 species (26 woody and 7 herbaceous) regenerated under eucalyptus stands in Malawi are used for 10 different purposes by the community. These regenerates are used as herbal medicines (20 species), as edible products (12 species), fuelwood (9 species) and fibre (8 species), while the plantation crop was harvested to supply regional and national wood demands. McNamara et al (2006) indicated that by using fast-growing exotics as facilitator crops for restoring degraded lands, not only are high-value native timber species regenerated but also the exotic nurse crops provide revenues that are sufficient to cover the cost of the restoration process. This is crucially important for east African countries like Ethiopia, which are not only facing severe wood products supply shortage, but also have a poor economy hindering engagement in costly and time-consuming restoration ventures. In Ethiopia existing plantation forests that facilitate ecosystem recovery are simultaneously supplying the local people with a number of products and services including fuelwood (Figure 9.6), both for sale or home consumption, forest grazing and opportunities for harvesting non-timber forest products such as honey. Future plantings for restoration of

**Figure 9.6** *Plantation forests are important for local livelihoods: Fuelwood collectors from a* Eucalyptus *plantation near Addis Ababa, Ethiopia*

*Source:* F. Bongers

degraded ecosystems need to be tuned to socio-economic needs and priorities of local people and national economies before passing to decisions on species selection, planting system and management plans.

## CONCLUSIONS

Lessons learnt from the experiences reviewed in this chapter clearly demonstrate that fast-growing tropical plantation forests when established on degraded lands are capable of fostering the return of diverse flora and fauna species and improving soil conditions. Some of these experiences provided strong evidence on the positive role of even the most contentious species, *Eucalyptus*, and are, therefore, particularly relevant to the '*Eucalyptus* controversy'. All forms of plantation forests, established on degraded sites with proper site–species match, can contribute to biodiversity restoration, although the intensity of fostering and thus effectiveness varies considerably based on the intrinsic species attributes such as crown structure, stand density (thin or overstocked), proximity to natural seed source, stand- or landscape-level structural diversity, stand age, plantation size, and stand management. The most important question in employing plantation forests as foster ecosystems is not whether a planted species is of exotic or indigenous origin but the intrinsic species behaviour such as canopy architecture, site/species match, stand management, and stand and/or landscape diversity. Applications of appropriate silvicultural management can induce any plantation forest ecosystem to be conducive to promote a high level of biodiversity restoration as well as soil fertility. The decision on origin of species (exotic versus indigenous) and mode of establishment (monoculture or mixed species) should be made rather based on the social, economic and political considerations within which the restoration projects are exercised.

To obtain good soil-fertility improvement on planted degraded sites, all management practices likely to decrease inputs, such as the burning of harvest residues, the harvest of twigs and leaves for fuel, and the harvest of logs with the bark, should be minimized (Brown et al, 1997; Lemenih et al, 2004a). Moreover, the choice of species should not be based only on growth and wood qualities, but also on litter quality and decomposability. Mixing of species is recommended. Simple guidelines (Box 9.1) for employing plantation forests in biodiversity restoration are based on a simple start (with only one to three species), whereafter complexity is introduced over time to achieve the desired ecosystem. Such a step-by-step approach will increase the acceptability of plantation forests that over time will develop into a more natural vegetation system capable of delivering a whole array of forest ecosystem functions.

---

### Box 9.1 Guideline for cost-effective ecosystem restoration using plantation forests as a tool

*Setting objectives*

Although the ultimate goal of restoration is clear – recovery of a resilient ecosystem – planting for ecosystem restoration can address multiple objectives. It can encompass the provision of wood and non-wood products and ecosystem services all together. As restoration operations are costly and time-consuming, it is recommended to incorporate other socio-economic gains while the ecosystem is restoring. Such intermediate objectives not only attract revenue streams that can partly or wholly cover the cost, but are also effective in winning popular participation. Therefore, considerations must be given to both the speed at which the restoration proceeds as well as intermediate products from the activity. Such objectives must be set with the participation of the beneficiaries.

*Which species to plant?*

Decisions should be made based on the site biophysical conditions and the need (objectives) of the landowner/community. For economic targets, it is recommended to select species that provide products with good market demand (timber or non-timber). Furthermore, fast growth, seed/seedling availability, management know-how and the community's need for forest products are important criteria to attend to. The following elements need careful consideration:

## 1 Site quality

- **On heavily degraded lands:** Start with monoculture or two to three mixed exotic high-value species (timber or non-timber) as these are generally easily and rapidly established. Then the target indigenous species can gradually be introduced either by natural recolonization over time or by artificial planting.
- **Relatively less degraded site:** Start with either exotic or indigenous species provided that available choices of species can satisfy the socioe-conomic needs and expectation of the landowner/community. However, if high-value indigenous species that are adapted to the open conditions of the degraded site are not available, consider either mixing high-value exotics with high-value indigenous or start with exotics and then introduce the indigenous shortly after a proper shade environment is established by the exotics.

## 2 Economic objectives

- Wood production on short rotation: species involved in this case should be coppicing type that rapidly resprout and provide the required shade for the regenerates or planted indigenous.
- Timber/industrial wood: coppice or non-coppice species of the desired timber quality, either exotic or indigenous, can be used.
- Diverse products: mixed species, exotic or indigenous or a mix of exotic and indigenous that provide the targeted products can be used. When considering mixed species it is essential to evaluate (i) the functional and structural complementarity of the species and (ii) the temporal diversity of products obtainable for addressing broad socio-economic demands, such as incorporating forage or fruit-yielding species. Furthermore, on degraded sites, mixing N-fixing species is highly recommended.

## 3 Social considerations

- **Private landowners:** Species that address the economic needs of the owner are selected, but within the general policy framework for forest development and ecological restoration. In poor economies, mixed species planting that provides various products (wood and non-wood) at various temporal scales are best choices as these could secure the livelihoods of the households. Nonetheless, based on market availability, favourability of price, policy environment, access to credit and other factors, private landowners can also chose to make monocultures. Thus it is essential to balance both ecological and household interests. Optimizing the economic return by also introducing NTFPs production programmes such as honey, mushrooms and livestock could increase success.
- **Community landowner:** Define clearly the community members and their territories in a participatory manner. Discuss targets (interests) and choose species based on participatory and transparent discussion. Mixed species planting is likely to address the desired diverse type of products to satisfy different groups and interests.

### Preparation for launching

Secure the restoration site, develop partnership with locals and plan activities in a participatory way. Produce a map of the area, and on the map show where to plant which species and what sizes, and at what stocking. Do this in a participatory manner. Start with sites close enough to remnant forest patches. If these are not available, start with sites having good stock of remnant solitary trees. Negotiate this.

## Site preparation for planting

For planting site preparations as little disturbance as possible should be made. It is recommended to use, for example, spot hoeing rather than tilling the entire area, though the choice could be determined based on labour and technology available. Probably, on heavily degraded sites, some kind of intensive site preparations may be needed such as soil and water conservation structures. Avoid intensive site preparations on relatively good quality sites.

## Planting

- If monoculture, plant at spacing usually used for timber plantation of the selected species;
- If mixed species, plant each species in alternate rows and at normal stocking for timber plantation.

## Weeding and early tending operations

- At early establishment age, weeding (spot weeding) should be practised to allow the planted seedlings to establish well and grow fast;
- Survival counts and replacement planting for dead seedlings should be done as is usual for timber planting.

## Thinning and pruning

Adaptive thinning and pruning is required. Monitor the dynamics of regenerates and the canopy density of the plantation canopy to design the required thinning and pruning intensity. Generally, a more frequent thinning regime than normally applied is likely to be needed to encourage more understorey. However, the intensity and frequency could differ based on site, plantation species, original stocking and rate of recolonization, thus constant monitoring of the understorey regeneration dynamics is a necessity. When many emergent seedlings are observed at the forest floor, lightening the canopy mildly to allow them to receive enough PAR is needed, and thus thinning or pruning. For light canopy species, thinning and pruning should not be heavy, while for evergreen and dense canopy species, thinning and pruning intensities and frequencies that constantly maintain at least 20–35 per cent of canopy openness should be applied as necessary.

## Protection

Exclusion of human interference during some period may strongly stimulate plantation development. Observations from Ethiopia, for instance, show strong negative effects of free forest grazing, litter racking and illegal cutting of the planted trees. There is no question that protection against such acts will improve performance of plantations as foster ecosystems and promote rapid recolonization. However, such exclusion is also a source of conflict between locals and plantation managers. Thus, based on the local situation, instigation of potential challenges and negotiation on such issues through a participatory approach is needed. For protection against weeds, pests and disease, use of chemicals of any type should be avoided.

### Length of rotation and harvesting

The length of rotation is usually set based on the products target and economic needs of the landowners. In all cases harvest must be made with as little disturbance as possible. Also for insuring good ecological requirements of the understorey regenerates, unless economic drivers are not forcing, it is better to gradually remove the plantation tree crops by progressive thinning.

### Replanting?

In short-rotation coppice management, the regenerates will gradually grow tall and vigorous during the successive harvest and regrowth, to a level where they themselves overtop and suppress the plantation tree crops. However, in long-rotation and non-coppice species, if the regenerates are not diverse and vigorous enough, it is recommended to replant, but at lower disturbance and lower density than originally planted. The enrichment planting density should be decided based on the understorey density and growth performance.

# NOTE

1 (Re)generation refer to the establishment of plant species either of the original ecosystem (regeneration) or new to the area (generation), while (re)introduction is used to refer to the colonization of animal populations of the original ecosystem (reintroduction) or new to the area species (introduction) under plantation forests.

# REFERENCES

Aide, T. M. and Cavelier, J. (1994) 'Barriers to lowland tropical forest restoration in the Sierra Nevada de Santa Marta, Colombia', *Restoration Ecology*, vol 2, pp219–229

Aide, T. M., Zimmerman, K. J., Herrera, L., Rosario, M. and Serrano, M. (1995) 'Forest recovery in abandoned tropical pastures in Puerto Rico', *Forest Ecology and Management*, vol 77, pp77–86

Alem, S. (2006) 'Regeneration of indigenous woody plants: Status of soil fertility and quality of coffee found in a *Eucalyptus grandis* plantation and the adjacent natural forest in south-western Ethiopia', MSc thesis, Addis Ababa University, Addis Ababa

Allen, R., Platt, K. and Wiser, K. (1995) 'Biodiversity in New Zealand plantations', *New Zealand Forestry*, vol 39, pp26–29

Anonymous (2003) Kenya Forest Department unpublished information, Nairobi, Kenya

Anonymous (2006) Eastern and Southern Africa Forest Investment Forum, 13–16 June, Pietermaritzburg, South Africa

Ashton, P. M. S., Gamage, S., Gunatilleke, I. A. U. N. and Gunatilleke, C. V. S. (1998) 'Using Caribbean pine to establish a mixed plantation: Testing effects of pine canopy removal on plantings of rain forest tree species', *Forest Ecology and Management*, vol 106, pp211–222

Ashton, M. S., Gunatilleke, C. V. S., Singhakumara, B. M. P. and Gunatilleke,
I. A. U. N. (2001) 'Restoration pathways for rain forest in southwest Sri Lanka:
A review of concepts and models', *Forest Ecology and Management*, vol 154,
pp409–430

Atkinson, D. (2000) 'Root characteristics: Why and what to measure', in A. L. Smit,
A. G. Bengough, M. Van Noordwijk, S. Pellerin and S. C. van De Gein (eds) *Root
Methods, a Hand Book*, Springer-Verlag, Berlin, Heidelberg and New York

Augusto, L., Ranger, J., Binkley, D. and Rothe, A. (2002) 'Impact of several common
tree species of European temperate forests on soil fertility', *Annals of Forest Science*,
vol 59, pp233–253

Baum, A. K., Haynes, J. K., Dillemuth, P. F. and Cronin, T. J. (2004) 'The matrix
enhances the effectiveness of corridors and stepping stones', *Ecology*, vol 85,
pp2671–2676

Bekele, M. (1992) 'Forest history of Ethiopia from early times to 1974', MSc thesis,
School of Agricultural and Forest Sciences, University College of North Wales,
Bangor, UK

Bernhard-Reversat, F. (ed) (2001) *Effect of Exotic Tree Plantations on Plant Diversity
and Biological Soil Fertility in the Congo Savanna: With Special Reference to
Eucalypts*, Center for International Forestry Research, Bogor, Indonesia

Binkley, D. and Giardina, C. (1998) 'Why do tree species affect soils? The warp and
woof if tree–soil interactions', *Biogeochemistry*, vol 42, pp89–106

Binkley, D. and Resh, S. C. (1999) 'Rapid changes in soils following *Eucalyptus*
afforestation in Hawaii', *Soil Science Society of America Journal*, vol 63, pp222–225

Binkley, D., Senock, R., Bird, S. and Cole, T. (2003) 'Twenty years of stand development in pure and mixed stands of *Eucalyptus saligna* and nitrogen-fixing *Facaltaria
mollucana*', *Forest Ecology and Management*, vol 182, pp93–102

Bone, R., Lawrence, M. and Magombo, Z. (1997) 'The effect of a *Eucalyptus camaldulensis* (Dehn.) plantation on native woodland recovery on Ulumba Mountain,
southern Malawi', *Forest Ecology and Management*, vol 99, pp83–99

Bosch, J. M. and Hewlett, J. D. (1982) 'A review of catchments experiments to determine the effect of vegetation changes on water yield and evapotranspiration', *Journal
of Hydrology*, vol 55, pp3–23

Brady, N. C. and Weil, R. R. (1999) *The Nature and Properties of Soils* (12th edition),
Prentice Hall, Upper Saddle River, NJ

Brosset, A. (2001) 'Effect of exotic tree plantations on vertebrate fauna', in F.
Bernhard-Reversat (ed) *Effect of Exotic Tree Plantations on Plant Diversity and
Biological Soil Fertility in the Congo Savanna: With Special Reference to Eucalypts*,
CIFOR, SMK Grafika Desa Putera, Indonesia, pp24–26

Brown, A. G., Nambiar, E. K. S. and Cossalter, C. (1997) 'Plantations for the tropics –
Their role, extent and nature', in E. K. S. Nambiar and A. G. Brown (eds)
*Management of Soil, Nutrients and Water in Tropical Plantation Forests*, ACIAR
Monograph No. 43, ACIAR, Canberra, pp3–23

Brown, S. and Lugo, A. E. (1994) 'Rehabilitation of tropical lands: A key to sustaining
development', *Restoration Ecology*, vol 2, pp97–111

Butler, R., Montagnini, F. and Arroyo, P. (2008) 'Woody understorey plant diversity in
pure and mixed native tree plantations at La Selva Biological Station, Costa Rica',
*Forest Ecology and Management*, vol 255, pp2251–2263

Calder, I. R., Hall, R. L. and Adlard, P. G. (eds) (1992) *Growth and Water Use of Forest
Plantations*, Wiley and Sons, New York

Carnevale, N. J. and Montagnini, F. (2002) 'Facilitating regeneration of secondary forests with the use of mixed and pure plantations of indigenous specie', *Forest Ecology and Management*, vol 163, pp217–227

Carnus, J. M., Parrotta, J., Brockerhoff, E., Arbez, M., Jactel, H., Kremer, A., Lamb, D., O'Hara, K. and Walters, B. (2003) 'Planted forests and biodiversity', IUFRO Occasional Paper, vol 15, pp33–49

Carnus, J. M, Parrotta, J., Brockerhoff, E. G., Arbez, M., Jactel, H., Kremer, A., Lamb, D., O´Hara, K. and Walters, B. (2006) 'Planted forests and biodiversity', *Journal of Forestry*, vol 104, pp65–77

Castellón, T. and Sieving, E. K. (2006) 'An experimental test of matrix permeability and corridor use by an endemic understory bird', *Conservation Biology*, vol 20, pp135–145

Chamshama, S. A. O. and Nwonwu, F. O. C. (2004) 'Forest plantations in Sub-Saharan Africa: Lessons learnt on sustainable forest management in Africa', AFORNET, AAS, FAO and KSLA

Chandrasekar-Rao, A. and Sunquist, M. E. (1996) 'Ecology of small mammals in tropical forest habitats of southern India', *Journal of Tropical Ecology*, vol 12, pp561–571

Chapman, C. A. and Chapman, L. J. (1996) 'Exotic tree plantations and the regeneration of natural forests in Kibale National Park, Uganda', *Biological Conservation*, vol 76, pp253–257

Chapman, C. A. and Chapman, L. J. (2004) 'Unfavorable successional pathways and the conservation value of logged tropical forest', *Biodiversity and Conservation*, vol 13, pp2089–2105

Chapman, C. A., Chapman, L. J., Kayfman, L. and Zanne, A. E. (1999) 'Potential causes of arrested succession in Kibale National Park, Ugnada: Growth and mortality of seedlings', *African Journal of Ecology*, vol 37, pp81–92

Chazdon, R. L. (2008) 'Beyond deforestation: Restoring forests and ecosystem services on degraded lands', *Science*, vol 320, pp1458–1460

Chen, X., Li, B. L. and Lin, Z. S. (2003) 'The acceleration of succession for the restoration of the mixed-broadleaved Korean pine forests in northeast China', *Forest Ecology and Management*, vol 177, pp503–514

Christian, D. P., Hoffman, W., Hanowski, J. M., Niemi, G. J. and Beyea, J. (1998) 'Bird and mammal diversity on woody biomass plantations in North America', *Biomass and Bioenergy*, vol 14, pp395–402

Corlett, R. T. (1999) 'Environmental forestry in Hong Kong: 1871–1997', *Forest Ecology and Management*, vol 116, pp93–105

Corley, J., Sackmann, P., Rusch, V., Bettinelli, J. and Paritsis, J. (2006) 'Effects of pine silviculture on the ant assemblages (Hymenoptera: Formicidae) of the Patagonian steppe', *Forest Ecology and Management*, vol 222, pp162–166

Cubina, A. and Aide, T. M. (2001) 'The effect of distance from forest edge on seed rain and soil seed bank in a tropical pasture', *Biotropica*, vol 33, pp260–267

Cummings, J. and Reid, N. (2008) 'Stand-level management of plantations to improve biodiversity values', *Biodiversity and Conservation*, vol 17, pp1187–1211

Cummings, J., Reid, N., Davies, I. and Grand, C. (2005) 'Adaptive restoration of sand-mined areas for biological conservation', *Journal of Applied Ecology*, vol 42, pp160–170

Cummings, J., Reid, N., Davies, I. and Grant, C. (2007) 'Experimental manipulation of restoration barriers in abandoned eucalypt plantations', *Restoration Ecology*, vol 15, pp156–167

Cusack, D. and Montagnini, F. (2001) 'The role of native species plantations in recovery of understory woody diversity in degraded pasturelands of Costa Rica', *Forest Ecology and Management*, vol 188, pp1–15

de la Cretaz, L. A. and Kelty, J. M. (2002) 'Development of tree regeneration in fern-dominated forest understories after reduction of deer browsing', *Ecological Restoration*, vol 10, pp416–426

Diaz, M., Carbonell, R., Santos, T. and Telleria, J. L. (1998) 'Breeding bird communities in pine plantations of the Spanish plateaux: Biogeography, landscape and vegetation effects', *Journal of Applied Ecology*, vol 35, pp562–574

Dobson, P. A., Bradshaw, A. D. and Baker, A. J. M. (1997) 'Hopes for the future: Restoration ecology and conservation biology', *Science*, vol 277, pp515–522

Donovan, T. M., Jones, P. W., Annand, E. M. and Thompson, F. R. III (1997) 'Variation in local-scale edge effects: Mechanisms and landscape context', *Ecology*, vol 78, pp2064–2075

Dowsett-Lemaire, F. and Dowsett, R. J. (1991) 'The avifauna of the Kouilou basin in Congo', in F. Dowsett-Lemaire and R. J. Dowsett (eds) *Flore et Faune du Kouilou et Leur Exploitation*, Touraco Research Report 4, Touraco, Liège, Belgium, pp189–239

Dozier, H., Gaffney, J. F., McDonald, S. K., Johnson, E. R. R. L. and Shilling, D. G. (1998) 'Cogongrass in the United States: History, ecology, impacts and management', *Weed Technology*, vol 12, pp737–743

Duncan, R. S. and Chapman, C. A. (2001) 'Seed dispersal and potential forest succession in abandoned agricultural in tropical Africa', *Ecological Applications*, vol 9, no 3, pp998–1008

Duncan, R. S. and Chapman, C. A. (2003) 'Consequences of plantation harvest during tropical forest restoration in Uganda', *Forest Ecology and Management*, vol 173, pp235–250

Elliot, C. (2003) 'WWF vision for planted forests', paper presented at the UNFF Intersessional Experts Meeting on the Role of Planted Forest in Sustainable Forest Management, Wellington, New Zealand, www.maf.govt.nz/mafnet/unff-planted-forestry-meeting/conference-papers/www-vision-for-planted-forests.htm, pp25–27

Erskine, D. P., Lamb, D. and Bristow, M. (2006) 'Tree species diversity and ecosystem function: Can tropical multi-species plantations generate greater productivity?', *Forest Ecology and Management*, vol 233, pp205–210

Evans, J. (1992) *Plantation Forestry in the Tropics. Tree Planting for Industrial, Social, Environmental and Agroforestry Purposes* (second edition), Clarendon Press, Oxford, UK

Evans, J. and Turnbull, J. W. (2004) *Plantation Forestry in the Tropics* (third edition), Oxford University Press, Oxford, UK

Faaborg, J., Brittingham, M., Donovan, T. and Blake, J. (1995) 'Habitat fragmentation in the temperate zone', in T. E. Martin and D. M. Finch (eds) *Ecology and Management of Neotropical Migratory Birds*, Oxford University Press, New York, pp357–380

Fang, W. and Peng, S. L. (1997) 'Development of species diversity in the restoration process of establishing a tropical man-made forest ecosystem in China', *Forest Ecology and Management*, vol 99, pp185–196

FAO (2001) 'Global Forest Resources Assessment 2000, Main Report', FAO Forestry Paper 140, Food and Agriculture Organization of the United Nations, Rome

FAO (2005) 'Global Forest Resources Assessment 2005', Food and Agricultural Organization of the United Nation, Rome

Fernandes, E. C. M., Motavalli, P. P., Castilla, C. and Mukurumbira, L. (1997) 'Management of soil organic matter dynamics in tropical land-use systems', *Geoderma*, vol 79, pp49–67

Fimbel, R. A. and Fimbel, C. C. (1996) 'The role of exotic conifer plantations in rehabilitating degraded tropical forest lands: A case study from the Kibale Forest in Uganda', *Forest Ecology and Management*, vol 81, pp215–226

Fisher, R. F. (1995) 'Amelioration of degraded rain forest soils by plantations of native trees', *Soil Science Society of America Journal*, vol 59, pp544 – 549

Garten, C. T. Jr. (2002) 'Soil carbon storage beneath recently established tree plantations in Tennessee and South Carolina, USA', *Biomass and Bioenergy*, vol 23, pp93–102

Geldenhuys, C. J. (1997) 'Native forest regeneration in pine and eucalyptus plantations in Northern Province, South Africa', Forest Ecology and Management, vol 99, pp101–115

George, L. T. and Zack, S. (2001) 'Spatial and temporal considerations in restoring habitat for wildlife, *Restoration Ecology*, vol 9, pp272–279

Gibbs, P. J., Marquez, C. and Sterling, J. E. (2008) 'The role of endangered species reintroduction in ecosystem restoration: Tortoise–cactus interactions on Espanola Island, Galapagos, *Restoration Ecology*, vol 16, pp88–93

Gibson, C. W. D. and Brown, V. K. (1992) 'Grazing and vegetation change: Deflected or modified succession?', *Journal of Applied Ecology*, vol 29, pp120–131

Glenday, J. (2006) 'Carbon storage and emissions offset potential in an East African tropical rainforest', *Forest Ecology Management*, vol 235, pp72–83

Gómez-Aparicio, L., Zavala, A. M., Bonet, J. F. and Zamora, R. (2009) 'Are pine plantations valid tools for restoring Mediterranean forests? An assessment along abiotic and biotic gradients', *Ecological Applications*, vol 19, pp2124–2141

Guariguata, M. R., Rheingans, R. and Montagnini, E. (1995) 'Early woody invasion under tree plantations in Costa Rica: Implications for forest restoration', *Restoration Ecology*, vol 3, pp252–260

Haggar, J., Wightman, K. and Fisher, R. (1997) 'The potential of plantations to foster woody regeneration within a deforested landscape in lowland Costa Rica', *Forest Ecology and Management*, vol 99, pp55–64

Hardwick, K., Healy, J., Elliott, S., Garwood, N. C. and Anusarnsunthorn, V. (1997) 'Understanding and assisting natural regeneration processes in degraded seasonal evergreen forests in northern Thailand', *Forest Ecology and Management*, vol 99, pp203–214

Harrington, A. C. (1999) 'Forests planted for ecosystem restoration or conservation', *New Forests*, vol 17, pp175–190

Harrington, R. A. and Ewel, J. J. (1997) 'Invasibility of tree plantations by native and non-indigenous plant species in Hawaii', *Forest Ecology and Management*, vol 99, pp153–162

Harvey, C. A. and Haber, W. A. (1999) 'Remnant trees and the conservation of biodiversity in Costa Rican pastures', *Agroforestry Systems*, vol 44, pp37–68

Hartley, M. J. (2002) 'Rationale and methods for conserving biodiversity in plantation forests', *Forest Ecology and Management*, vol 155, pp81–95

Hayes, F. E. and Samad, I. (1998) 'Diversity, abundance and seasonality of birds in a Caribbean pine plantation and native-broadleaved forest at Trinidad, West Indies', *Bird Conservation International*, vol 8, pp67–89

Helles, F. and Linddal, M. (1996) 'Afforestation experiences in the Nordic countries', Nord 1996:15, Nordic Council of Ministers, Copenhagen, Denmark

Hinde, J. R., Corti, R. G., Fanning, E. and Jenkins, K. B. R. (2001) 'Large mammals in miombo woodland, evergreen forest and a young teak (*Tectona grandis*) plantation in Kilombero Valley, Tanzania', *African Journal of Ecology*, vol 39, pp318–321

Hobbs, R. J. (2004) 'Restoration ecology; the challenge of social values and expectations', *Frontiers in Ecology and the Environment*, vol 2, pp43–44

Horsley, B. S, Stout, L. S. and Decalesta, S. D. (2003) 'White-tailed deer impact on the vegetation dynamics of a northern hardwood forest', *Ecological Applications*, vol 13, no 1, pp98–118

Howe, H. F. and Smallwood, J. (1982) 'Ecology of seed dispersal', *Annual Review of Ecology and Systematics*, vol 13, pp201–228

Hunter, I. R., Hobley, M. and Smale, P. (1998) 'Afforestation of degraded land – Pyrrhic victory over economic, social and ecological reality?', *Forest Ecology and Management*, vol 10, pp97–106

Huttel, C. and Loumeto, J. L. (2001) 'Effect of exotic tree plantations and site management on plant diversity', in F. Bernhard-Reversat (ed) *Effect of Exotic Tree Plantations on Plant Diversity and Biological Soil Fertility in the Congo Savanna: With Special Reference to Eucalypts*, CIFOR, SMK Grafika Desa Putera, Bogor, Indonesia, pp14–23

Igarashi, T. and Kiyono, Y. (2008) 'The potential of hinoki (*Chamaecyparis obtusa* [Sieb. et Zucc.] Endlicher) plantation forests for the restoration of the original plant community in Japan', *Forest Ecology and Management*, vol 255, pp183–192

Inman, M. F., Wentworth, R. T., Groom, M., Brownie, C. and Lea, R. (2007) 'Using artificial canopy gaps to restore Puerto Rican parrot (*Amazona vittata*) habitat in tropical timber plantations', *Forest Ecology and Management*, vol 243, pp169–177

Jaiyeoba, A. I. (2001) 'Soil Rehabilitation through afforestation: Evaluation of the performance of eucalyptus and pine plantations in a Nigeria savanna environment', *Land Degradation & Development*, vol 12, pp183–194

Jenkins, R. K. B., Roettcher, K. and Corti, G. (2003) 'The influence of stand age on wildlife habitat use in exotic teak tree *Tectona grandis* plantations', *Biodiversity and Conservation*, vol 12, pp975–990

Jenny, H. (1941) *Factors of Soil Formation*, McGraw-Hill, New York, NY

Jim, C. Y. (2001) 'Ecological and landscape rehabilitation of a quarry site in Hong Kong', *Restoration Ecology*, vol 1, pp85–94

Johnson, G. A. M. and Freedman, B. (2002) 'Breeding birds in forestry plantations and natural forest in the vicinity of Fundy National Park, New Brunswick', *Canadian Field-Naturalist*, vol 116, pp475–487

Kamo, K., Vacharangkura, T., Tiyanon, S., Viriyabuncha, C., Nimpila, S. and Doangsrisen, B. (2002) 'Plant species diversity in tropical planted forests and implication for restoration of forest ecosystems in Sakaerat, northeastern Thailand', *JARQ – Japan Agricultural Research Quarterly*, vol 36, pp111–118

Katyal, J. C., Rao, N. H. and Reddy, M. N. (2001) 'Critical aspects of organic matter management in the tropics: The example of India', *Nutrient Cycling in Agroecosystems*, vol 61, pp71–88

Keenan, R., Lamb, D., Woldring, O., Irvine, T. and Jensen, R. (1997) 'Restoration of plant biodiversity beneath tropical tree plantations in northern Australia', *Forest Ecology and Management*, vol 99, pp117–131

Keenan, R. J., Lamb, D., Parrotta, J. and Kikkawa, J. (1999) 'Ecosystem management in tropical timber plantations: Satisfying economic, conservation, and social objectives, *Journal of Sustainable Forestry*, vol 9, pp117–134

Kelly, E. F., Chadwick, O. A. and Hilinski, T. E. (1998) 'The effect of plants on mineral weathering', *Biogeochemistry*, vol 42, pp21–53

Kelty, M. J. (2006) 'The role of species mixtures in plantation forestry', *Forest Ecology and Management*, vol 233, pp195–204

Kenya MENR (1994) 'Kenya Forestry Master Plan', Ministry of Environment and Natural Resources, Nairobi, Kenya

Kwok, H. K., and Corlett, R. T. (2000) 'The bird communities of a natural secondary forest and a *Lophostemon confertus* plantation in Hong Kong', *Forest Ecology and Management*, vol 130, pp227–234

Kuusipalo, J., Goran, A., Jafarsidik, Y., Otsamo, A., Tuomela, K. and Vuokko, R. (1995) 'Restoration of natural vegetation in degraded *Imperata cylindrica* grassland: Understory development in forest plantations', *Journal of Vegetation Science*, vol 6, pp205–210

Lamb, D. (1998) 'Large-scale ecological restoration of degraded tropical forest lands: The potential role of timber plantations', *Restoration Ecology*, vol 6, pp271–279

Lamb, D. (2003) 'Is it possible to reforest degraded tropical lands to achieve economic and also biodiversity benefits?', in H. C. Sim, S. Appanah and P. B. Durst (eds) *Bringing Back the Forests: Policies and Practices for Degraded Lands and Forests. Proceedings of an International Conference*, 7–10 October 2002, FAO Regional Office for Asia and the Pacific, Kuala Lumpur, Malaysia, pp17–25

Lamb, D., Erskine, P. D. and Parrotta, J. A. (2005) 'Restoration of degraded forest landscapes', *Science*, vol 310, pp1628–1632

Laycock, W. A. (1995) 'New perspectives on ecological condition of rangelands: Can state and transition or other models better define condition and diversity?', in L. Montes and G. E. Olivia (eds) *Proceedings of the International Workshop on Plant Genetic Resources, Desertification and Sustainability*, INTA-EEA, Rio Gallegos, Argentina

Lee, W. S. E., Hau, C. H. B. and Corlett, T. R. (2005) 'Natural regeneration in exotic tree plantations in Hong Kong, China', *Forest Ecology and Management*, vol 212, pp358–366

Lemenih, M. (2004) 'Effects of land use changes on soil quality and native flora degradation and restoration in the highlands of Ethiopia. Implications for sustainable land management', *Acta Universitatis Agriculturae Sueciae, Silvestria*, vol 306, PhD thesis, Swedish University of Agricultural Sciences, Department of Forest Soils, Upsala, Sweden

Lemenih, M. (2006) 'Expediting ecological restoration with the help of foster tree plantations in Ethiopia', *Journal of the Drylands*, vol 1, pp72–84

Lemenih, M. (2008) 'Current and prospective economic contributions of the forestry sector in Ethiopia', in T. Hechett and N. Aklilu (eds) *Proceeding of a Workshop on 'Ethiopian Forestry at Crossroads: On the Need for Strong Institutions'*, Addis Ababa, Ethiopia, pp59–82

Lemenih, M. and Teketay, D. (2004a) 'Potentials of plantation forests in fostering the restoration of native flora and fauna at degraded sites in Ethiopia: A review', *Ethiopian Journal of Biological Society*, vol 3, pp81–111

Lemenih, M. and Teketay, D. (2004b) 'Constraints and opportunities associated with restoration of native forest flora in the degraded highlands of Ethiopia: A review', *SINET: Ethiopian Journal of Science*, vol 27, pp75–90

Lemenih, M. and Teketay, D. (2005) 'Effect of prior land use on the recolonization of native plants under plantation forests in Ethiopia', *Forest Ecology and Management*, vol 218, pp60–73

Lemenih, M. and Teketay, D. (2006) 'Changes in soil seed bank composition and density following deforestation and subsequent cultivation of dry Afromontane forest in Ethiopia', *Journal of Tropical Ecology*, vol 47, pp1–12

Lemenih, M., Gidyelew, T. and Teketay, D. (2004a) 'Effects of canopy cover and understorey environment of tree plantations on species richness, density and sizes of colonizing native woody species in southern Ethiopia', *Forest Ecology and Management*, vol 194, pp1–10

Lemenih, M., Lemma, B. and Teketay, D. (2004b) 'Changes in soil carbon and total nitrogen following reforestation of previously cultivated land in the highlands of Ethiopia', *SINET: Ethiopian Journal of Science*, vol 28, no 2, pp99–108

Lemenih, M., Karltun, E. and Olsson, M. (2004c) 'Comparison of soil attributes under *Cupressus lusitanica* and *Eucalyptus saligna* established on abandoned farmlands with continuously cropped farmlands and natural forest in Ethiopia', *Forest Ecology and Management*, vol 195, pp57–67

Lemenih, M., Tolera, M. and Karltun, E. (2008) 'Deforestation: Impact on soil quality, biodiversity and livelihoods in the highlands of Ethiopia', in I. B. Sanchez and C. L. Alonso (eds) *Deforestation Research Progress*, Nova Science Publishers, Hauppage, NY, pp22–39

Lindenmayer, D. B. (2002) *Plantation Design and Biodiversity Conservation*, Rural Industries Research and Development Corporation, Canberra, Australia

Lindenmayer, D. B. and Hobbs, R. J. (2004) 'Fauna conservation in Australian plantation forests – A review', *Biological Conservation*, vol 119, pp151–168

Lindenmayer, D. B., Cunningham, R. B. and Pope, M. L. (1999) 'A large-scale "experiment" to examine the effects of landscape context and habitat fragmentation on mammals', *Biological Conservation*, vol 88, pp387–403

Lonsdale, W. M. (1999) 'Global patterns of plant invasions and the concept of invisibility', *Ecology*, vol 80, pp1522–1536

Loubana, P. M. and Reversat, G. (2001) 'Effect of exotic tree plantations on free living and plant parasitic soil nematodes and population changes with eucalypt hybrids and plantation age', in F. Bernhard-Reversat (ed) *Effect of Exotic Tree Plantations on Plant Diversity and Biological Soil Fertility in the Congo Savanna: With Special Reference to Eucalypts*, Center for International Forestry Research (CIFOR), Bogor, Indonesia, pp43–48

Loumeto, J. J. and Bernhard-Reversat, F. (2001) 'Soil fertility changes with eucalypt hybrids and plantation age: Soil organic matter', in F. Bernhard-Reversat (ed) *Effect of Exotic Tree Plantations on Plant Diversity and Biological Soil Fertility in the Congo Savanna: With Special Reference to Eucalypts*, CIFOR, SMK Grafika Desa Putera, Bogor, Indonesia, pp35–40

Loumeto, J. J. and Huttel, C. (1997) 'Understory vegetation in fast growing tree plantations on savanna soils in Congo', *Forest Ecology and Management*, vol 99, pp65–81

Lugo, A. E. (1992) 'Tree plantations for rehabilitating damaged lands in the tropics', in M. K. Wali (ed) *Environmental Rehabilitation*, vol 2, SPB Academic Publishing, The Hague, pp247–255

Lugo, A. E. (1997) 'The apparent paradox of reestablishing species richness on degraded lands with tree monocultures', *Forest Ecology and Management*, vol 99, pp9–19

Lugo, A. E., Parrotta, J. A. and Brown, S. (1993) 'Loss in species caused by tropical deforestation and their recovery through management', *Ambio*, vol 22, pp107–109

MacArthur, R. H. (1972) *Geographical Ecology: Patterns in the Distribution of Species*, Harper and Row, New York

Mboukou-Kimbatsa, I. M. C. and Bernhard-Reversat, F. (2001) 'Effect of exotic tree plantations on invertebrate soil macrofauna', in F. Bernhard-Reversat (ed) *Effect of Exotic Tree Plantations on Plant Diversity and Biological Soil Fertility in the Congo Savanna: With Special Reference to Eucalypts*, CIFOR, SMK Grafika Desa Putera, Bogor, Indonesia, pp51–57

McCulloch, J. S. G. and Robinson, M. (1993) 'History of forest hydrology', *Journal of Hydrology*, vol 150, pp189–216

McInvar, J. and Starr, L. (2001) 'Restoration of degraded lands in the interior Columbia River basin: Passive vs. active approaches', *Forest Ecology and Management*, vol 153, pp15–28

McNamara, S., Duong, V. T., Erskine, P. D., Lamb, D., Yates, D. and Brown, S. (2006) 'Rehabilitating degraded forest land in central Vietnam with mixed native species plantings', *Forest Ecology and Management*, vol 233, pp358–365

Mekonnen, Z., Kassa, H., Lemenih, M. and Campbell, B. (2007) 'The role and management of eucalyptus in Lode Hetosa district, central ethiopia', *Forests, Trees and Livelihoods*, vol 17, pp309–323

Mengak, M. T. and Guynn, D. C. Jr. (2003) 'Small mammal microhabitat use on young loblolly pine regeneration areas', *Forest Ecology and Management*, vol 173, pp309–317

Mengak, M. T., Guynn, D. C. Jr. and Van Lear, D. H. (1989) 'Ecological implications of loblolly pine regeneration for small mammal communities', *Forest Science*, vol 35, pp503–514

Miller, R. M and Jastrow, J. D. (1992) 'The application of VA mycorrhizae to ecosystem restoration and reclamation', in M. F. Allen (ed) *Mycorrhizal Functioning*, Chapman and Hall, New York, pp438–467

Moges, Y. (1998) 'The role of exotic plantation forests in fostering the regeneration of native trees in an Afromontane forest area in Ethiopia', MSc thesis, Wageningen Agricultural University, Wageningen, The Netherlands

Montagnini, F. (2000) 'Accumulation in above-ground biomass and soil storage of mineral nutrients in pure and mixed plantations in a humid tropical lowland', *Forest Ecology and Management*, vol 134, pp257–270

Montagnini, F. (2001) 'Strategies for the recovery of degraded ecosystems: Experiences from Latin America', *Interciencia*, vol 26, pp498–503

Montagnini, F. and Jordan, C. F. (eds) (2005) *Tropical Forest Ecology: The Basis for Conservation and Management*, Springer, Berlin, Germany

Nair, C. T. S. (2006) 'What is the future for African forests and forestry?', *International Forestry Review*, vol 8, pp4–13

Nepstad, D. C., Uhl, C. and Serrao, E. A. S. (1991) 'Recuperation of a degraded Amazonian landscape: Forest recovery and agricultural restoration', *Ambio*, vol 20, pp248–255

Newmark, W. (1993) 'Tropical forest fragmentation and the local extinction of understory birds in the eastern Usambara mountains, Tanzania', *Conservation Biology*, vol 5, pp67–73

Nichols, J. D. and Carpenter, F. L. (2006) 'Interplanting *Inga edulis* with *Terminalia amazonia* yields nitrogen benefits to the timber tree', *Forest Ecology and Management*, vol 233, pp344–351

Oberhauser, U. (1997) 'Secondary forest regeneration beneath pine (*Pinus kesiya*) plantations in the northern Thai highlands: A chronosequence study', *Forest Ecology and Management*, vol 99, pp171–183

Olsson, M. (2001) 'Do trees improve soil?', *Currents*, Swedish University of Agricultural Sciences, vol 25/26, pp31–34

Omiti, J. M., Parton, K. A., Sinden, J. A. and Ehui, S. K. (1999) 'Monitoring changes in land-use practices following agrarian de-collectivisation in Ethiopia', *Agriculture, Ecosystem and Environment*, vol 72, pp111–118

Otsamo, R. (1998) 'Effect of nurse tree species on early growth of *Anisotepa marginata* Korth. (Diptercarpaceae) on an *Imperata cylindrical* (L.) Beauv. grassland site in South Kalimantan, Indonesia', *Forest Ecology and Management*, vol 105, pp303–311

Otsamo, R., (2000) 'Secondary forest regeneration under fast-growing forest plantations on degraded *Imperata cylindrica* grasslands', *New Forests*, vol 19, pp69–93

Paritsis, J. and Aizen, M. A. (2008) 'Effects of exotic conifer plantations on the biodiversity of understorey plants, epigeal beetles and birds in *Nothofagus dombeyi* forests', *Forest Ecology and Management*, vol 255, pp1575–1583

Parrotta, J. A. (1992) 'The role of plantation forests in rehabilitating degraded tropical ecosystems', *Agriculture, Ecosystem and Environment*, vol 41, pp115–133

Parrotta, J. A. (1993) 'Secondary forest regeneration on degraded tropical lands: The role of plantations as "foster ecosystem"', in H. Lieth and M. Lohman (eds) *Restoration of Tropical Forest Ecosystems*, Kluwer Publishers, Dordrecht, The Netherlands, pp63–73

Parrotta, J. A. (1999) 'Productivity, nutrient cycling and succession in single- and mixed-species plantations of *Casuarina equisetifolia*, *Eucalyptus robusta*, and *Leucaena leucocephala* in Puerto Rico', *Forest Ecology and Management*, vol 124, pp45–77

Parrotta, J. A., Turnbull, W. J. and Jones, N. (1997) 'Catalysing native forest regeneration on degraded tropical lands', *Forest Ecology and Management*, vol 99, pp1–7

Paul, E. A. and Clark, F. E. (1996) *Soil Microbiology and Biochemistry*, Academic Press, San Diego, CA

Paul, K. I., Polglase, P. J., Nyakuengama, J. G. and Khanna, P. K. (2002) 'Change in soil carbon following afforestation', *Forest Ecology and Management*, vol 168, pp241–257

Perrow, M. R. and Anthony, J. D. (eds) (2002) *Handbook of Ecological Restoration, Vol 1: Principles of Restoration*, Cambridge University Press, Cambridge, UK

Pohjonen, V. and Pukkala, T. (1988) 'Profitability of growing *Eucalyptus globulus* plantations the Ethiopian highlands', *Silva Fennica*, vol 22, pp307–321

Pohjonen, V. and Pukkala, T. (1990) '*Eucalyptus globulus* in Ethiopian forestry', *Forest Ecology and Management*, vol 36, pp19–31

Polglase, P. J., Paul, K. I., Khanna, P. K., Nyakungana, J. G., O'Connell, A. M., Grove, T. S. and Battaglia, M. (2000) 'Change in soil carbon following afforestation or reforestation: Review of experimental evidence and development of a conceptual framework', NCAS Technical Report No. 20, Australian Greenhouse Office, Canberra, www.greenhouse.gov.au/ncas/files/publications.html

Post, M. W. and Kwon, C. K. (2000) 'Soil carbon sequestration and land-use change: Processes and potential', *Global Change Biology*, vol 6, pp317–327

Powers, J. S., Haggar, J. P. and Fisher, R. F. (1997) 'The effect of overstorey composition on understorey woody regeneration and species richness in 7-year-old plantations in Costa Rica', *Forest Ecology and Management*, vol 99, pp43–54

Raulund-Rasmussen, K., Borggaard, O. K., Hansen, H. C. B. and Olsson, M. (1998) 'Effect of natural organic soil solutes on weathering rates of soil minerals', *European Journal of Soil Science*, vol 49, no 3, pp397–406

Rubio, A., Gavilan, R. and Escudero, A. (1999) 'Are soil characteristics and understorey composition controlled by forest management?', *Forest Ecology and Management*, vol 113, pp191–200

Ruiz-Jaen, C. M. and Aide, T. M. (2005) 'Restoration success: How is it being measured?', *Restoration Ecology*, vol 13, pp569–577

Sarmiento, F. O. (1997) 'Arrested succession in pastures hinders regeneration of tropandean forests and shreds mountain landscapes', *Environmental Conservation*, vol 24, pp14–23

Sayer, J., Chokkalingam, U. and Poulsen, J. (2004) 'The restoration of forest biodiversity and ecological values', *Forest Ecology and Management*, vol 201, pp3–13

Scherr, S., White, A., Khare, A., Inbar, M. and Molar, A. (2004) 'For services rendered: The current status and future potential of markets for the ecosystem services provided by forests', ITTO Technical Series No. 21, International Tropical Timber Organization (ITTO), Yokohama, Japan

Sedjo, R. A. and Botkin, D. (1997) 'Forest plantations to spare natural forests', *Environment*, vol 39, pp14–20

Senbeta, F. and Teketay, D. (2001) 'Regeneration of indigenous woody species under the canopy of tree plantations in Central Ethiopia', *Tropical Ecology*, vol 42, pp175–185

Senbeta, F., Teketay, D. and Näslund, B. Å. (2002a) 'Native woody species regeneration in exotic tree plantations at Munessa-Shashamane forest, Southern Ethiopia', *New Forests*, vol 24, pp131–145

Senbeta, F., Beck, E. and Luttge, U. (2002b) 'Exotic trees as nurse-trees for the regeneration of natural tropical forests', *Trees*, vol 16, pp245–249

Shono, K., Davies, J. S. and Kheng, Y. C. (2006) 'Regeneration of native plant species in restored forests on degraded lands in Singapore', *Forest Ecology and Management*, vol 237, pp574–582

Silver, W. L., Ostertag, R. and Lugo, A. E. (2000) 'The potential for carbon sequestration through reforestation of abandoned tropical agricultural and pasture lands', *Restoration Ecology*, vol 8, pp394–407

Smith, P., Powlson, D. S., Glendining, M. J. and Smith, J. U. (1997) 'Potential for carbon sequestration in European soils: Preliminary estimates for five scenarios using results from long-term experiments', *Global Change Biology*, vol 3, pp67–79

Stangel, P. J. (1991) 'Plant nutrients in sustainable land management systems', presentation during International Workshop on Evaluation for Sustainable Land Management in the Developing World, Chiangrai, Thailand

Stevenson, F. J. (1982) *Humus Chemistry. Genesis, Composition, Reactions*, John-Wiley and Sons, New York

Sturm, K. and Apel, U. (2006) 'Forest restoration through nature-oriented reforestation practices in North Eastern Vietnam', *International Forestry Review*, vol 8, pp350–360

Sweeney, B. W., Czapka, S. J. and Yerkes, T. (2002) 'Riparian forest restoration: Increasing success by reducing plant competition and herbivory', *Restoration Ecology*, vol 10, pp392–400

Teketay, D. (1996) 'Seed ecology and regeneration in dry afro-montane forests of Ethiopia', Doctoral thesis, Swedish University of Agricultural Sciences, Umeå, Sweden

Teketay, D. and Granstrom, A. (1995) 'Soil seed banks in dry Afromontane forests of Ethiopia', *Journal of Vegetation Science*, vol 6, pp777–786

Teklu, G. (2003) *Expanse of Plantation Forest in Ethiopia (An Outcome of More Than Half a Century's Effort)*, Resources Development and Regulatory Department (NRM&RD) under the Ethiopian Ministry of Agriculture (MoA), Addis Ababa

Trouve, C., Mariotti, A., Schwartz, D. and Guillet, B. (1994) 'Soil organic carbon dynamics under *Eucalyptus* and *Pinus* planted on savannas in the Congo', *Soil Biology and Biochemistry*, vol 26, pp287–295

Tucker, N. I. J. and Murphy, T. M. (1997) 'The effects of ecological rehabilitation on vegetation recruitment: Some observations from the wet tropics of North Queensland', *Forest Ecology and Management*, vol 99, pp133–152

Turner, J. C., Gerwin, J. A. and Lancia, R. A. (2002) 'Influences of hardwood stand area and adjacency on breeding birds in an intensively managed pine landscape', *Forest Science*, vol 48, pp323–330

Tutin, C. E. G., White, L. J. T. and Mackanga-Missandzou, A. (1997) 'The use by rain forest mammals of natural forest fragments in an equatorial African savanna', *Conservation Biology*, vol 5, pp1190–1303

Tyynelä, T. M. (2001) 'Species diversity in *Eucalyptus camaldulensis* woodlots and miombo woodland in northeastern Zimbabwe', *New Forests*, vol 22, pp239–257

UNFF (2003) 'The role of planted forests in sustainable forest management. Report of the UNFF inter-sessional experts meeting', Wellington, New Zealand, 25–27 March, www.maf.govt.nz/mafnet/unff-planted-forestry-meeting/report-of-unff-meeting-nz.pdf, accessed 8 May 2010

van Ruremonde, R. H. A. C. and Kalkhoven, J. T. R. (1991) 'Effects of woodland isolation on the dispersion of plants with fleshy fruits', *Journal of Vegetation Science*, vol 2, pp377–384

Victor, G. D. (2003) 'Forest plantations and a vision for restoring the forests', keynote speech at 'The Role of Planted Forests in Sustainable Forest Management', an intersessional experts meeting for the UN Forum on Forests, convened by the Ministry of Agriculture and Forestry, Wellington, New Zealand, 24–28 March, www.maf.govt.nz/mafnet/unff-planted-forestry-meeting/index.htm

Vieira, D. and Scariot, A. (2006) 'Principles of natural regeneration of tropical dry forests for restoration', *Restoration Ecology*, vol 14, pp11–20

Wassie, A., Sterck, F. J., Teketay, D. and Bongers, F. (2009a) 'Tree regeneration in church forests of Ethiopia: Effects of microsites and management', *Biotropica*, vol 41, no 1, pp110–119

Wassie A., Sterck, F. J., Teketay, D. and Bongers, F. (2009b) 'Effect of livestock exclusion on tree regeneration in church forests of Ethiopia', *Forest Ecology and Management*, vol 257, no 3, pp765–772

Woldu, K. (1999) 'Natural regeneration of *Juniperus procera* (Mocht.) in *Eucalyptus globulus* (Labill) plantation at Entoto mountain, central Ethiopia', Ethiopian MSc thesis work, Report No. 1999–43, Skinnskatteberg, Sweden

Weins, J. A. (1992) 'Ecological flows across landscape boundaries: A conceptual overview', in A. J. Hansen and F. di Castri (eds) *Landscape Boundaries: Consequences for Biotic Diversity and Ecological Flows*, Springer-Verlag, Berlin, pp218–235

Whisenant, S. G. (1999) *Repairing Damaged Wildlands: A Process-Orientated, Landscape-Scale Approach*, Cambridge University Press, Cambridge, UK

Wunderle, J. M. (1997) 'The role of animal seed dispersal in accelerating native forest regeneration on degraded tropical lands', *Forest Ecology and Management*, vol 99, pp223–235

Yirdaw, E. (2001) 'Diversity of naturally regenerated native woody species in forest plantations in the Ethiopian highlands', *New Forests*, vol 22, pp159–177

Yirdaw, E. (2002) 'Restoration of the native woody-species diversity, using plantation species as foster trees, in the degraded highlands of Ethiopia', PhD dissertation,

Viikki Tropical Resources Institute, University of Helsinki, Helsinki

Yirdaw, E. and Luukkanen, O. (2003) 'Indigenous woody species diversity in Eucalyptus globulus Labill. plantations in the Ethiopian highlands', *Biodiversity and Conservation*, vol 12, pp567–582

Zamora, C. O. and Montagnini, F. (2007) 'Seed rain and seed dispersal agents in pure and mixed plantations of native trees and abandoned pastures at La Selva Biological Station, Costa Rica', *Restoration Ecology*, vol 15, pp453–461

Zanne, A. E., Keith, B., Chapman, C. A. and Chapman, L. J. (2001) 'Protecting terrestrial mammal communities: Potential role of pine plantations', *African Journal of Ecology*, vol 39, pp399–401

Zerfu, H. (2002) 'Ecological impact evaluation of eucalyptus plantations in comparison with agricultural and grazing land-use types in the highlands of Ethiopia', PhD dissertation, Institute of Forest Ecology, Vienna University of Agricultural Sciences, Vienna

Zobrist, K. (2005) 'Increasing biodiversity in intensively managed loblolly pine plantations: A literature review', RTI Fact Sheet No. 37, University of Washington, Seattle, WA

Zurita, G. A., Rey, N., Varela, D. M., Villagra, M. and Bellocq, M. I. (2006) 'Conversion of Atlantic Forest into native and exotic tree plantations: Effects on bird communities from the local and regional perspectives', *Forest Ecology and Management*, vol 235, pp164–173

# 10

# Future Options for *Maesopsis*: Agroforestry Asset or Conservation Catastrophe?

*John B. Hall*

## INTRODUCTION

*Maesopsis eminii* Engl. (Rhamnaceae) is a widespread, generally short-lived, pioneer species of Africa's equatorial rainforests, with a relatively wide and open crown, bearing simple alternate leaves on the young branches at the periphery. In the eastern part of the range, which includes much of Uganda, northwestern Tanzania and the most westerly part of Kenya, where heights of 25–35m are commonly attained, it is the main indigenous timber tree in volume production terms. The straight, unbuttressed, self-pruned bole 20m or more long and 70–80cm diameter at breast height (DBH), coupled with fast growth and suitability as a light hardwood for packaging and indoor use, account for extensive use in plantation forestry, both within and beyond the natural range.

In the broad context of agroforestry as understood today, the species has, for generations, been a key pole and timber element. Lately, this forestry importance has also led to formalized agroforestry tree roles, the longest-standing being in intercropping (*taungya*) and as a shade tree, with over 60 years of application. Overall, agroforestry interest in the species reflects potential to contribute services, often combined with supplies of products, or cash revenue at household or cottage enterprise level. Today, agroforestry use is a major aspect of *Maesopsis* management, and the species is widely acknowledged as an agroforestry tree, both in Africa and in other regions of the tropics, often in circumstances where it is an element of resource diversification and the rehabilitation of land that has been degraded.

This traditional forestry significance has resulted in several forestry treatments of the species, notably those of Normand (1935), Eggeling and Harris (1939), Fenton et al (1977) and Mugasha (1981), and its silviculture is well established, guiding its use in agroforestry for both service and product benefits. Currently overshadowing forestry interest in the species, however, are invasiveness concerns, expressed by a vigorous and effective environmental lobby. A debate centres on the two perspectives, with forestry seeking to promote use of *Maesopsis* as an agroforestry tree and the environmental lobby forcefully urging its elimination from the areas to which it has been introduced.

These contrasting views of *Maesopsis* are of considerable significance for the future management of the species, particularly within east Africa. Both views are examined here through a review and evaluation of past experience and assumptions.

## THE AGROFORESTRY EXPERIENCE

Service use is currently the major agroforestry role of *Maesopsis*, usually combined with generating harvestable products. In addition, the potential of combination with cereals and herbaceous legumes as an intercrop has been explored. Also of interest are situations where by-products are extracted or cultured using the species.

### Service roles

There are early reports of use as a shade tree in coffee plantations in Congo (Anon, 1947) and Uganda (Eggeling and Harris, 1939). After World War II, the practice was extended to the coffee-growing areas of southern India (Radcliffe, 1985; Sreenivasan and Dharmaraj, 1991) and, later, Indonesia (Silitonga et al, 1987) and was also adopted in cocoa plantations in Ghana (Benstead, 1954). In India, use as shade for coffee and cardamom is widespread. In Africa, use is reported as shade for bananas in Kenya (Oduol and Akunda, 1989) and as a retained, spontaneously regenerating, home-garden border shade tree, potentially yielding marketable timber, in Tanzania (Rugalema et al, 1994a and b).

The principal attraction of *Maesopsis* as a shade tree for perennial plantation crops is that, as a timber tree, it diversifies benefits from the land. Despite the necessity either to plant at wide intervals (12m × 12–15m – Sreenivasan and Dhamaraj, 1991) or to apply drastic thinning (Radcliffe, 1985), even without crown contact *Maesopsis* exhibits self-pruning behaviour, and forms a single straight trunk. The wide, open crown casts a modest and uniform shade. Being used predominantly in perennial plantation crop areas, where dry season conditions are not excessively severe, leaf fall is never complete (Radcliffe,

1985). Additional positive characteristics of the species as a shade tree are the small size of shed branches, the rapid decomposition of its leaf litter (Radcliffe, 1985; Macfadyen, 1989) and the continuing reputation for freedom from major pests.

Sreenivasan and Dhamaraj (1991) report mean annual height and diameter increments of over 2m and 5cm respectively for the initial 10 years in coffee plantations in Karnataka, India. Radcliffe (1985), however, for older trees (21 years), indicates considerably slower growth from the same part of India – mean annual height increments of 0.7m and diameter increments of 2.3cm. Combined with self-pruning, the better growth rates result in a bole free of branches for upwards of 10m in less than ten years. Impact on the shaded crop appears to vary. Thomas (1940) credits *Maesopsis* shade with a beneficial effect on coffee. Silitonga et al (1987), with reference to Bondowoso, Indonesia, present some evidence that when combined with a coffee crop the growth rate of *Maesopsis* is reduced, but under circumstances where a coffee understorey was established in a timber plantation. To improve light and humidity conditions for the coffee, after the pre-flowering coffee pruning, home-garden *Maesopsis* crowns may be subject to some lopping (Rugalema et al, 1994a).

The fast growth and good survival on what are often considered poor soils have aroused interest in a further service use of *Maesopsis* – as a potential tool for soil conservation and improvement in humid tropical lowlands (Mwihomeke, 1989; Schabel and Latiff, 1997). Initiatives involving such use of the species combined with useful product generation have been taken in Rwanda (Egger, 1986) and Indonesia (Widiarti and Arasjid, 1987). For reinforcing banks and in terrace creation and consolidation, Egger (1986) names *Maesopsis* and eight other tree and shrub species as effective in trials of mixed age and species planting at 200–400 individuals per hectare. Gintings and Yusuf (1983) in Java, Indonesia, concluded that forest cover incorporating *Maesopsis* administered better protection against erosion (negligible runoff and erosion) than areas under agricultural crops. At Kadipaten, also in Indonesia, Anwar et al (1989) found that a mixture of planted trees including *Maesopsis*, on slopes up to 30 per cent, restricted erosion losses over six months (236mm rainfall) to 0.03t/ha, and runoff to 0.7mm but was less effective than a shrub fallow. In a third Indonesian investigation, at Cijambu in West Java, Purwanto and Gintings (1989) found higher water infiltration rates (23cm/h) in a stand of *Maesopsis* 3.5 years old than where a crop of cabbage was grown (9cm/h). Comparing soil characteristics (top metre of soil) under a pure *Maesopsis* plantation with those beneath a plantation under which coffee had been planted, after an unspecified period, at Bondowoso, Indonesia, Silitonga et al (1987) noted a range of differences. The wide soil horizon reported, however, detracts from the informativeness, and the differences (more exchangeable Mg and a more acidic soil reaction with coffee present) that were significant may reflect site differences pre-dating the study.

# Products

## Leaf material

*Maesopsis* leaves have been investigated for mulch and fodder use. Drechsel and Reck (1998) give elemental concentrations in *Maesopsis* prunings, presumably mainly of leaf matter, gathered in southern Rwanda for composting in the dry season (dry weight basis) as: N, 3.6 per cent; P, 0.23 per cent; K, 0.6 per cent; Ca, 0.8 per cent; Mg, 0.38 per cent. The nitrogen level is high. Phosphorus, calcium and magnesium levels are not unusual but the potassium level is low – perhaps arising from translocation at the end of the growing season, possibly with the cumulative impact of leaching in throughfall contributing.

References to the fodder suitability of the species are mostly made in passing, without supporting or contextual detail (Mwihomeke, 1989; Schabel and Latiff, 1997), although Rugalema et al (1994b) refer to use specifically as goat fodder in the Bukoba region of Tanzania. *Maesopsis* leaves obtained by lopping are an established livestock forage in Java (Mahyuddin et al, 1988). In the only detailed fodder study, these workers reported in vitro evaluations of feed quality for the species. Analysis of the leaves (season and leaf age not specified) indicated unremarkable leaf dry matter proportions of carbohydrate (45 per cent), neutral detergent fibre (20 per cent) and fat (4 per cent), a low level of ash (5 per cent) and a high level of crude protein (26 per cent). Estimates were also made of lignin (5.4 per cent of dry matter), total phenols (2.4 per cent of dry matter), condensed tannin (vanillin-HCl method, 5.6 per cent) and total tannin by pepsin precipitation (0.9 per cent). Intraruminal (*Bos indica* L.) dry matter disappearance for fresh *Maesopsis* leaves was determined as 90 per cent, indicating a potentially excellent feed material.

## On-farm timber, poles and fuelwood

Where used as plantation shade (Radcliffe, 1985) or retained in home gardens as an indigenous, spontaneously regenerating tree (Rugalema et al, 1994a), *Maesopsis* is always appreciated as a timber source, growth to timber size in just 20–30 years accounting for this appeal. Consideration of timber yield is reflected in management to favour the early growth of shade trees over perennial plantation crops, with a grass-free area maintained around each tree in its first year (Sreenivasan and Dhamaraj, 1991). To minimize harvesting complications, trees are grown to timber size in areas easily accessed for harvesting and felled at 25–30 years of age, before the crown attains a spread making it vulnerable to wind throw. These measures enable closer control over exploitation and limit disturbance to the shaded crop. A *Maesopsis* timber harvest is intended as the main benefit from formal intercropping; this has prompted some research into appropriate tree-management measures (Okorio et al, 1994). Individual tree fertilization (20g calcium ammonium nitrate per tree, at planting) and replacement of early mortality loss are intended to ensure full stocking of vigorous young trees, while the lowest one-third of the tree crown is pruned to reduce shading effects on the arable crop when the trees are only a few metres tall.

Poles can be generated in agroforestry situations in 3–8 years according to Peden et al (1996) and Schabel and Latiff (1997) or 10–12 years (Radcliffe, 1985). For Uganda, Peden et al (1996) describe experimental agroforestry pole production plantings at 2m intervals in lines longitudinally bisecting plots 8m wide and 12m long. At Bushenyi (1600m; 1200mm mean annual rainfall), Uganda, after 41 months, 73 per cent (11 of 15) of the *Maesopsis* individuals produced poles at least 2m long, and with a distal diameter of 20mm, in three years. At a second, higher and drier, location (Kachwekano – 2000m; 1000mm mean annual rainfall) only 20 per cent (3 of 15) produced such poles in three years. Pole quality, as well as quantity, contrasted sharply between sites. Mean pole length and proximal diameter at Bushenyi were 4.0m and 126mm, respectively, while mean taper was a fall in diameter of 26mm/m (proximal to distal). Corresponding values at Kachwekano were 2.2m and 78mm, with similar taper. The best poles produced at Bushenyi were of *Eucalyptus grandis* Maiden (mean length 12.8m; mean taper 15mm/m) and *Casuarina equisetifolia* L. (5.9m; 15mm/m), while the worst were from *Maesopsis* and *Alnus acuminata* Kunth (2.8m; 25mm/m). The best poles produced at Kachwekano were of *Eucalyptus grandis* (10.9m; 14mm/m) and *Casuarina cunninghamiana* Miq. (4.9m; 13mm/m), while the worst were from *Maesopsis* and *Cedrela odorata* L. (2.1m; 25mm/m).

Along roadsides and field boundaries in Java, Indonesia, *Maesopsis* is commonly planted as a source of fuelwood (Smiet, 1990), and in Tanzania fuelwood is obtained from the species, where it is retained in home gardens (Schabel and Latiff, 1997). In experimental trial plantings for fuelwood at a standard spacing (2m × 2m) in Java, Indonesia (Widiarti and Alrasjid, 1987), over the initial eight months the *Maesopsis* plants which survived (64 per cent) reached mean heights of about 0.7m.

## *Maesopsis* and arable crop combinations

When the two are combined, compatibility depends on the extent of positive or negative influence of *Maesopsis* on arable crop production. Initiatives at Butare, Rwanda (Egli and Pietrowicz, 1990), and at Kabanyolo, Uganda (Okorio et al, 1994), provide information on *Maesopsis* in combination with several arable crops. In Rwanda, the arable crops (maize, soya and haricot beans, and sweet potatoes) were underplanted in a *Maesopsis* stand 34 years old (dominant individuals 23.5m tall and 34cm diameter at ground height, DGH), at a spacing equivalent to 130 per hectare, and in stands of nine other tree species. In Uganda, arable crops (maize and local beans) were grown during a four-year experimental period starting when rows of *Maesopsis* (of two provenances) and six other tree species were established in the field. In this experiment, where tree spacing was equivalent to 830 per hectare, the trees eventually reached a mean height of around 7.5m and a mean root collar diameter of around 15cm.

**Table 10.1** *Arable crop yields relative to control in association with* Maesopsis eminii* *at Butare, Rwanda*

| Cropping season and crop | Control yield (kg/ha) | Relative yield in light shade (light intensity 82% of control value) | Relative yield in heavy shade (light intensity 45% of control value) |
|---|---|---|---|
| Maize (end of year) | 970–2290 | 57–67% | 36–52% |
| Maize (mid-year) | 237–900 | 27–65% | 21–26% |
| Haricot beans (end of year) | 860–1080 | 48–55% | 38–59% |
| Soya beans (mid-year) | 260–540 | 54–87% | 55–62% |
| Sweet potatoes (mid-year) | 580–2700 | 35–73% | 25–28% |

*Note:* * 34 years old; mean dominant height 23.5m; mean dominant DBH 34cm.
*Source:* Egli and Pietrowicz (1990)

Over a series of cropping seasons, control plot (no trees) crop yields at Butare were 240–2200kg/ha, 860–1080kg/ha and 260–542kg/ha seed dry weights for maize, haricot beans and soya beans respectively. For sweet potatoes, control plot tuber harvests (fresh weight) were 578–2703kg/ha. Under *Maesopsis* (Table 10.1) and *Grevillea robusta* A. Cunn., yields were reduced markedly, and generally to a greater extent than under *Acacia elata* Benth., *Acrocarpus fraxinifolius* Arn., *Casuarina littoralis* Salisb., *Toona sinensis* (A.Juss.) M.Roem., *Croton megalocarpus* Hutch., *Entandrophragma excelsum* (Dawe and Sprague) Sprague, *Podocarpus falcatus* (Thunb.) Mirb. and *Polyscias fulva* (Hiern) Harms. Possibly, low availability of phosphorus found in the soil beneath *Toona sinensis, Grevillea robusta* and *Maesopsis* (2–16ppm for the 0–20cm horizon; 1–17ppm for the 20–40cm horizon) contributed to low crop yields. It was not established whether the presence of the trees (for 33–40 years) had reduced phosphorus levels or if they reflected environmental variation present before the trees were planted. Surprisingly, the *Maesopsis* stocking was the lowest of any of the species compared, the basal area (12m²/ha) was among the lowest and most of the other trees cast heavier shade. The most shade-tolerant arable crops grown at Butare were the two types of bean. With all the arable crops there was generally a more severe yield reduction under heavy shade.

Over a series of cropping seasons, control plot (no trees present) crop yields at Kabanyolo were 6–1258kg/ha seed dry weights for beans and 2399–3572kg/ha for maize grains. Arable crop yields from harvests during the first year were not affected by the *Maesopsis* trees, but from an age of 18 months, by which stage the *Maesopsis* trees were ca. 3.2m tall, yields of associated beans and maize were depressed. At all assessments from the age of 24 months, both maize and beans yields under *Maesopsis* were lower than under any of the other tree species (*Alnus acuminata*, *Casuarina equisetifolia*, *Cordia africana* Lam., *Cupressus lusitanica* Mill., *Markhamia lutea* (Benth.) K.Schum. and *Melia azedarach* L.) included in the study. The final crop of maize, with *Maesopsis* trees 42 months old and ca. 7.0m tall, yielded only 41 per cent (Wundaji provenance) and 52 per cent (Kakamega provenance) of control plot level (3346kg/ha). The final bean crop, with *Maesopsis* trees 48 months old and ca. 7.5m tall, yielded only 22 per cent (Wundaji provenance) and 28 per cent (Kakamega provenance) of control plot level (301kg/ha).

In the Kabanyolo experiment, the possibilities of tree impact on associated crops changing with separation distance, and whether installation of a barrier to tree root spread modified any pattern, were also considered. Separation distances were examined on a crop row by crop row basis outwards to 300cm from lines of *Maesopsis* trees, intervals being 75cm for maize and 60cm for beans. Maize yields progressively increased from 20 per cent of control plot value (3010–3190kg/ha) at 75cm to 92 per cent at 300cm, rising above 50 per cent from 225cm. Beans yields progressively increased from 25 per cent of control plot value (420–450kg/ha) at 60cm to 70 per cent at 300cm. For both crops, consistently lower yields were recorded with the Wundaji provenance; with this provenance bean yield was below 50 per cent of control plot value even at 240cm. Over the timescale of the study, at every distance bean yields under *Maesopsis* were lower than under the other tree species, where yields had mostly risen to 75 per cent of control plot values at 240cm distance. With maize, reductions were more pronounced under *Maesopsis* than under other species to a distance of 150cm; beyond this distance, yields were similar under *Cordia africana*, *Maesopsis* and *Melia azedarach* and were in excess of 50 per cent of control plot values for all tree species included.

The influence of the root barrier was investigated with maize when the *Maesopsis* trees were 42 months old and 7m tall, and with beans when the trees were 48 months old and 7.5m tall. Confining tree root spread, thereby neutralizing below-ground tree–arable crop competition, resulted in enhanced arable crop yields at all distances studied. It was concluded that only close to the trees (75cm distance) did *Maesopsis* shade markedly reduce maize yields relative to control plot values (3150–3390kg/ha), and that root competition accounted for the low yields (always <50 per cent of control plot values) from 150–300cm distance. The root barrier had a more marked affect on the maize yield under *Maesopsis*, *Markhamia lutea* and *Melia azedarach* than under the potentially nitrogen-fixing *Casuarina equisetifolia*. Shade effects on beans emerged as extending further outwards and, while progressively declining, persisted as an important impact even at 240cm distance. With shading and root competition effects combined, bean yields barely reached 50 per cent of control plot yields (260–340kg/ha), even at 300cm distance. The presence of the root barrier affected beans yields similarly under both *Maesopsis* and *Markhamia lutea* and had a less pronounced impact on yields under *Casuarina equisetifolia* and *Melia azedarach*.

# By-products from agroforestry with *Maesopsis*

Spontaneously regenerating or intentionally planted *Maesopsis* can create opportunities to harvest or produce dependent or specialized products in the form of mushrooms, non-wood phytochemicals and medicinal materials, and oil and gum derived from the fruit.

**Mushroom culture.** Suprapti (1987) reports cultivation of oyster mushrooms (*Pleurotus ostreatus* (Fr.) Kummer) in Indonesia on sawdust of *Maesopsis* and 10 other species. Mushrooms were harvested daily after fruit body growth and the average yield was 0.38kg/kg dry substrate. The highest yield (0.66kg/kg dry substrate) was recorded on rubber sawdust. In another study (Suprapti, 1989), also in Indonesia (Bogor), *Pleurotus flabellatus* (Berk. and Br.) Sacc. mushrooms were cultivated on logs inclined at 60° using bamboo supports under *Altingia excelsa* Noroña shade. Over a four-month period, a yield of 103kg/m$^3$ was harvested from logs of *Maesopsis*. There were much lower yields from four other species: *Piper aduncum* L. (14kg/m$^3$), *Schima wallichii* (DC.) Korth. (5kg/m$^3$), *Paraserianthes falcataria* (L.) Nielsen (4kg/m$^3$) and *Castanopsis argentea* (Blume) A.DC. (2kg/m$^3$) and nothing from *Lithocarpus sundaicus* (Blume) Rehder and *Altingia excelsa*.

**Non-wood phytochemicals and medicinal materials.** Throughout the natural range medicinal values, most often based on use of the bark, are attributed to *Maesopsis*. Little has been formally reported on specific chemicals that might be active in the effectiveness of the treatments, but the Rhamnaceae as a family is noteworthy for the presence of anthraquinone glycosides. In *Maesopsis* saponins and alkaloids (0.3 per cent) were found in bark and roots from Congo (Bouquet, 1969), and three anthraquinones (cynodontin, chrysophanol and one unidentified) were detected in root bark from Nigeria (Ekpa et al, 1985). Neither alkaloids nor saponin were found in leaf material from Congo, although the presence of tannin was confirmed. Both saponins and tannin were detected in leaves from Côte d'Ivoire (Bouquet and Debray, 1974).

The secondary compounds recorded from *Maesopsis* offer phytochemical explanations for findings in two studies involving interactions with other species. Molluscidal effects (against *Bulinus globosus* Morelet) of preparations from *Maesopsis* were demonstrated experimentally in Nigeria (Adewunmi and Sofowora, 1980). Mortality was achieved with both root-based (10 of 10 snails exposed over 24 hours to a 100ppm concentration of root extract in water) and stem-based (2 of 10 snails exposed over 24 hours to a 100 ppm concentration of stem extract in water) preparations, but one using a leaf extract was ineffective. The effects of the root and stem preparations are likely to reflect the presence of saponins. Laine et al (1985) suggested saponins or tannins explained antifungal activity of the bark in Congo. Further support for the view that secondary compounds in *Maesopsis* give the species medicinal value emerges in the frequency with which it is used as a purgative (and the inclusion in the treatment of measures to limit the reportedly violent purgative action – Burkill, 1997). In fact, most uses reported in the literature concern conditions affecting the alimentary tract, including intestinal worms. The active elements are generally administered as bark-based medicine: in soaked, macerated, infusion or decoction form, often variously combined with such materials as sugar-cane juice or palm wine. Less often leaves are the basis of a decoction. Use as an abortifacient (Cameroon – Noumi and Tchakonang, 2001) and to accelerate labour (Congo – Bouquet, 1969) also suggests active compounds stimulating strong reactions from the treated party. Urino-genital complaints are also treated, with both leaf decoctions and root/stem bark preparations.

**Oil and gum.** In one of the oldest published references to *Maesopsis*, Pierre (1897) drew attention to the oily, edible cotyledons within the seed. Almost a century passed before the seed oil (of trees grown in Karnataka, India) was characterized qualitatively and estimates were made of the quantity present (Prabhu and Theagarajan, 1986). Extracted from the woody endocarp, the seed contains 40–45 per cent, by weight, of a light brown oil. The oil is rich in oleic acid, which contributes nearly half of the fatty acid total. Defatted seed is protein-rich (48–51 per cent), suggesting animal feed potential.

Barminas and Eremosele (2002), using kernels marketed at Yola, Nigeria, have recently undertaken a preliminary examination of the suitability of gum from *Maesopsis* as a hydrocolloid for use in the food and pharmaceutical sectors. Samples of gum (comminuted kernels, with seeds) were found to include water-soluble (polymer) and swellable insoluble fractions. In water, the gum (at a concentration of 0.5–2 per cent weight/volume) displayed the pseudoplastic behaviour of a soft gel, with high viscosity at low shear rates and vice versa, indicating possible usage as a viscosity builder and stabilizer. At 25°C, with a 2 per cent (weight/volume) gum concentration, there was no evidence that dilute solutions (<1.5mmol) of sodium chloride affected viscosity, suggesting potential as a pharmaceutical suspending and thickening agent, but viscosity was progressively reduced as temperature was raised to 50°C.

# THE CONSERVATION PERSPECTIVE: *MAESOPSIS* AS INVASIVE AND ALIEN

Easy establishment, good survival, fast growth and relative freedom from diseases and pests contribute to favourable opinions of *Maesopsis* as an agroforestry tree. However, these attributes also underlie an aggressive invasive tendency under the conditions prevailing in the East Usambara Mountains, Tanzania. The ability to regenerate spontaneously and spread beyond points of introduction was first recognized in Tanzania's East Usambara Mountains over 75 years ago – in the notes Peter Greenway provided with a specimen (Greenway 922, K) he collected at Amani in 1928. Published comments about this spread appeared a few years later (Moreau, 1935), while a picture of subsequent actions and events involving *Maesopsis* in the East Usambaras, and their possible implications, emerges from forestry and conservation documents.

## Concerns in the East Usambara Mountains

The *Maesopsis* invasion of the East Usambara rainforest in Tanzania has been considered one of the ecological disasters of the 1980s and a consequence has been a widely publicized assertion (see, for example, Cronk and Fuller, 1995) that *Maesopsis* illustrates that some alien tree species can aggressively invade closed tropical forest subject only to natural disturbance. In East African terms, it was feared that the ecological integrity of the East Usambara forest ecosystem, one of the most valued natural forest ecosystems in the region, would be irreversibly damaged, negating its significance as a conservation area. The explosive spread of *Maesopsis* in the East Usambaras from the 1960s is not in dispute. As published reports and explanations of this process have paid little attention to the autecology of the species, or whether situations encountered suggest any exceptional or unique qualities, emphasis in this evaluation is on a broad context.

## *The* Maesopsis *timeline*

A timeline for *Maesopsis* in the East Usambara Mountains indicates phases of initial introduction, spontaneous spread from this and deliberate planting in additional areas (Table 10.2). The introduction of *Maesopsis* to the Usambara Mountains took place within 30 years of the first botanical collection being made, by Franz Stuhlmann in 1891 (Stuhlmann 971, M). It is probable that this was a direct consequence of the potential highlighted for the species by Engler (1906). There appears to be no formal record of introduction to the East Usambaras but initiatives at Amani presumably paralleled efforts to grow the tree in the West Usambaras (Siebenlist, 1914), and a date of 1913 is widely accepted as the date of introduction at Amani (Wood, 1966).

Post-World War II planting began on a small scale, to establish the species in felling gaps, in 1957, nursery activity beginning in the previous year. From 1963, however, the scale of planting was greatly expanded, with increasing tendency to clear residual forest after timber was extracted and replace it with plantations. From 1962 to 1968 pure stands of *Maesopsis* were established. From 1968 to 1977, mixtures combining *Maesopsis* as a fast-growing nurse species with the slower-growing, but more valuable, *Cephalosphaera usambarensis* (Warb.) Warb. were used instead. Planting *Maesopsis* in quantity ceased in 1977 as a result of financial constraints. However, by this time the trees planted from 1957 to around 1970 were already fruiting and attracting the attention of dispersal agents. As logging was continuing, and creating large gaps, more stands of *Maesopsis* arose spontaneously in newly logged areas.

Realization that the spontaneous spread that had been attracting comment over the preceding 60 years or so might become an invasion problem was first publicized in 1982 (Rodgers and Homewood, 1982). At the time concern was limited and Rodgers and Homewood also suggested *Maesopsis* should continue to be used for plantations in the area. As late as July 1985 (Buck et al, 1985), the possibility that *Maesopsis* might significantly alter the ecological balance of the forests was not recognized, although logging was becoming increasingly intensive and destructive. In October of the same year, however, *Maesopsis* spread was underlined as a significant factor in changing forest composition (Mustonen and Räsänen, 1985). This prompted attention to the general *Maesopsis* situation in the East Usambaras, a concern reflected in contributions

**Table 10.2** *A* Maesopsis eminii *timeline, with particular reference to the presence of the species in the East Usambara Mountains, Tanzania*

| Year | Maesopsis *event* | *Reference* |
|---|---|---|
| 1891 | First botanical collection (Bukoba area, Tanzania) | |
| 1906 | Forestry potential highlighted | Engler (1906) |
| ca. 1913 | Introduction to East Usambaras (Amani Biological Agricultural Institute) | Wood (1966) |
| 1928 | First report of spontaneous regeneration – attributed to hornbills | notes with Greenway 922 (K) |
| 1930 | Seed produced from trees at Amani available for distribution | Anon (1930) |
| 1935 | Indication ('several miles') of extent of spread by dispersers | Moreau (1935) |
| 1956 | Plantings in East Usambara lowlands (Longuza) | Sangster (1957) |
| 1963–1968 | Pure stands planted on a large scale | Geddes (1998) |
| 1968–1978 | Extensive use as a nurse crop for plantings of *Cephalosphaera usambarensis* | Geddes (1998) |
| 1981 | Forestry planting in East Usambaras terminated | Geddes (1998) |
| 1982 | Initial speculation that species had become invasive | Rodgers and Homewood (1982) |
| 1985–1986 | Serious concerns expressed about invasiveness | |
| 1987 | IUCN review of situation | Hamilton and Bensted-Smith (1989) |
| 1994 (April–June) | Formal regeneration and *Maesopsis* survey in forest, with sampling at 112 points and a total of 14,000 observations (all species) | Geddes (1998) |
| 1994 (December) | Reconnaissance reassessment of situation eight years after cessation of commercial logging | Hall (1995) |
| 1997 (May) | Amani Nature Reserve gazetted | |
| 1998 | Formal survey of floristic composition of *Maesopsis* plantation understorey, with sampling of sapling and lower canopy individuals at 152 sampling points and a total of ca. 4900 observations (all species) | Viisteensaari et al (2000) |

to Hamilton and Bensted-Smith's (1989) book, particularly in Binggeli's (1989) contribution.

## Evaluation of the situation in the 1980s

Attention to possible site impact from *Maesopsis* developed from impressions gained as the IUCN programme started in the East Usambaras in 1986. A series of exploratory studies resulted, primarily evaluating conditions associated with stands of *Maesopsis*. Central to the perception that *Maesopsis* invasion had brought persisting and intensifying damage to the East Usambara forests was the presence of site conditions contrasting with those in uninvaded natural forest. Contrasts were attributed to *Maesopsis* (Binggeli and Hamilton, 1993). Additional field studies subsequently extended insight into the ecosystem impact ten years after the IUCN programme activities (Hall, 1995; Geddes, 1998; Viisteensaari et al, 2000).

231

Wherever *Maesopsis* is present in quantity, there has been major impact on site conditions, through land clearance/preparation or tree fall events, which has led to mineral soil exposure and localized loss of litter cover. Typically, the consequence of such events (Lundgren, 1980) is a phase of rapid change, as a large mass of litter is deposited but swiftly decomposed, as microbial activity responds to the higher temperature of exposure and abundant substrate material. In conditions like those in the East Usambara forests, decomposition releases a pulse of nutrients within an ecological system with weak nutrient conservation capacity. If clearance has involved burning, the nutrient pulse is greater, as woody material is reduced to ash, and is evident earlier. Under poor management practices, the ultimate result is loss of both organic matter and nutrient reserves from the system. In the context of the East Usambaras, interest centres on three issues. One is the extent to which observations that were made reflect pre-disturbance site characteristics. The second is whether a *Maesopsis* cover is in any way unusual compared with other covers. And the third issue is whether restoration of the original soil–litter–vegetation conditions can be expected after the major disturbance of heavy exploitation of the forest.

## Maesopsis *and site changes*

Binggeli and Hamilton (1993) concluded that a series of 'major changes' had affected site conditions:

* loss of the mat of superficial tree roots;
* reduction of litter quantity and higher proportion of its woody component;
* reduction of topsoil organic matter content;
* reduction of soil fauna diversity;
* loss of distinctive mycorrhizal rootlets;
* reduction of topsoil acidity; and
* increase in soil surface exposure, runoff and erosion.

**Tree root mat.** Taylor (1989) noted a contrast between the Kwamkoro natural forest and the stands of *Maesopsis* at Amani in the restriction of a distinct superficial root mat to the former. He suggested that a superficial root mat previously present at Amani had been lost as a *Maesopsis* cover developed. However, deterioration of soil conditions has been widely noted in vigorously growing 5–10-year-old plantations of fast-growing tree species in the tropics (Lundgren, 1980) and it is questionable whether it is particularly attributable to *Maesopsis*. In tropical plantation forestry rapid litter decomposition (Zobel et al, 1987) and paucity of roots close to the surface (Bowyer, 2001), as noted by Taylor for *Maesopsis*, have often been reported. Further, in the absence of repeated fires or litter gathering for fuel or mulching material, changes in soil conditions and soil surface cover due to plantation establishment are reversed as time elapses (Evans, 1999).

**Litter quantity and composition.** Litter under *Maesopsis* is rich in shed branchlets, as the species is self-pruning. In stands of *Maesopsis* most fallen leaves will be of this species. Binggeli and Hamilton (1993) suggested rapid litter decomposition of *Maesopsis* using litter bags. However, the study did not cover a full annual cycle, no litter turnover coefficients were determined, and no estimates were presented of the standing crop of foliage and how this changes through an annual cycle or whether it is similar to that of the natural forest. Experience elsewhere indicates that seasonal variation in both litter fall and litter standing crop may be considerable, particularly when attention is restricted to particular species. Even from species-rich forest, variation within the year may be as great as from 100 to 800g/m$^2$ in dry weight terms (Swift and Anderson, 1989). Species with leaves of a quality less rapidly broken down, and varying leaf-fall cycles, can be expected to contribute to the litter standing crop under forest cover, damping temporal variations. From September to November the monthly rainfall of the East Usambaras tends to be high, favouring detritivore activity, and the standing crop of litter predominantly of a readily decomposed species would be low. The possibility that *Maesopsis* illustrates this effect was examined by estimating litter cover in 112 plots (in 59 of which *Maesopsis* contributed to the canopy) in the Kwamkoro Forest Reserve (Geddes, 1998). Where *Maesopsis* was present as a canopy tree (individuals ⩾20cm DBH), there was a significantly (P<0.01) higher proportion (28 per cent vs. 18 per cent) of exposed mineral soil. However, there was no correlation between the proportion of exposed mineral soil and the contribution of *Maesopsis* to the basal area present. Mineral soil exposure may have resulted from logging operations before *Maesopsis* invasion.

**Topsoil organic matter.** Hamilton (1989b) recorded organic carbon values in the uppermost 2cm of topsoil for twelve sampling points across the interface of the oldest (ca. 1913) *Maesopsis* plantation with adjacent forest at Amani. Organic carbon percentages under *Maesopsis* plantation were as high as 10 per cent, and mostly ⩾5 per cent. Three of the five values for samples from the adjacent forest were also in the 5–10 per cent range, one was 20 per cent and the last was approximately 12 per cent but regarded as atypical on the basis of other attributes. Macfadyan (1989) estimated organic horizon thicknesses (F and H layers) at intervals between individuals of *Allanblackia stuhlmannii* (Engl.) Engl. and invading *Maesopsis* trees at Amani, and in tree fall gaps at Kwamkoro. At Amani, organic horizons, particularly H layers, were thinner at transect positions close to *Maesopsis* trees, and at Kwamkoro portions of gap areas had organic horizons <5cm thick.

   Hamilton's organic carbon percentages for the uppermost 2cm cannot be compared directly with other topsoil values from the East Usambaras, but the values of 5–10 per cent are not particularly low: Hamilton (1989a) found values of 2.8 per cent and 6 per cent for the top 5cm under forest. It is not possible to rule out persisting effects from past destruction of the H layer and, therefore, loss of pools of organic matter having a residence time of decades. Mineral soil exposure may have resulted from operations preparatory to planting or from clearance predisposing sites to *Maesopsis* invasion where Macfadyan's observations were made, within the Amani Institute residential area. The original disturbance is also considered the most likely explanation for the poor expression of F and H layers in the tree fall gaps at Kwamkoro (Macfadyan, 1989).

The rapid decomposition of *Maesopsis* leaf litter may influence how rapidly topsoil organic matter pools are recreated. The suspicion (Macfadyan, 1989; Binggeli and Hamilton, 1993) that *Maesopsis* leaves decompose rapidly is consistent with field observations made during the IUCN studies. Rates of leaf litter decomposition vary widely among species because the material varies in quality. High-quality leaf litter decomposes rapidly and *Maesopsis* seems to produce such litter, as do many other species, including several important tropical plantation trees. Macfadyan noted similarities in decomposition rate between *Maesopsis* and *Chrysophyllum gorungosanum* Engl., an associate indigenous to the East Usambaras. Leaf litter breakdown in *Gmelina arborea* L. occurs readily (Evans, 1984), and in *Tectona grandis* L.f. takes less than six months (Egunjobi, 1974). Lugo (1992) recorded, over a year, more leaf litter disappearing than accumulating in plantations of *Swietenia macrophylla* King, and more litter disappearance than accumulation in several samples of secondary forest, the source habitat for most fast-growing tropical plantation species.

**Soil and litter fauna diversity.** Evaluation of soil fauna data from the East Usambaras is constrained by confounded site variables and restriction to single sampling exercises. Macfadyan (1989) recorded observations of soil invertebrates while comparing soils under forest and under *Maesopsis* stands. The results were inconclusive and Macfadyan attributes this to an inappropriate sampling time – the end of the dry season.

Taylor (1989) found that earthworms were more numerous in *Maesopsis* stands (Amani) than in submontane forest (Kwamkoro). Of the two types of worms found in the study, only the smaller one was present under *Maesopsis*. Under forest, this species was found in smaller numbers and accompanied by a few individuals of the larger type. A direct *Maesopsis* leaf litter effect on the earthworm population is questionable: tropical forest earthworm populations do not appear to share the litter-moving habits of their temperate equivalents (Swift and Anderson, 1989). With the information reported, any *Maesopsis* effect is inseparable from a soil reaction effect, since pH was higher under *Maesopsis* stands than under the forest.

Mahunka (1989) collected arthropods from forest at Amani and Kwamsambia and from *Maesopsis* stands at Kwamkoro and Amani to explore the general character of the arthropod fauna. The soil arthropod fauna under *Maesopsis* stands was more uniform than under forest but the total number of individuals was not lower. Comparative studies of soil arthropod populations in tropical forests and plantations replacing them have been too few to gauge the implications of the greater uniformity, although with uniform litter composition in plantations uniformity can be expected. However, changes in decomposer communities apply generally for fast-growing plantation species. Lundgren (1980) reports that under many commonly used plantation trees there is a low activity of decomposers, as maximum above-ground biomass production is reached in the first rotation. Mahunka's *Maesopsis* sites represent this stage.

Tattersfield (1996; Tattersfield et al, 2001) found that the *Maesopsis* plantations in Kakamega forest, Kenya, had a somewhat impoverished mollusc fauna compared to the remnant rainforest. It should be noted, however, that *Maesopsis* is naturally indigenous at this locality, and local germplasm appears to have been among that used for the plantations. Nevertheless, more of the local rainforest mollusc fauna was present than under plantations of *Bischofia javanica* Blume and *Pinus* sp. and represented a substantial proportion of that recorded for the rainforest.

**Loss of distinctive mycorrhizal rootlets.** Hamilton (1989b) draws attention to the occurrence of mycorrhizal rootlets under natural forest cover and their absence from the surface soil under *Maesopsis*. The rootlets, described as 'chubby', are presumably ectomycorrhizal. Ubiquitous presence of these roots in the natural forest has not yet been demonstrated, but would be at variance with a claim that ectomycorrhiza tend to be abundant only locally (Bagyaraj, 1989). Absence from areas supporting a *Maesopsis* cover thus may not be evidence of suppression of ectomycorrhizal fungi by *Maesopsis*.

**Soil acidity.** Tropical moist forest ecosystems display systematic soil variation along catenas in addition to considerable non-directional heterogeneity over distances as short as a few centimetres. These variations probably explain pH differences (Hamilton, 1989a and b; Taylor, 1989) between soils under *Maesopsis* and under forest. It has been established that soil reaction is less acidic in low sites, where bases accumulate from laterally percolating drainage water, than in higher positions (Nye, 1955). Within the *Maesopsis* plantation, Taylor recognizes lower, middle and upper sampling zones but the whole plantation is contained in the lower part of a valley. In Taylor's forest site, several kilometres away, in less dissected terrain with more gentle slopes suggesting generally higher catenary positions, pH values were lower – as would be expected even if both sites were under natural forest cover. The second study comparing soil reaction under forest and under *Maesopsis* (Hamilton, 1989b) illustrates small-scale pH variation in a constant catenary position and uniform gradient. Hamilton infers that higher pH is linked to litter cover richer in *Maesopsis* foliage, but the trend is weak and was not demonstrated statistically. There is evidence from the moist tropics of topsoil pH rises for a few years where monocultures of other plantation tree species were established (Chijioke, 1980), but the general picture (Evans and Turnbull, 2004) is that tropical tree plantations have no consistent effect on soil reaction. The claim that *Maesopsis* changes topsoil pH in the East Usambaras awaits confirmation.

**Increased soil surface exposure, runoff and erosion.** Taylor (1989) draws attention to a more open *Maesopsis* canopy (compared with the natural forest), a large part of the rain tending to fall uninterrupted to the ground, over much of which there is no litter cover. Binggeli and Hamilton (1993) report that soil erosion is evident under *Maesopsis* and that exposed and eroding red upper soil may be readily observed. It is implied that *Maesopsis* exacerbates these effects by forming a relatively open canopy and by not generating a continuous layer of leaf litter able to absorb throughfall and reduce runoff. Plantation eucalypts are noteworthy for narrow crowns allowing considerable direct throughfall in stands, especially when the trees are young (Evans and Turnbull, 2004), but plantation species in general achieve early canopy closure at normal spacing. With *Maesopsis*, there is no evidence that narrow crowns leave an unusually high proportion of the ground surface unprotected. In typical planted stands, or invasive concentrations, *Maesopsis* rapidly develops a wide crown and is not associated with unusually high throughfall. Where erosion and runoff are concerned, there is wide acceptance that choice of species has a negligible impact on the levels recorded in comparison with the impact of management actions. Undesirably high runoff quantity and rates of erosion (as suggested by Binggeli and Hamilton for areas of the East Usambaras supporting *Maesopsis*) are universally acknowledged to result from harvesting damage and poor plantation site preparation (denuding the ground surface of protective litter and herbaceous growth) or management shortcomings (unregulated livestock access, litter gathering and uncontrolled surface fires). In the East Usambaras, logging and other forest clearance activity can adequately explain soil exposure and subsequent runoff and erosion.

## Maesopsis *and ecosystem change*

In addition to site modifications, Hamilton and Bensted-Smith (1989) presume two principal ecosystem changes accompanying an increasing prominence of *Maesopsis*:

1    floristic distinctness of *Maesopsis*-dominated forest; and
2    a deflected successional path.

These inferences were based on extensive opportunistic observations made in the East Usambaras in 1986–1987 and limited systematic observations in *Maesopsis* stands reported by Binggeli (1989). Subsequent woody species data sets from Kwamkoro, based on 112 samples (Geddes, 1998) and 152 three-plot clusters in *Maesopsis* plantations 19–35 years old (Viisteensaari et al, 2000), allow further consideration of ecosystem change associated with *Maesopsis*.

**Floristic distinctness of *Maesopsis*-dominated forest.** Vegetation succession after major forest disturbance typically involves phases during which a fast-growing, light-demanding species dominates the canopy and is associated with other light-demanding species. For the East Usambaras, Pocs (1989) links strong representation of *Maesopsis* with increased biomass and diversity of light-demanding coarse herbs and climbers. In addition, Binggeli and Hamilton

(1993) report pure patches of *Maesopsis* canopy where understorey tends to be of light-demanding secondary forest tree species, but they also express unease about two other floristic contrasts between the natural forest and areas with abundant *Maesopsis* which could have longer-term impact.

The first of these contrasts is the development of strong representation of other exotic species in areas with *Maesopsis*. Binggeli and Hamilton (1993) list the common occurrence of the exotic species *Clidemia hirta* (L.) D.Don, *Lantana camara* L. and *Rubus rosifolius* Sm. in the shrub layer of forest where *Maesopsis* has become established as a major ecosystem change. Geddes (1989) found no relationship between *Maesopsis* and *Lantana camara*, while *Clidemia hirta* was more prominent in forest with *Maesopsis* as part of the canopy. As Geddes found no positive relation between abundance of *Clidemia hirta* and *Maesopsis*, however, he concluded that both species were independently responding to creation of conditions allowing establishment through forest disturbance.

The second floristic contrast of concern is the reduced representation in *Maesopsis*-rich forest of the many endemic/near-endemic tree taxa in the East Usambara forests (Iversen, 1991). Most of these are mature forest elements; a few are pioneers. Binggeli (1989) suggests *Maesopsis* displaces *Allanblackia stuhlmannii* (Engl.) Engl., *Alsodeiopsis schumannii* (Engl.) Engl., *Anisophyllea obtusifolia* Engl. and Brehmer, *Englerodendron usambarense* Harms, *Greenwayodendron suaveolens* (Engl. and Diels) Verdc. subsp. *usambaricum* Verdc., *Isoberlinia scheffleri* (Harms) Greenway, *Morinda asteroscepa* K.Schum. and *Uvariodendron* spp. Geddes (1998) quantitatively confirmed that, regardless of species, there were fewer regenerating individuals in total under a *Maesopsis* canopy. More of these individuals were of species other than *Maesopsis*, and they were mainly young plants of large forest species (particularly of *Cephalosphaera usambarensis* (Warb.) Warb.). Geddes also found that regenerates of the near-endemic species *Allanblackia stuhlmannii* declined while those of *Cephalosphaera usambarensis* increased with increasing numbers of *Maesopsis* in the canopy.

Viisteensaari et al (2000) also show that endemic/near-endemic tree taxa regenerate successfully under the canopy of *Maesopsis* plantations in the East Usambaras. Sapling numbers (individuals ⩾1m tall and <5cm DBH) of *Allanblackia stuhlmannii*, *Beilschmiedia kweo* (Mildbr.) Robyns and Wilczek and *Greenwayodendron suaveolens* subsp. *usambaricum* increased with increasing *Maesopsis* plantation age, and so did numbers of lower canopy individuals (5<20cm DBH) of *Greenwayodendron suaveolens* subsp. *usambaricum*.

**Deflected successional path.** Binggeli (1989) inferred a 'worst case' future scenario from the dominance of *Maesopsis* in invaded areas of forest. As a consequence of explosive spread into tree fall gaps exceeding ca. 300m² in extent, and the sparse representation of primary tree species beneath the *Maesopsis* canopy, little prospect of natural forest reoccupying the sites over a long period – perhaps as long as 200 years – was predicted. Within areas of *Maesopsis* plantation there was also a lack of regenerating primary tree species but a sub-canopy composed mainly of light-demanding pioneers. At rotation-end harvesting of the *Maesopsis*, Binggeli (1989) foresaw development of secondary forest largely devoid of primary species. These predictions are not supported by later studies, however. Although determining the relationship of *Maesopsis* with the regeneration of primary forest trees is rendered difficult by the low densities of individuals, overall and at species level, Geddes (1998) and Viisteensaari et al (2000) have quantitatively evaluated tree regeneration under *Maesopsis* canopies in invasive and plantation conditions respectively. In invaded forest at Kwamkoro, Geddes revealed that the regeneration of several large or relatively large non-pioneer forest trees, including *Cephalosphaera usambarensis*, *Myrianthus holstii* Engl. and *Strombosia scheffleri* Engl., was associated with *Maesopsis*. Also at Kwamkoro, Viisteensaari et al (2000) found numbers of lower canopy individuals of *Greenwayodendron suaveolens* subsp. *usambaricum*, *Parinari excelsa* Sabine and *Strombosia scheffleri* rose with increasing plantation age. Observations on saplings indicated that numbers of *Allanblackia stuhlmannii* and *Beilschmiedia kweo* also increased with plantation age.

## Experience in other areas

*Maesopsis* has been introduced in at least 12 countries outside the natural range (Table 10.3) and taken beyond the eastern and southern limits of the range in others (Democratic Republic of Congo, Kenya, Tanzania, Uganda and Zambia). Published reference to spontaneous regeneration after introduction to areas away from the natural range is very fragmentary. Regeneration may have occurred after seeds have fallen or have been dropped (in India – Sreenivasan and Dharmaraj, 1991; in the Solomon Islands – Marten, 1980). Spontaneous spread after introduction has also been found (for example in Kenya – Faden et al, 1988; in Rwanda – Troupin, 1982). Apart from in the East Usambara Mountains, however, such spread has been viewed with major concern only in Puerto Rico and on Pemba Island in Tanzania. Francis and Liogier (1991) report rapid spread and abundant reproduction of *Maesopsis* in Puerto Rico, resulting in the species becoming 'common to abundant', where it has spread from a planted source of <100ha, and they predict that *Maesopsis* will continue to spread on the island. Abdullah et al (1996) draw attention to the success of *Maesopsis* in colonizing open areas in Pemba's Ngezi forest, seedlings developing from seed dispersed from stands planted in the 1970s and 1980s.

Various eradication or control measures, which could be considered anywhere *Maesopsis* is spreading, have been suggested (Binggeli, 1989; Geddes, 1998). The individuals to be targeted include both those in existing plantations and those that have become established singly or in abundance in natural forest or on public or estate land. Developed into plans for action, control measures include actions to kill the individual trees, actions to avoid creating sites predisposed to *Maesopsis* invasion, and actions to reduce the prospects of *Maesopsis* populations recovering or being renewed.

**Table 10.3** *Circumstances of introductions of* Maesopsis eminii
*to countries outside the natural range*

| Country | Year introduced | Circumstance | Reference |
|---|---|---|---|
| Australia | no details available | no details available | Richardson et al (2000) |
| Brazil | ca. 1958 | small-scale trial | Dubois et al (1966) |
| Costa Rica | by 1962 | trial | Fenton et al (1977) |
| Fiji | ca. 1964 | plantations | Fenton et al (1977) |
| India | 1952 | for coffee shade | Radcliffe (1985) |
| Indonesia – from east Africa | by 1939 | plantations | Fenton et al (1977) |
| Malawi | by 1962 | trial | Streets (1962) |
| Malaysia (Peninsula) – from east Africa via Indonesia | 1952 | plantations | Streets (1962) |
| Malaysia (Sabah) | ca. 1958 | trial | Willan (1966) |
| Rwanda – from Tanzania and Uganda | 1948 | plantations | Kalinganire (1989)* |
| Samoa | before 1977 | trial | Fenton et al (1977) |
| Solomon Islands – from Uganda | ca. 1962 | trial | Fenton et al (1977) |
| US (Hawaii) | by 1977 | trial | Fenton et al (1977) |
| US (Puerto Rico) | ca. 1930; | trial | P. Binggeli (pers. comm.) |
| | ca. 1960 (reintroduction) | adaptability trial | Francis (1988) |

*Note:* * Kalinganire (1989) also makes the intriguing suggestion that *Maesopsis* was introduced to Rwanda (Gisaka) from Tanzania (Bukoba) some 500 years ago!

## DISCUSSION

Much of the attention that has been paid to *Maesopsis* in the agroforestry context is superficial and there are few documented experimental studies or analyses of detailed descriptive or monitoring data. Most opinions of whether land is better protected, or soil conditions improved, by a tree cover of *Maesopsis* than by herbaceous agricultural crops are speculative and systematic studies are yet to be reported. Nevertheless, use as a shade tree for plantation crops is generally described in favourable terms and positive hydrological effects have been noted, although these probably apply to trees in general and are not evidence of any special benefit arising from use of *Maesopsis*.

Despite assumptions that it will generate a locally improved microclimate for associated herbaceous crops (Oduol and Akunda, 1989), experience suggests negative impacts tend to be stronger, bearing out the reservations farmers express. The fast growth, the broad but relatively dense canopy and the concentration of feeder roots in the upper 30cm of the soil profile (Okorio et al, 1994) make *Maesopsis* highly competitive in intercropping situations, especially with light-demanding annual crops. It must be concluded that *Maesopsis* is not a particularly compatible tree for intercropping arrangements.

In terms of quality, no *Maesopsis* product has any exceptional attraction, as far as is known at present, and some have limitations. The fast growth rate that has encouraged planting with use as fuelwood in view results in wood of relatively low density (ca. $0.45kg/m^3$, air dried) and an inferior fuel. The poles of *Maesopsis* harvested at 41 months from the experiment of Peden et al (1996) fell far short of commercial needs, which are for lengths of at least 7m and much less taper ($\leq$6mm/m). Perhaps because the growing period was short, however, the other 14 species included in that study also failed to meet the expected market quality, although seldom by such a wide margin. Timber production through intercropping with arable crops has been advocated (Oduol and Akunda, 1989), but is not popular where the species is unfamiliar (Mwihomeke, 1989); the long-term acceptability to farmers of intercropping with such a fast-growing and wide-crowned tree is yet to be confirmed. As a fodder source, a more encouraging view of *Maesopsis* has emerged where it has been introduced in Indonesia (Java) and there is some analytical support for this (Mahyuddin et al, 1988).

Despite the limitations of *Maesopsis* products, an important consideration is the ease and speed with which they can be obtained and this convenience already explains some of the local interest in this species. In the future, this interest can be expected to grow and diversify as local familiarity with *Maesopsis* increases. Initiatives whereby the species supplies serviceable and useful products, including mushrooms as a by-product, and simultaneously helps reverse land degradation, seem destined to spread.

The degraded site conditions often associated with *Maesopsis* are not caused by the species, but it exhibits a range of reproductive characteristics which, together with its fast growth, have enabled the development of vigorous populations in the East Usambaras. Characteristics associated with colonizing behaviour are much in evidence, notably:

- early onset of reproductive activity;
- unspecialized flower structure;
- more or less continual reproductive activity;
- seeds easily dispersed by large generalist frugivores; and
- considerable longevity.

The collective effect of these characteristics is a more or less continuous supply of diaspores, compensating for the relatively short viability of individual seeds and the consequent lack of potential to stock a soil seedbank. The continual supply of new diaspores also makes fresh seed available for dispersal to suitable regeneration sites as they arise naturally through forest tree fall or from exploitation or other anthropic disturbance. Suitable regeneration sites typically are large ($>300m^2$) natural tree fall gaps (Binggeli, 1989), but such gaps remain suitable for establishment only briefly. Establishment is most likely when the diaspore (the regurgitated or evacuated stone from the fruit) contacts a moist soil surface free of loose litter (70 per cent of the area disturbed in uprooting – Binggeli, 1989). The large gap size ensures that, despite the extended period over which the slow germination takes place, full light still penetrates to the region of the gap with soil conditions (moist and relatively humus-rich) most favourable for establishment. Humus-rich soil, the substrate for germination, offsets some of the nutrient supply limitations of the mineral soil. In addition, the large, oil-rich seed contains reserves that can be utilized during the germination phase. Gaps providing these conditions, often for more extended periods of time, also arise frequently during logging, and comparable conditions occur along roadsides and at interfaces of forest with agricultural land.

Once *Maesopsis* has become established in a site, all stages in the life cycle can potentially contribute to a retention process, from incoming seed rain to stumps of adult individuals which produce new shoots. Incoming seed rain poses a threat only if the seed is dispersed to exposed soil – under a recently formed canopy gap or where a canopy gap forms over the seedling within a few months of germination. A conservation concern forcefully expressed by Binggeli (1989) and Binggeli and Hamilton (1993) is the ability of *Maesopsis* to retain sites through the appearance of a new generation of individuals or replacement of crowns from epicormic shoots. Other reports (Geddes, 1998; Viisteensaari et al, 2000) acknowledge these possibilities, but as less certain events. Provided full exposure to light continues after germination, height growth and crown development are rapid. Branches are carried at a wide angle and a self-pruning habit frees the trees of those at the crown base to which resources captured by the better-illuminated part would otherwise be diverted.

In well-illuminated sites with freshly exposed soil, *Maesopsis* readily becomes established from any incoming seed rain and a stand of the species may develop. This is equally a feature of many other pioneer tree species spontaneously forming stands which replace a natural closed humid forest ecosystem. No evidence suggests that in the East Usambaras *Maesopsis* is more likely than other broadleaved plantation species to suppress or prevent the regeneration of other trees. In the absence of disturbance, there are no indications that sites under *Maesopsis* dominance will be retained for a second generation, and no reason to doubt that a diverse forest with endemic/near-endemic and primary forest trees will develop through natural successional processes. *Maesopsis* does not retain dominance over a site after natural death of mature individuals or clearance action which preserves sufficient understorey. However, site retention is possible if the understorey is lost and no measures are taken to promote replacement by other species.

# CONCLUSIONS

*Maesopsis* is not a great agroforestry asset. However, it has potential for use as an agroforestry tree where it can serve an effective service role, and the tree's fast growth, predictable form, and resilience to pests and disease make management for timber attractive. Potential in intercropping arrangements with annual crops is limited. The conservation impact of *Maesopsis* as an aggressive invasive is not becoming catastrophic and has been overstated. While there is a potential invasion threat where *Maesopsis* is introduced outside its natural range, this should not disqualify forestry or agroforestry use of the species. Reports of

*Maesopsis* becoming established in large numbers in natural forest refer to observations made after exploitation has created invasible sites. Despite assertions otherwise, there is no evidence that *Maesopsis* will invade intact closed tropical forest which is not subject to severe disturbance. There is no basis for claims that *Maesopsis* predisposes sites to invasion by other exotic species. Where other exotic species become established, this reflects the underlying site conditions rather than the presence of *Maesopsis*.

# REFERENCES

Abdullah, H. S., Ali, M. S. and Kurikka, T. (1996) 'Ngezi Forest Reserve management plan', *Zanzibar Forestry Development Project Technical Paper*, vol 31, pp1–67

Adewunmi, C. O. and Sofowora, E. A. (1980) 'Preliminary screening of some plant extracts for molluscicidal activity', *Planta Medica*, vol 39, pp57–65

Anon (1930) 'Plant and seed list', East African Agricultural Institute Station, Amani, Tanganyika Territory

Anon (1947) 'Expérience sur les essences d'ombrage (1935): Rapport pour les exercises 1944 et 1945', Institut National pour l'Étude Agronomique du Congo Belge, Brussels, p49

Anwar, C., Baheramsyah, K. and Hamzah, Z. (1989) 'Efektivitas semak dan agroforestry di desa Kadipaten (Sub DAS Citanduy Hulu) dalam memperkecil aliran permukaan dan erosi', *Buletin Penelitian Hutan*, vol 511, pp1–8

Bagyaraj, D. J. (1989) 'Mycorrhizas', in H. Lieth and M. J. A. Werger (eds) *Tropical Rain Forest Ecosystems: Biogeographical and Ecological Studies*, Elsevier, Amsterdam, pp537–546

Barminas, J. T. and Eremosele, I. C. (2002) 'Rheological properties and potential industrial application of konkoli (*Maesopsis eminii*) seed gum', in P. A. Williams and G. O. Phillips (eds) *Gums and Stabilizers for the Food Industry*, 11, Royal Society of Chemistry, Cambridge, UK, pp306–310

Benstead, R. J. (1954) 'Timber trees suitable for cocoa shade' in *Proceedings of the West African Cocoa Research Conference*, West African Cocoa Research Institute, Tafo, Ghana, pp95–98

Binggeli, P. (1989) 'The ecology of *Maesopsis* invasion and dynamics of the evergreen forest of the East Usambaras, and their implications for forest conservation and forestry practices', in A. C. Hamilton and R. Bensted-Smith (eds) *Forest Conservation in the East Usambara Mountains, Tanzania*, International Union for the Conservation of Nature and Natural Resources, Gland, Switzerland, pp269–300

Binggeli, P. and Hamilton, A. C. (1993) 'Biological invasion by *Maesopsis eminii* in the East Usambara forests, Tanzania', *Opera Botanica*, vol 121, pp229–235

Bouquet, A. (1969) 'Féticheurs et médecines traditionelles du Congo (Brazzaville)', *Mémoires de l'Office de la Recherche Scientifique et Technique d'Outre-Mer*, vol 36, pp1–282

Bouquet, A. and Debray, M. (1974) 'Plantes médicinales de la Côte d'Ivoire', *Travaux et Documents de l'Office de la Recherche Scientifique et Technique d'Outre-Mer*, vol 32, pp1–232

Bowyer, J. L. (2001) 'Environmental implications of wood production in intensively managed plantations', *Wood and Fiber Science*, vol 33, pp318–333

Buck, L., Hall, J., Speich, A. and Stocking, M. (1985) *Agricultural Development and*

*Environmental Conservation in the East Usambara Mountains, Tanga Region, Tanzania*, Mission Report, International Union for the Conservation of Nature and Natural Resources Regional Office, Nairobi

Burkill, H. M. (1997) *The Useful Plants of West Tropical Africa. Vol 4, Families M–R* (second edition), Royal Botanic Gardens, Kew, UK

Chijioke, E. O. (1980) 'Impact on soils of fast-growing species in the lowland humid tropics', *FAO Forestry Paper*, vol 21, pp1–111

Cronk, Q. C. B. and Fuller, S. L. (1995) *Plant Invaders: The Threat to Natural Ecosystems Worldwide*, Chapman and Hall, London

Drechsel, P. and Reck, B. (1998) 'Composted shrub-prunings and other organic manures for smallholder farming systems in southern Rwanda', *Agroforestry Systems*, vol 39, pp1–12

Dubois, J., Hallewas, P. H. and Knowles, O. H. (1966) 'The role of the lower Brazilian Amazon as a source of wood products', in *Proceedings of the Sixth World Forestry Congress*, Madrid, vol 3, pp3213–3223

Eggeling, W. J. and Harris, C. M. (1939) *Forest Trees and Timbers of the British Empire. IV. Fifteen Uganda Timbers*, Clarendon Press, Oxford, UK

Egger, K. (1986) 'L'intensification écologique: Conservation (LAE) et amelioration des sols tropicaux par les systèmes agro-sylvo-pastoraux', *Collection Documents Systèmes Agraires*, vol 6, pp129–135

Egli, A. and Pietrowicz, P. (1990) L'associations des arbres et des cultures vivrières', L'Arboretum de Ruhande, Institut des Sciences Agronomiques du Rwanda, Butare, Rwanda

Egunjobi, J. K. (1974) 'Litter fall and mineralization in a teak, *Tectona grandis*, stand', *Oikos*, vol 25, pp222–226

Ekpa, O., Anam, E. and Vethaviyasar, N. (1985) 'Anthraquinones from the root bark of *Maesopsis eminii*', *Planta Medica*, vol 6, pp528

Engler, A. (1906) 'Über *Maesopsis Eminii* Engl., einen wichtigen Waldbaum des nordwestlichen Deutsch-Ostafrika, und die Notwendigkeit einer gründlichen forstbotanischen Erforschung der Wälder dieses Gebietes', *Notizblatt des Königlich Botanischen Gartens und Museums zu Berlin*, vol 38, pp239–242

Evans, J. (1984) 'Maintaining and improving the productivity of tropical and subtropical plantations', in D. C. Grey, A. P. G. Schönau, C. J. Schutz and A. van Laar (eds) *Proceedings of the Symposium on Site and Productivity of Fast Growing Plantations*, South African Forest Research Institute, Pretoria, pp893–905

Evans, J. (1999) *Sustainability of Forest Plantations: A Review of Evidence Concerning the Narrow-Sense Sustainability of Planted Forests*, United Kingdom Department for International Development, London

Evans, J. and Turnbull, J. W. (2004) *Plantation Forestry in the Tropics: The Role, Silviculture and Use of Planted Forests for Industrial, Social, Environmental and Agroforestry Purposes* (third edition), Clarendon Press, Oxford, UK

Faden, R. B., Beentje, H. J. and Nyakundi, D. O. (1988) 'Checklist of the forest species', *Utafiti*, vol 1, pp43–66

Fenton, R., Roper, R. E. and Watt, G. R. (1977) *Lowland Tropical Hardwoods: An Annotated Bibliography of Selected Species with Plantation Potential*, External Aid Division, Ministry of Foreign Affairs, Wellington, ppME1–24

Francis, J. K. (1988) '*Maesopsis eminii* Engl., musizi: Rhamnaceae, buckthorn family', Research Note SO-ITF-SM-8, United States Department of Agriculture Forest Service Southern Forest Experiment Station, New Orleans, pp1–4

Francis, J. K. and Liogier, H. A. (1991) 'Naturalized exotic trees in Puerto Rico', USDA

Forest Service Southern Experiment Station General Technical Report SO-82, pp1–12

Geddes, N. R. (1998) '*Maesopsis* invasion of the tropical forest in the East Usambara mountains, Tanzania', MPhil thesis, University of Wales, Bangor, UK

Gintings, A. N. and Yusuf, H. (1983) 'Aliran permukaan dan erosi tanah pada lahan beberapa jenis tanaman pangan dan hutan di Waspada-Garut', *Laporan, Pusat Penelitian dan Pengembangan Hutan*, vol 412–414, pp12–26

Hall, J. B. (1995) '*Maesopsis eminii* and its status in the East Usambara Mountains', *East Usambara Catchment Forest Project Technical Paper*, vol 13, pp1–39

Hamilton, A. C. (1989a) 'Soils', in A. C. Hamilton and R. Bensted-Smith (eds) *Forest Conservation in the East Usambara Mountains, Tanzania*, International Union for the Conservation of Nature and Natural Resources, Gland, Switzerland, pp87–95

Hamilton, A. C. (1989b) 'Some effects of *Maesopsis* on litter and soil on the East Usambaras', in A. C. Hamilton and R. Bensted-Smith (eds) *Forest Conservation in the East Usambara Mountains, Tanzania*, International Union for the Conservation of Nature and Natural Resources, Gland, Switzerland, pp330–332

Hamilton, A. C. and Bensted-Smith, R. (1989) *Forest Conservation in the East Usambara Mountains, Tanzania*, International Union for the Conservation of Nature and Natural Resources, Gland, Switzerland

Iversen, S. T. (1991) 'The Usambara Mountains, NE Tanzania: Phytogeography of the vascular plant flora', *Acta Universitatis Upsaliensis Symbolae Botanicae Upsaliensis*, vol 29, no 3, pp1–234

Kalinganire, A. (1989) '*Maesopsis eminii*, essence forestière et agroforestière à croissance rapide: Arbre d'avenir pour le Rwanda', *Bulletin Agricole du Rwanda*, vol 22, pp1–4

Laine, C., Baiakina, J., Vaquette, J., Chaumont, J. P. and Simeray, J. (1985) 'Activité antifongique d'écorces de troncs de sept phanérogames congolaises', *Plantes Médicinales et Phytothérapie*, vol 19, pp75–83

Lugo, A. E. (1992) 'Comparison of tropical tree plantations with secondary forests of similar age', *Ecological Monographs*, vol 62, pp1–41

Lundgren, B. (1980) 'Plantation forestry in tropical countries – Physical and biological potentials and risks', *Rural Development Studies*, vol 8, pp1–134, Swedish University of Agricultural Sciences, Uppsala, Sweden

Macfadyan, A. (1989) 'A brief study of the relationships between *Maesopsis* and some soil properties in the East Usambaras', in A. C. Hamilton and R. Bensted-Smith (eds) *Forest Conservation in the East Usambara Mountains, Tanzania*, International Union for the Conservation of Nature and Natural Resources, Gland, Switzerland, pp333–343

Mahunka, S. (1989) 'Preliminary study of the soil fauna of primary and secondary submontane rain forests on the East Usambaras', in A. C. Hamilton and R. Bensted-Smith (eds) *Forest Conservation in the East Usambara Mountains, Tanzania*, International Union for the Conservation of Nature and Natural Resources, Gland, Switzerland, pp345–346

Mahyuddin, P., Little, D. A. and Lowry, J. B. (1988) 'Drying treatment drastically affects feed evaluation and feed quality with certain tropical forage species', *Animal Feed Science and Technology*, vol 22, pp69–78

Marten, K. D. (1980) 'A summary of the performance of the major plantation species in divisional trials plots', Solomon Islands Forestry Department Research Report S/1/80, Ministry of Forest, Environment and Conservation, Honiara, Solomon Islands, pp1–7

Moreau, R. E. (1935) 'The breeding biology of certain East African hornbills', *Journal of the East African Natural History Society*, vol 13, pp1–28

Mugasha, A. G. (1981) 'The silviculture of Tanzanian indigenous tree species. II. *Maesopsis eminii*', *Tanzania Silviculture Technical Note (New Series)*, vol 52, pp1–14

Mustonen, P. and Räsänen, P. K. (1985) 'The Amani Forest Planning Project: Project document', FINNIDA/Finnmap Oy/Silvestria, Helsinki

Mwihomeke, S. T. (1989) 'Comparative growth and productivity of potential tree species (exotic and indigenous) at the Lushoto Arboretum, in the West Usambara Mountains, Tanzania', in *Trees for Development in Sub-Saharan Africa*, International Foundation for Science, Stockholm, pp77–87

Normand, D. (1935) 'Sur le *Maesopsis* de l'ouest africain et la bois de nkanguele', *Revue Internationale de Botanique Appliquée et d'Agriculture Tropicale*, vol 15, pp252–263

Noumi, E. and Tchakonang, N. Y. C. (2001) 'Plants used as abortifacients in the Sangmelina region of southern Cameroon', *Journal of Ethnopharmacology*, vol 76, pp263–268

Nye, P. H. (1955) 'Some soil forming processes in the humid tropics. III. Laboratory studies on the development of a typical catena over granitic gneiss', *Journal of Soil Science*, vol 6, pp63–72

Oduol, P. A. and Akunda, E. W. (1989) 'Tropical rainforest trees with agroforestry potential', in *Trees for Development in Sub-Saharan Africa*, International Foundation for Science, Stockholm, pp49–56

Okorio, J., Byenkya, S., Wajja, N. and Peden, D. (1994) 'Comparative performance of seventeen upperstorey tree species associated with crops in the highlands of Uganda', *Agroforestry Systems*, vol 26, pp185–203

Peden, D. G., Okorio, J. and Wajja-Musukwe, N. (1996) 'Commercial pole production in linear agroforestry systems', *Agroforestry Systems*, vol 33, pp177–186

Pierre, L. (1897) 'Plantes du Gabon. Sur le genre *Karlea* de la tribu des Zyzyphées des Rhamnacées', *Bulletin Mensuel de la Société Linnéenne de Paris*, vol 159, pp1270–1272

Pocs, T. (1989) 'A preliminary study of the undergrowth of primary and secondary submontane rainforests in the East Usambara Mountains, with notes on epiphytes', in A. C. Hamilton and R. Bensted-Smith (eds) *Forest Conservation in the East Usambara Mountains, Tanzania*, International Union for the Conservation of Nature and Natural Resources, Gland, Switzerland, pp301–306

Prabhu, V. V. and Theagarajan, K. S. (1986) 'Studies on the fatty oil of *Maesopsis eminii* seed', *Van Vigyan*, vol 24, nos 3–4, pp116–117

Purwanto, I. and Gintings, A. N. (1989) 'Pengaruh berbagai jenis vegetasi terhadap kapasitas infiltrasi tanah di Cijambu, Sumedang, Jawa Barat', *Buletin Penelitian Hutan*, vol 518, pp13–22

Radcliffe, R. (1985) 'A note on the cultivation and growth of *Maesopsis eminii* Engl. (Musizi)', *Indian Forester*, vol 111, pp760–764

Richardson, J. E., Fay, M. F., Cronk, Q. C. B., Bowman, D. and Chase, M. W. (2000) 'A phylogenetic analysis of Rhamnaceae using *rbcL* and *trnL-F* plastid DNA sequences', *American Journal of Botany*, vol 87, pp1309–1324

Rodgers, W. A. and Homewood, K. M. (1982) 'Species richness and endemism in the Usambara mountain forests', *Biological Journal of the Linnean Society*, vol 18, pp197–242

Rugalema, G. H., Okting'ati, A. and Johnsen, F. H. (1994a) 'The homegarden agroforestry system of Bukoba district, north-western Tanzania. 1: Farming system analysis', *Agroforestry Systems*, vol 26, pp53–64

Rugalema, G. H., Johnsen, F. H., Okting'ati, A. and Minjas, A. (1994b) 'The home garden agroforestry system of Bukoba district, north-western Tanzania. 3: An economic appraisal of possible solutions to falling productivity', *Agroforestry Systems*, vol 28, pp227–236

Sangster, R. G. (1957) *Annual Report of the Forest Department, Tanganyika, 1960*, Government Printer, Dar es Salaam

Schabel, H. G. and Latiff, A. (1997) '*Maesopsis eminii* Engler', in I. F. Hanum and L. J. G. van der Maesen (eds) *Auxilliary Plants. Plant Resources of South-East Asia*, Backhuys, Leiden, The Netherlands, pp184–187

Siebenlist, T. (1914) *Forstwirtschaft in Deutsch-Ostafrika*, Paul Parey, Berlin

Silitonga, H., Hamzah, Z. and Manan, S. (1987) 'The KPH Bondowso agroforestry model in East Java. Possible uses of coffee plants (*Coffea robusta* Lind.) as an auxiliary forest crop and as a source of income for local people', *Duta Rimba*, vol 13, pp79–80

Smiet, A. C. (1990) 'Agro-forestry and fuel-wood in Java', *Environmental Conservation*, vol 17, pp235–238

Sreenivasan, M. S. and Dharmaraj, P. S. (1991) '*Maesopsis eminii* Engl. – A fast growing shade tree for coffee', *Indian Coffee*, vol 55, no 7, pp17–20

Streets, R. J. (1962) *Exotic Trees in the British Commonwealth*, Clarendon Press, Oxford, UK

Suprapti, S. (1987) 'Pemanfaatan limbah industri penggergajian untuk media tumbuh jamur tiram putih', *Duta Rimba*, vol 13, nos 87–88, pp38–40

Suprapti, S. (1989) 'Pembudidayaan jamur tiran (*Pleurotus flabellatus*) padu tujuh jenis kayu', *Duta Rimba*, vol 15, nos 109–110, pp13–17

Swift, M. J. and Anderson, J. M. (1989) 'Decomposition', in H. Lieth and M. J. A. Werger (eds) *Tropical Rain Forest Ecosystems: Biogeographical and Ecological Studies*, Elsevier, Amsterdam, pp547–569

Tattersfield, P. (1996) 'Local patterns of land snail diversity in a Kenyan rain forest', *Malacologia*, vol 38, pp161–180

Tattersfield, P., Seddon, M. B. and Lange, C. N. (2001) 'Land-snail faunas in indigenous rainforest and commercial forestry plantations in Kakamega Forest, western Kenya', *Biodiversity and Conservation*, vol 10, pp1809–1829

Taylor, D. (1989) 'Root distribution in relation to vegetation and soil type in the forests of the East Usambaras', in A. C. Hamilton and R. Bensted-Smith (eds) *Forest Conservation in the East Usambara Mountains, Tanzania*, International Union for the Conservation of Nature and Natural Resources, Gland, Switzerland, pp313–329

Thomas, A. S. (1940) 'Cover crops and shade trees: Shade trees', in J. D. Tothill (ed) *Agriculture in Uganda*, Oxford University Press, London, pp480–484

Troupin, G. (1982) *Flore des Plantes Ligneuses de Rwanda*, Institut National de Recherche Scientifique, Butare

Viisteensaari, J., Johansson, S., Kaarakka, V. and Luukkanen, O. (2000) 'Is the alien tree species *Maesopsis eminii* Engl. (Rhamnaceae) a threat to tropical forest conservation in the East Usambaras, Tanzania?', *Environmental Conservation*, vol 27, pp76–81

Widiarti, A. and Alrasjid, H. (1987) 'Penanaman introduksi jenis pohon kayu baker di lahan kritis Paseh dan Kadipaten', *Buletin Penelitian Hutan*, vol 488, pp1–17

Willan, R. L. (1966) *Fast Growing Tropical Trees*, Commonwealth Forestry Institute, Oxford, UK

Wood, P. J. (1966) 'A guide to some German forestry plantations in Tanga Region', *Tanzania Notes and Records*, vol 66, pp203–206

Zobel, B., Wyk, G. van and Stahl, P. (1987) *Growing Exotic Forests*, Wiley, New York

# 11

# Single-Tree Management Models: *Maesopsis eminii*

*Thomas Buchholz, Timm Tennigkeit*
*and Axel Weinreich*

## INTRODUCTION

Single-tree management models aim to maximize individual tree performance and can be adapted to different management regimes, like commercial tree growing and agroforestry. In general these models are designed to help forest and plantation managers to maximize stand value by predicting relationships between bole diameter and tree height as timber value generally increases in an exponential fashion related to bole diameter. This is especially the case for coniferous tree species (Uzoh and Oliver, 2008; Vargas-Larreta et al, 2009). Such height/diameter models are generally used to maximixe diameter growth using easily available variables (for example from exisiting inventories) such as tree height and DBH data sets. However, their predictive capability is hampered by the generally weak relationship of tree height and diameter development observed for most forest stands and species, and recently stand density functions have been included in the models (Condés and Sterba, 2008).

For most deciduous species, however, single-tree management models focused on height–diameter relationships produce poor results, especially when besides bole diameter (branch-free) bole length should be maximixed for optimum revenue as well. In such cases models including stand density factors generally have a higher prediction power. Therefore better results are achieved when data are also available on individual crown cover area (expressed by crown width), growing space and self-pruning dynamics.[1] Inclusion of these variables into the models has resulted in very accurate and predictive single-tree management models, particularly for deciduous trees of the temperate and tropical zone. For instance, Schröder et al (2007) developed a model with a strong focus on the height–diameter relationship but also incorporating stand density functions in uneven-aged and mixed-species stands. Some of the most elaborate and precise single-tree management models for decidous trees in Europe prediciting bole diameter and length are those from Nutto (1999), Hein (2003), Hein and Spiecker (2008), Schmidt and Kändler (2009), Spiecker (2009) and Weinreich, 2001. Examples from the tropics (Dawkins, 1963a and b; Ohland, 1998; Tennigkeit and Weinreich, 2000) cover a range from even-aged monocultures to mixed-species, uneven-aged stands.

## The situation of the forest sector in Uganda and the significance of *Maesopsis eminii*

Today forests cover 21 per cent of Uganda's area. This figure reflects a long history of deforestation. From 1970 to 1990 forest cover was reduced by 50 per cent (FAO, 1995). The actual reduction rate per year is presumed to be 2 per cent (FRA, 2000). Most of the natural forests in Uganda have disappeared or are highly degraded (SPGS, 2009). The remaining timber plantation area is small after years of political instability and to a large extent in dire condition. Owing to the steadily improving political and economic situation, the demand for construction timber as well as timber prices are increasing. Since the mid-1990s there have been growing efforts to boost domestic timber production. Exotic timber species like *Eucalyptus grandis* or *Pinus caribaea* are often preferred by plantation investors over indigenous species due to the superior management knowledge on these species gathered across the globe.

*Maesopsis eminii* (henceforth referred to as *Maesopsis*) is a native Ugandan hardwood species with superior timber and growth characteristics (Dawkins, 1963a; Kriek, 1970; Kingston, 1974; Hall, Chapter 10 of this volume). Its timber is a light but mechanically loadable hardwood, similar in weight and strength to *Pinus sylvestris* (Anon, 1953). It is easy to handle and well suited for construction. It is able to produce good veneer and, therefore, plywood because of its straight, branchless bole with a medium, even texture and density, cylindrical shape, and absence of buttresses (Tack, 1956).

Its ecological status as a colonizing tree indicates its fast growth, especially in sapling stages when competing with climbers. Its strong self-pruning tendency to producing long branch-free boles makes it a promising species for natural forest management, plantations and agroforestry systems. These growth patterns diminish the need for cost-intensive weeding and pruning activities.

Naturally, *Maesopsis* occurs in east, central and west Africa from Liberia to Uganda to south Angola, both in rainforest and riverine forest (Katende, 1995; Binggeli, 1997). In Uganda *Maesopsis* occurs between altitudes of 700m and 1500m with a mean annual rainfall of 1500–2000mm (Hall, 1995).

Outside of its natural range, *Maesopsis* is able to grow in Fiji, India, Malaysia and Indonesia. *Maesopsis* plantations in those areas are desirable mainly due to its suitability as shade tree for coffee or as pulpwood (Earl, 1968; Binggeli, 1997). In its natural range, a maximum tree height of 40m and a DBH of 120cm can be reached. An exploitable DBH of 50cm can be reached at rotation cycles of 25 to 30 years according to Dupuy and Mille (1998), though other sources argue that 30 to 35 years is required for that DBH, while poles can be harvested in 5–8 years. In a study by Kriek (1970) *Maesopsis* reached a DBH of 30cm within 5.5 years under optimal conditions and average height of 16m under optimal humidity conditions. Pesme (Dupuy and Mille, 1998) observed a rapid growth of 2.5–3m per year. Tack (1956) reported mature trees with branchless boles of 9–20m.

These timber qualities and growth charateristics have long been recognized and *Maesopsis* has a track record as an important indigenous forest plantation species in Uganda. *Maesopsis* plantations which sum up to several hundred hectares in Uganda (especially in Mubende and Bukaleba districts) have been established by the public and private sectors in order to grow high-quality hardwood, as its multipurpose fast-growing timber is highly desired and prized. Most existing plantations of this species were established in the 1970s and only a few new ones have been planted since 1995. In addition, enrichment plantings in degraded forests are carried out using *Maesopsis*. *Maesopsis* is further widely cultivated by small-scale farmers. In this context, it is grown to supply fuel and poles for building as well as to take advantage of its agroforestry compatibility (light shade, low crop competition) and its wide distribution of seeds and seedlings.

Early plantation forest management aimed at maximizing volume production at stand level. The first yield table for *Maesopsis* in Uganda was developed by Kingston (1974). Based on different site conditions, his model aimed for volume maximization on a stand level rather than producing quality timber in a short to medium term. For instance, no focus was given to prevent bending of boles due to light competition, to which *Maesopsis* is highly susceptible (Buchholz, 2003) and the beneficial characteristic of *Maesopsis* to develop branch-free boles through natural pruning (Francis, 1997). Management regimes for *Maesopsis* have been recommending a high tree density. As a result, rotation periods of 25 to 55 years were proposed for *Maesopsis* plantations, with a final harvesting DBH of 50 to 60cm.

In contrast, modern silvicultural concepts focus on the production of quality timber (maximizing bole length and diameter) in short rotation periods to increase return on investments rather than maximum volume production over longer time periods. However, such a management model for *Maesopsis* is absent to date. Meanwhile, there is growing recognition of the silvicultural potential of *Maesopsis* in its native range and beyond.

## Study objectives and approach

The main objective of this study was to develop a single-tree management model and a management strategy for *Maesopsis* optimizing the single-tree growth in order to maximize the economic value of a plantation by reducing the rotation period and improving timber quality. A second objective was to use this model to compare growth rates and the economic performance of *Maesopsis* with *Pinus caribaea*, the main timber plantation species in Uganda; and to suggest strategies to further enhance domestic timber production capacity in Uganda.

The development of a single-tree management model as well as a comparison of the growth and economic performance of *Maesopsis* with exotic tree species requires subsequent model development and assessment steps. To be able to develop a complete single-tree management model, we developed preliminary models for (a) site index curves, (b) DBH development, (c) crown width and (d) bole length. The single-tree management model developed here is based on site index and diameter growth curves, the strong correlation between DBH and crown projection area, and a bole length model. This model predicts the necessary growing space at a predefined rotation period and harvesting diameter. The management strategy implied in the model is that, first, the crown develops under minimum competition (similar to a solitary tree, no canopy closure). Accordingly, the diameter development reflects mainly site conditions and age. And second, a sufficient branch-free bole length (referred to as 'bole length' in this study) is crucial for high-quality timber production and depends on tree genetics and light conditions, in other words crown length.

# MATERIAL AND METHODS

## Site

The climate where *Maesopsis* prevails is characterized by a bimodal rainfall pattern, with rains from March to April and from September to November. Naturally, *Maesopsis* occurs between altitudes of 700 and 1500m asl, with mean annual rainfall of 1500–2000mm (Hall, 1995). Study sites were chosen all over the natural range of *Maesopsis* in Uganda.

The survey of *Maesopsis* trees was conducted between November 2002 and February 2003. This primary data contains 29 stands at 8 sites, totalling 389 trees. The stands (Figure 11.1 and Table 11.1) were established both naturally and artificially, growing in mixed stands and even-aged plantations with different stand densities. There was no observable damage to the trees or signs of senescence. Records of stand age were available for most stands.

Additionally, growth records for *Maesopsis* were used from previous studies, after undergoing a plausibility check. We used data from Kriek taken between 1967 and 1968 in Uganda (according to Fenton et al, 1977) and data from

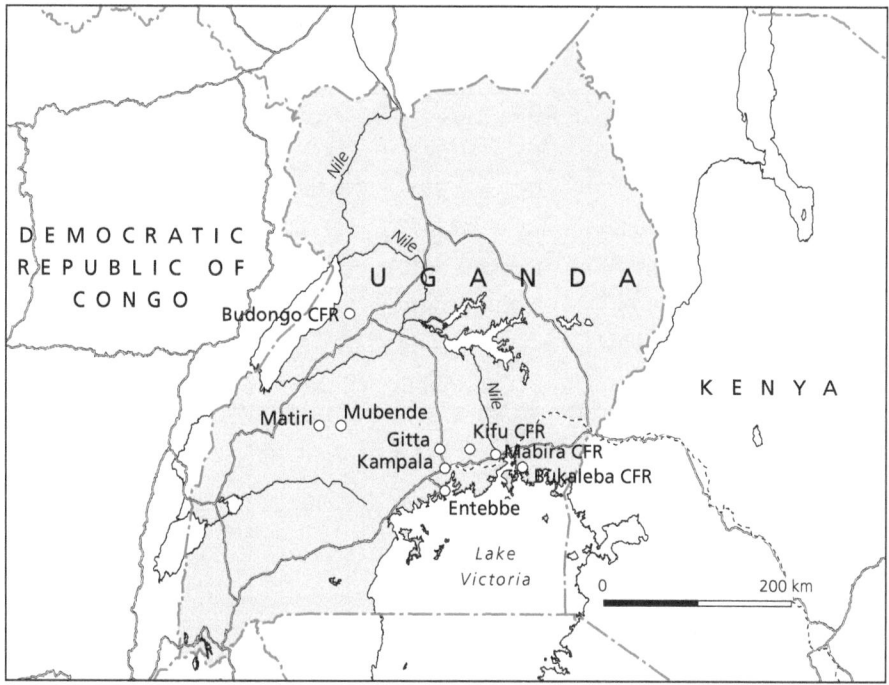

**Figure 11.1** *Location of the eight study sites in Uganda: Entebbe, Matiri, Mubende, Budongo, Mabira, Kifu, Bukaleba and Kampala*

Morton (1975, according to Francis, 1997) and Warsopranoto et al (1966, according to Fenton et al, 1977) taken in Malaysia.

## Preliminary models for a single-tree management model

**Site index model** curves are based on the stand dominant tree height ($H_{dom}$), the average height of the 20 per cent largest diameter trees of a stand, in other words trees growing under a minimum of competition (Kramer and Akça, 1995). The age of 10 years served as the base, in other words the site index curves were labelled according to the assumed dominant tree height of the stands at the age of 10 years. A combination of two types of equations was used to describe the height growth of *Maesopsis*. The Chapman-Richards equation (eq 1) is well adapted to simulate the growth of pioneer trees as its shape is characterized by an early inflection point in growth, simulating an early culmination in tree height (see Alder et al, 2003; Van Laar and Akça, 1997).

$$h = b_0 \times (1 - \exp)-b_1 \times A))^{b_2} \text{ (Chapman-Richards)} \qquad \text{(eq 1)}$$

where $h$ is the height, $b_0$, $b_1$, and $b_2$ are parameters, and $A$ is the age.

**11.1** *Overview of primary data collected in Uganda*

| MAT (°C) | No. of stands | Sample size | Stand density (N/ha) | Age (yrs) | Dominant tree height (m) | DBH$_{dom}$ (cm) | Mean DBH (cm) | Mean bole height (m) |
|---|---|---|---|---|---|---|---|---|
| 21.5 | 2 | 32–30 | 239–768 | 6–14 | 12.2–17.7 | 17–30 | 13–24 | 7.9–11.7 |
| 20.3 | 1 | 31 | 96 | 37 | 34.3 | 57 | 41 | 18.7 |
| 20.3 | 1 | 27 | 54 | 37 | 33.1 | 55 | 44 | 15.7 |
| 22.7 | 2 | 15–30 | 150–212 | 4–8 | 13.6–18.3 | 27–34 | 22–30 | 5.9–10.9 |
| 22.1 | 1 | 29 | 88 | 38 | 36.0 | 66 | 49 | 20.0 |
| 20.7 | 6 | 5–24 | 54–78 | 1.5–9 | 16.1–19.3 | 12–41 | 9–37 | 2.4–7.5 |
| 22.1 | 2 | 29–31 | 131–296 | 4–5 | 12.4–12.9 | 21–25 | 17–21 | 5.8–4.8 |
| 21.7 | 1 | 29 | 179 | 8 | 18.7 | 30 | 24 | 10.4 |

Accordingly, site index curves (eq 2) were calculated as:

$$h = SI / (1 - \exp(-b_1 \times A_i))^{b_2} \times 1 - \exp(-b_1 \times A_j))^{b_2} \qquad \text{(eq 2)}$$

where $SI$ is the site index, $A_i$ the site index base age and $A_j$ the actual age.

The Chapman-Richards equation described the height growth of *Maesopsis* well up to an age of 10 years. In order to identify the regression parameter, all stands younger than 25 years were included in the analysis. However, due to missing stand records corresponding to ages of 10 to 30 years, this type of equation did not satisfactorily describe the height growth of *Maesopsis* when exceeding the age of 10 years. Therefore, the Schumacher equation (eq 3) (Van Laar and Akça, 1997) was used to describe height growth development of *Maesopsis* stands past an age of 10 years. All stands above an age of 5 years were used to identify the regression parameter. Overlapping of the data set to calculate the parameters for the two different equations resulted in a smooth interface of the curves at age 10.

$$h = \exp(b_3 + b_4 \times (1/A) \qquad \text{(eq 3)}$$

where $b_3$ and $b_4$ are parameters and $A$ is the age.

Accordingly, site index curves (eq 4) were calculated as:

$$h = \exp(\ln(SI) + b_4 \times (1/A_j - 1/A_i) \qquad \text{(eq 4)}$$

**The diameter model.** The management model aims to maximize DBH growth. There was a lack of data about DBH development in time series of *Maesopsis*, with no records of individuals growing without competition. Therefore, the DBH development of dominant trees (DBH$_{dom}$), or false time series taken from the primary data, were used to develop a diameter growth model (eq 5).

$$DBH_{dom} = b_5 + (b_6 \times \ln(A)) \times b_7 \times SI \qquad \text{(eq 5)}$$

where $b_5$, $b_6$ and $b_7$ are parameters, $A$ is the age and $SI$ the site index.

**The crown width model** was intended to confirm the model developed for *Maesopsis* by Dawkins (1963a) in order to predict a tree's canopy surface depending on a desired DBH. The idea of this model is that the part of the tree crown directly exposed to the light is the photosynthetically most active area. As a result, the crown projection area is a good indicator of expected diameter growth. The relationship between available crown projection area and diameter growth was determined by Duchaufour (1903). In his model, site and age were not included as factors. The crown projection area (CPA) for each tree was calculated following Kramer and Akça (1995; eq 6):

$$CPA = (r_1+r_2)/2)^2 + ((r_2+r_3)/2)^2 + ((r_3+r_4)/2)^2 + ((r_4+r_1)/2)^2) \times \pi/4$$

$$\text{(eq 6)}$$

where $r_1$, $r_2$, $r_3$ and $r_4$ are crown radii measured in 90° angles.

**Bole length model.** A sufficient free (unbranched) bole length is necessary for high-quality timber production. Branches influence knottiness and therefore timber quality. The best timber quality for any construction purpose requires a long, branch-free cylindrical bole with sufficient diameter. A tree's natural bole length is influenced by its genetically determined self-pruning habit and by neighbouring trees that compete for light. With increasing shade of the developing crown or increased light competition with neighbouring vegetation, branches located at the crown base die and break off. This results in branch-free boles favourable for timber purposes. In the case of *Maesopsis*, this self-pruning tendency is exceptionally strong, as the self-pruning process takes place under very light shade compared to other tree species (Buchholz, 2003). Subsequently, bole length is affected by stand density development (the intensity of competition for each tree). Unfortunately, competition information was not available for any of the stands and therefore could not be considered. However, the DBH can be used as a substitute indicator for competition (Nutto, 1999). High competition depresses diameter growth. Therefore, trees with the same height but different competition history will differ in DBH. Subsequently, for dominant trees we investigated their bole length in relation to height, age and DBH.

**Taper functions.** Taper functions of a tree describe the form of the bole as its diameter recedes with increasing height. Having a taper function available therefore allows us to calculate volume based on the single measurement of the DBH. Absence of taper functions valid for *Maesopsis* led to the application of a modified BRINK function and the associated parameters for *Fagus sylvatica* (Nagel et al, 2003). The function was used to calculate tree volume and products (assortments). The same function was applied to *Pinus caribaea*, and supplemented with a PAIN function calculating the individual tree taper (applied parameter calculated for *Pinus sylvestris*; Nagel et al, 2003).

Having species-specific parameters available for *Maesopsis* of the above-described functions of bole length, bole diameter and crown width based on height and age allows us to model the growth of a single tree when grown under the absence of competition for light.

# MODULES FOR A SINGLE-TREE MANAGEMENT MODEL – RESULTS AND VALIDATION

## Site index model

Equations i and ii in Table 11.2 describe the site index curves for trees younger and older than 10 years respectively. Parameters were generated via a non-linear regression procedure using equations 1 and 3 respectively. In other words, equation i has to be inserted into equation 2, equation ii into equation 4 and for DBH modelling equation iii into equation 5. The regression functions do not

**Table 11.2** *Equations used in the single-tree management model for* Maesopsis

| Model description | Equation | N | $R^2$ | Equation no. |
|---|---|---|---|---|
| Site index model <10 yrs | $h=20.3505\times(1-\exp(-0.1932\times A))^{1.1833}$ | 150 | 0.50 | i |
| Site index model >10 yrs | $h=\exp(3.6163+-6.5225\times(1/A)$ | 130 | 0.69 | ii |
| Diameter growth model | $DBH_{dom}=-1.1146+$ $(15.2004\times\ln(A))\times0.0535\times SI$ | 13 | 0.92 | iii |
| Crown width model | $CW=2.1789+0.2329\times DBH$ | 361 | 0.89 | iv |
| Bole length model | $bl=0.4983\times th$ | 66 | 0.8462 | v |

*Notes:* Equation i and ii: where h is the dominant tree height in m and *A* is the age in years.
Equation iii: where $DBH_{dom}$ is given in cm, *A* is the age in years and *SI* is the site index.
Equation iv: where *CW* is the crown width in m and *DBH* is given in cm.
Equation v: where *bl* is the bole length in m and *th* the tree height in m.

take site differences into account and explain about 50 per cent to 70 per cent of the variance.

Figure 11.2a visualizes the shape of the resulting site index curves when applying the parameters in the respective site index equations 2 and 4. The lines represent the assumed growth of the stands' dominant tree height and serve as site index.

The dominant tree height development was predicted for stands with a dominant tree height of 5, 10, 15, 20 and 25 metres at an age of 10 years, which is the base age (vertical line). It is remarkable that primary data records from Uganda show similar vigorous growth to the stands recorded in Malaysia, while older records of *Maesopsis* stands in Uganda tend to demonstrate slower height growth.

The site index SI 25 approaches a dominant tree height of around 40m, which is in line with general maximum tree height recorded in other studies (Hall, 1995; Bingelli, 1997). The dotted line in Figure 11.2b represents the site

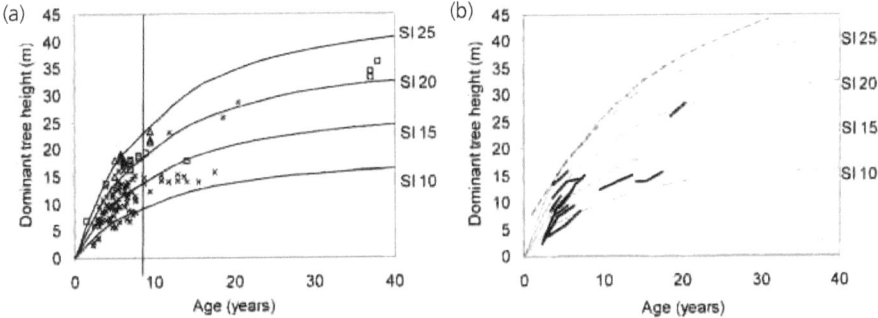

**Figure 11.2** *Site index curves for* Maesopsis eminii

*Notes:* (a) Dominant tree height in relation to age for plots with data from own field research (squares) and literature data from Uganda (crosses, data from Kriek, 1967 and 1968) and Malaysia (triangles, data from Warsopranoto, 1966, and Morten, 1975). The curves show the relation for sites with a different site index. The vertical line marks the base age which is used to name the site index curves according to the height reached at that age.
(b) The curves of (a) with additionally the highest broken line indicating the SI 25 curve of Kingston (1974) and with recorded time series; data as in (a).

index SI 25 modelled following Kingston (1974). The equation is a three polynomial function resulting in unrealistic height in the initial growth that already exceeds the general maximum tree height of 40m at the age of 25 years. In order to validate the shape of the site index curves, they were compared with the real dominant tree height development of recorded stands. The bold lines represent time series of stands recorded by Kriek (according to Fenton et al, 1977) in Uganda. The shape of the lines representing the time series is comparable in shape with the site index curves and does simulate tree height growth better than the site index equations of Kingston (1974).

## Diameter growth model

Diameter development depends on site, age and the crown development potential, in other words is expressed by the stand density or canopy space. Actual canopy closure ranged between 73 per cent (in general, 80 per cent is considered as canopy closure in a forest) and 146 per cent (significant overlap of canopies), with an average of 100 per cent. Stand densities varied from 13 to 780 trees/ha at an age of 5 to 38 years. As no time series were available, this data set was inappropriate to identify any relation to diameter development directly.

Age and site index clearly affect the development of the $DBH_{dom}$ (Figure 11.3a), with higher site indices resulting in thicker trees. The lines in Figure 11.3a show the $DBH_{dom}$ development on sites with SI of 20 and 25 (equation iii). A $DBH_{dom}$ in the range of 48cm to 60cm can be expected at an age of 20 years for *Maesopsis* stands classified with a site index above SI 20.

Equation iii shown in Table 11.2 is derived from equation 5 and describes the development of DBH depending on age and site index. Contrary to expectations, canopy closure did not affect $DBH_{dom}$. (Figure 11.3b). Our model developed for a site index of 25 SI predicts higher $DBH_{dom}$ for all ages than the Kingston (1974) model (Figure 11.3c). The difference between the curves might be explained by a difference in stand density development, but this variable was not recorded for the data set used by Kingston.

## Crown width model

The crown width model relates crown width to DBH and allows us to calculate a stand's maximum density for a given DBH and canopy closure (Figure 11.4a). The relation between crown width and DBH was linear. Age did not have a significant effect on the predictive capacity (results not shown).

The crown width model developed by Dawkins (1963a) (CW = 0.85+25.15×DBH; with CW and DBH in the same units) results generally in a lower stand density for a given DBH compared to our model. Assuming a canopy closure of 80 per cent and a DBH of 50cm, our crown width model predicts a stocking density of 53 trees/ha, while the Dawkins (1963a) model predicts a stocking density of 57 trees/ha. This difference may be caused by

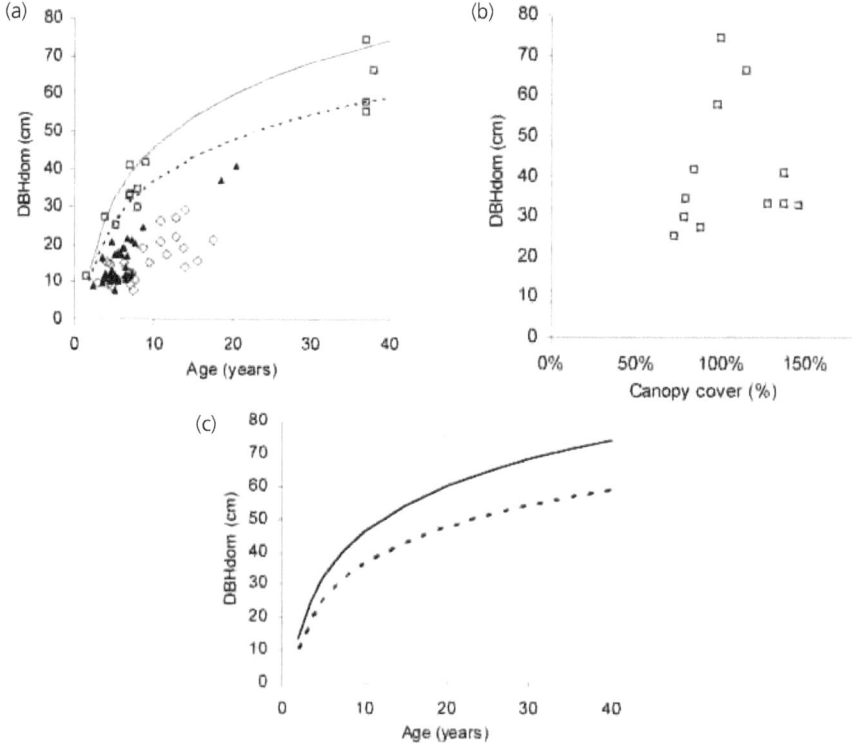

**Figure 11.3** *Relationships of* Maesopsis eminii *dominant diameter with age and canopy cover*

*Notes:* (a) DBH$_{dom}$ in relation to age and for different SI classes – diamonds indicate SI between 0 and 15, triangles SI 16–20 and squares SI >20. The unbroken line indicates the development of dominant diameter on sites with SI of 25; the broken line on sites with SI of 20.
(b) DBH$_{dom}$ in relation to canopy closure for sites with a SI above 20.
(c) DBH$_{dom}$ development models for sites with a SI of 25 (continuous line) in comparison with Kingston (1974) (broken line).

different approaches to calculate the crown projection area, but Dawkins did not describe his method. Concerning the influence of age on the model, Spiecker (2009) found a strong influence of age on the predictive ability of the crown width model for oak ssp. Ohland (1998), however, also found that the age only slightly improved the power of his crown width model for *Gmelina arborea*.

## Bole length model

Development of a branch-free bole length is predicted to be influenced by the tree height and the stand density. We related bole height using a stepwise multi-

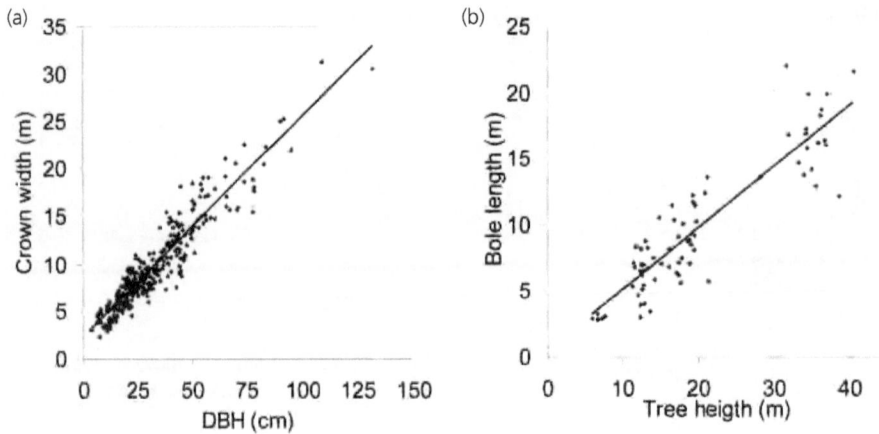

**Figure 11.4** *Tree allometric relationships for* Maesopsis eminii

*Notes:* (a) Crown width to diameter at breast height (DBH): CW = 2.18 + 0.23 × DBH, $R^2$ = 0.88.
(b) Bole length to tree height: BL= 0.498 × TH, $R^2$ =0.81.

ple linear regression model to the independent variables height, age and DBH for dominant trees. Tree height was the only significant influence on bole height of dominant trees. Therefore, bole length for stands can be modelled by the linear equation v shown in Table 11.2 and describing the trend line in Figure 11.4b. This model implicates that a dominant tree's bole length is approximately 50 per cent of the whole tree height and is nearly independent of stand density.

The data set used for developing the model was taken from Buchholz (2003). To develop the model, mean values of stands were used. To validate this model, data recorded from solitary trees in this original data set were used. These trees had a mean relative bole length of 40 per cent, indicating the distinctive self-pruning habit of *Maesopsis*. Consequently, even slight competition can lead to a mean relative bole length of 50 per cent derived from the model. The so-called rule (cited in Kingston, 1974) that bole length is independent of height when exceeding an age of 10 thus is not supported.

# OUTPUT OF THE SINGLE-TREE MANAGEMENT MODEL

## Management strategy

Based on the preliminary models developed above (Table 11.2) to correlate height and age growth with diameter, bole length and crown width development, we were able to develop a single-tree management model. However, as this is just a decision tool to pursue a given purpose, we still had to define a management strategy for which the model should be used. This model is based on the following strategic objectives in order to be in line with the aims mentioned: optimizing the single-tree growth in order to maximize the economic value of a plantation by reducing the rotation period and improving timber quality.

## Objectives of the single-tree management model for *Maesopsis*

- The model is restricted to sites classified with an site index between 20 and 25 SI.
- The rotation period is 20 years.
- Stand density is managed so as to avoid canopy cover exceeding 80 per cent; Only at the end of the rotation is canopy cover allowed to close to approximately 100 per cent.
- Establishment follows a 'twin-planting' system. For each future tree, two trees are planted 2m apart. After 1 to 2 years a sapling selection (quality and height) is executed and the weaker individual of the twins is removed. When choosing the initial square spacing, it has to be considered that the final square spacing must allow unrestricted growing space within the planned rotation period.
- Thinning allow canopy closure to be reached at an early time in stand life, which is desirable to reduce light and nutrient competition of the stand with ground vegetation. Having a high initial density and then removing parts of the stand also allows for quality selection. In this management strategy for *Maesopsis*, we recommend one thinning executed when a crown cover of 80 per cent is reached the first time – 80 per cent is considered to be the starting point of competition for light among trees, reducing single-tree growth. As *Maesopsis* is highly susceptible to heliographic stem bending caused by irregular light competition (Buchholz, 2003) we recommend a regular thinning irrespective of tree quality. This thinning removes every second diagonal row, or 50 per cent of the trees. This will lead to a final square spacing, enhancing fast growth of straight, cylindrical boles.

## Yield table

With the above-mentioned models and strategic objectives, the tree management model for *Maesopsis* was developed (Table 11.3). The assumption is a stand establishment with a density of 200 individuals per hectare, leading to a final density at time of harvest of 50 individuals per hectare. The density change is established through thinning, which in our example with an SI of 25 is applied at stand age of five years (80 per cent canopy closure).

The example presented (Table 11.3) leads to a mean annual increment (MAI) of 13.9m³/ha/yr at an age of 20 years. Reducing the rotation period to 12 years would result in a mean DBH of 50cm and a MAI of 14m³/ha/yr at the end of the rotation period. A minimum rotation period of 8 years results in

Table 11.3 *Single-tree management model for* Maesopsis eminii *for 25 SI*

| Activity | Age (yr) | Height (m) | DBH (cm) | Bole length (m) | Stand density (N/ha) | Square spacing (m) | Large log[a] (m³) | Small log[b] (m³) | Ind. wood[c] (m³) | Fuel wood[d] (m³) | MAI[e] m³/ha/ yr |
|---|---|---|---|---|---|---|---|---|---|---|---|
| Establishment | 0 | – | – | – | 200 | 10 | – | – | – | – | – |
| Thinning | 5 | 17.1 | 34 | 8.5 | 100 | 10 | 0 | 27 | 6 | 1 | 13.4 |
| – | 10 | 25 | 46 | 12.5 | 50 | 14 | – | – | – | – | 13.6 |
| – | 15 | 31.1 | 54 | 15.5 | 50 | 14 | – | – | – | – | 14.1 |
| Harvest | 20 | 34.6 | 59 | 17.3 | 50 | 14 | 184 | 0 | 50 | 9 | 13.9 |

*Notes:* a Top diameter min. 35cm; min. length 4m; branch-free bole log; 20 per cent saw logs, 80 per cent veneer logs;
b Top diameter min. 20cm; min. length 4m; branch-free bole log;
c Top diameter min. 10cm; min. length 2m; industrial wood;
d Top diameter min. 5cm;
e MAI = mean annual increment.

trees with a mean DBH of 41cm and a MAI of 17m³/ha/yr. For stands classified 20 SI, a DBH of 48cm and a MAI of 11m³/ha/yr would be reached at an age of 20 years.

The only comparable yield tables for *Maesopsis* were developed by Kingston (1974). The output of his yield table for a 25 SI at an age of 20 years was compared with the output of the single-tree management model for the corresponding site index. Coming up with a basal area close to our findings, he recommended a high final stand density (73 vs. 50 trees/ha), consequently resulting in a lower $DBH_{dom}$ than computed by our model (46 vs. 60cm).

The single-tree management model presented in this study was calculated based on a regular square spacing throughout the whole rotation period. This is based on the insight that *Maesopsis* develops bent boles under irregular light competition. However, the intensity of a bole bending reaction in relation to increasing competition (crown cover or irregularity of competition) could not yet be modelled (Buchholz, 2003). When regular spacing is not necessary to grow straight boles, we recommend to thin twice. This allows the harvest of more stems while increasing the chance to harvest more valuable products at the second thinning.

## Economic competitiveness of *Maesopsis eminii* based on a single-tree management model approach

One study objective was to prove the economic competitiveness of *Maesopsis* against *Pinus caribaea* to prove its competitiveness against exotic and well-known tropical timber species. Therefore, an economic analysis was conducted, comparing the output of the single-tree management model of *Maesopsis* introduced above with a production model for *Pinus caribaea* developed by Alder et al (2003) (Table 11.4). The pine production approach is characterized by a high stand density and two thinnings, each removing 33 per cent of standing trees.

In order to estimate the product output, taper functions were applied. In order to get comparable results, optimal site conditions for both tree species were assumed, leading to a 25 SI for *Maesopsis* and 20 SI for *Pinus caribaea* respectively. Timber prices are based on 2004 roadside prices in Uganda.

The lower volume output of the *Maesopsis* model compared to the *Pinus caribaea* model (Table 11.4, row 11) was balanced by a higher production of high-quality timber, for example for veneer (Table 11.4, row 13), thus resulting in higher revenue.

The analysis of the cost allocation produced comparable results for both cases, that is harvest activities claimed around 55 per cent of total expenditures, establishment/early tending activities around 6 per cent. Remarkable differences occurred only in the fixed expenditures like land lease and management thinning activities.

Considering financial aspects, the first and only thinning had a positive cash flow in the case of *Maesopsis* while the cash-flow of the first *Pinus caribaea* thinning was negative. The internal rate of return was slightly higher in the case of *Maesopsis* than in the case of *Pinus caribaea* (Table 11.4, row 19 – 11 per cent and 10 per cent respectively). Additionally, the investment costs for a one-hectare *Maesopsis* plantation were less than half of the investment costs required for the establishment of one hectare according to the *Pinus caribaea* plantation model (row 20). Production costs per cubic metre of timber were slightly higher in the *Maesopsis* model than in the *Pinus caribaea* model (6 per cent – Table 11.4, row 24). However, compared to the *Pinus caribaea* model, *Maesopsis* generated 12 per cent higher revenue per cubic metre of timber sold (Table 11.4, row 23) due to its high ratio of better-prized products.

The comparison was only calculated for plantations established on optimal sites, as it is difficult to predict relative growth difference of both species on poorer sites, in other words which site index should be applied. Another restriction was that taper functions were unavailable for either of the two species.

Considering financial aspects, harvest costs and product prices are difficult to validate. For both species, the same product prices and relatively high harvest costs were assumed, although the total volume production is different and influences these figures as well. Additionally, increasing timber dimensions tend to correspond to reduced harvesting costs per cubic metre of timber harvested. However, for both models the same harvest costs were assumed, although the *Maesopsis* model predicted bigger timber dimensions than the *Pinus caribaea* model for all cutting activities. Therefore, the harvest costs for the *Maesopsis* model might be overestimated.

There are additional considerations that make *Maesopsis* especially attractive for smallholders. These aspects are not considered in the economic analysis and favour the *Maesopsis* model over *Pinus caribaea*:

- The risks (fire and health hazards) and resulting cost for a *Pinus caribaea* plantation are relatively high compared to a wide-spaced *Maesopsis* plantation. Fire lines reducing the total area productivity are not necessary with the *Maesopsis* model.

- In the case of the *Maesopsis* model, the rotation period can be reduced to 12 years while still producing harvestable timber (ca. DBH 50cm). This might be an interesting option as it results in an earlier cash flow.
- A controlled *taungya* system may improve farm economics.
- The thinning itself generates a remarkable cash flow for *Maesopsis*.

With regard to environmental issues, negative impact on biodiversity, soil and water is assumed to be less in the case of *Maesopsis*.

# CONCLUSION

The study objective was to develop a single-tree management model for *Maesopsis* aiming at high-quality timber production in a short period and to prove its competitiveness compared to exotic timber species. We showed that *Maesopsis* has a high potential to produce high-quality timber in short rotation periods.

In comparison to *Pinus caribaea* as an example of a promoted exotic species in the tropics, the investigated species has lower area but higher individual tree productivity. In terms of high-quality timber and high rates of return, Maesopsis is performing better than *Pinus caribaea*, at least on good forest sites. In addition, *Maesopsis* is an excellent agroforestry tree (see also Hall, Chapter 10 of this volume).

The single-tree management model for *Maesopsis* is highly recommendable to smallholders as it is characterized by:

1 low initial investment costs;
2 the potential for a multiple land-use form (for example agroforestry); and
3 revenue in the foreseeable future with minimum rotation periods of 10 years on good sites.

For commercial forestry, the single-tree management model for *Maesopsis* is an interesting option:

1 on good sites;
2 if area productivity maximization is not an objective (land is not a limiting factor);
3 where high-quality timber is priced adequately on the market; and
4 where a high internal rate of return is desired.

If the low area productivity of *Maesopsis* is seen as a constraint in a pure plantation concept, it might be compensated with a two-storey/mixed-species model.

**Table 11.4** *Economic analysis of input and output for a one-hectare plantation of* Maesopsis eminii *vs.* Pinus caribaea, *rotation period 20 years*

| Row | Input Parameter | Unit | Maesopsis | Pinus caribaea |
|---|---|---|---|---|
| 1 | Interest rate | % | 10 | 10 |
| 2 | Land lease | €/ha | 5 | 5 |
| 3 | Management costs[a] | €/ha | 25 | 25 |
| 4 | Price per seedling | €/seedling | 0.10 | 0.10 |
| 5 | Labour establishment/early tending[b] | days/ha | 45 | 55 |
| 6 | Labour pruning | days/ha | - | 30 |
| 7 | Harvest costs thinnings[c] | €/m³ | 10 | 10 |
| 8 | Harvest costs final clear cut[c] | €/m³ | 7 | 7 |
| 9 | Harvest losses | % | 20 | 20 |
| 10 | Labour costs | €/day | 1.5 | 1.5 |
| 11 | Total volume | m³/ha | 277 | 760 |
| 12 | Mean annual increment | m³/ha/20 yr | 14 | 38 |

| **Product volumes and price assumptions[d]** | | **m³/ha** | **€/m³** | **m³/ha** | **€/m³** |
|---|---|---|---|---|---|
| 13 | Veneer | 147 | 30 | 43 | 30 |
| 14 | Large sawlog | 37 | 22 | 173 | 22 |
| 15 | Small sawlog | 27 | 20 | 341 | 20 |
| 16 | Industrial wood | 55 | 3 | 40 | 3 |
| 17 | Fuelwood | 10 | 2 | 142 | 2 |

| **Investment analysis** | | | | |
|---|---|---|---|---|
| 18 | Net present value | € | 27 | −36 |
| 19 | Internal rate of return | % | 11 | 10 |
| 20 | Investment costs (first 2 yrs) | € | 209 | 548 |
| 21 | Total revenue | € | 4184 | 11,494 |
| 22 | Total costs | € | 2413 | 7087 |
| 23 | Revenues per m³ timber | €/m³ | 24.7 | 22 |
| 24 | Costs per m³ timber[e] | €/m³ | 14.3 | 13 |

*Note:* a Including fire control, inventories, etc;
b Including site preparation, planting, weeding and selection in the sapling stage;
c Including logging trail construction, machinery and labour;
d Price including harvest (at roadside);
e Considering total volume of veneer and saw logs.

# NOTE

1   Self-pruning can be described as the dying off of lower branches of a tree crown due to restricted light in the lower part of the crown. This process can be triggered by shading of lower branches through higher positioned branches of the same tree but is in general most influenced by competition for light amongst neighbouring trees. Tree density in a stand can therefore be used to steer self-pruning dynamics to maximize branch-free bole sections which commonly reap higher market prices.

# REFERENCES

Alder, D., Drichi, P. and Elungat, E. (2003) 'Yields of eucalyptus and Caribbean pine in Uganda', consultancy report, www.bio-met.co.uk/pdf/uymdoc.pdf, accessed 11 November 2009

Anon (1953) 'Silviculture of species *Maesopsis eminii*', Uganda Forest Department, Kampala, Uganda

Bingelli, P. (1997) www.members.lycos.co.uk/WoodyPlantEcology/docs/web-sp8.htm, accessed 11 November 2009

Buchholz, T. (2003) 'Silvicultural potential of *Maesopsis eminii* – A study on tree quality', Waldbau Institut Albert-Ludwigs Universität, Freiburg, Germany, www.waldbau.uni-freiburg.de/ITOO/I-TOO-files/WP/I-TOO%20WP12.pdf, accessed 11 November 2009

Condés, S. and Sterba, H. (2008) 'Comparing an individual tree growth model for *Pinus halepensis* Mill. in the Spanish region of Murcia with yield tables gained from the same area', *European Journal of Forest Research*, vol 127, no 3, pp253–261

Dawkins, H. C. (1963a) 'Observations on crown-diameter, stocking, silvicultural requirements and possible yield of *Maesopsis*', Technical Note No. 114/63, Uganda Forest Department, Kampala, Uganda

Dawkins, H. C. (1963b) 'Crown diameters: Their relation to bole diameter in tropical forest trees', *The Commonwealth Forestry Review*, vol 42, no 2, pp318–333

Duchaufour, A. (1903) 'L'aménagement de la forêt de Compiègne', *Revue des Eaux et Forêts*, vol 42, pp65–78

Dupuy, B. and Mille, G. (1998) 'Timber plantations in the humid tropics of Africa', FAO Forestry Paper No. 98, FAO, Rome

Earl, D. E. (1968) 'Latest techniques in the treatment of natural high forest in south Mengo district', Ninth Commonwealth Forestry Conference, New Delhi

FAO (1995) 'NFAP National Forestry Action Programmes Update 32', FAO, Rome

Fenton, R., Roper, R. E. and Watt, G. R. (1977) 'Lowland tropical hardwoods – *Maesopsis eminii*', External Aid Division, Ministry of foreign Affairs, Wellington, New Zealand

FRA (2000) www.fao.org/forestry/fo/fra/main/index.jsp, accessed 11 November 2009

Francis, J. K. (1997) '*Maesopsis eminii* Engl. Musizi', USDA, www.fs.fed.us/global/iitf/Maesopsiseminii.pdf, accessed 11 November 2009

Hall, J. B. (1995) '*Maesopsis eminii* and its status in the East Usambara Mountains', Technical Paper No. 13, Ministry of Tourism, Natural Resources and Environment, Tanzania

Hein, S. (2003) 'Zur Steuerung von Astreinigung und Dickenwachstum bei Esche (*Fraxinus excelsior* L.) und Bergahorn (*Acer pseudoplatanus* L.)', dissertation, Albert-Ludwigs-Universität Freiburg, Freiburg, Germany

Hein, S. and Spiecker, H. (2008) 'Crown and tree allometry of open-grown ash (*Fraxinus excelsior* L.) and sycamore (*Acer pseudoplatanus* L.)', *Agroforestry Systems*, vol 73, pp205–218

Katende, A. B. (1995) *Useful Trees and Shrubs for Uganda*, Regional Land Management Unit, RELMA/Sida, Nairobi

Kingston, B. (1974) 'Growth and yield of *Maesopsis eminii* in Uganda', Technical Note No. 200/74, Uganda Forest Department, Kampala, Uganda

Kramer, H. and Akça, A. (1995) *Leitfaden zur Waldmeßlehre*, J. D. Sauerländer's Verlag, Frankfurt am Main, Germany

Kriek, W. (1967) 'Preliminary report on underplanting trials in tropical high forests', Technical Note No. 158/68, Uganda Forest Department

Kriek, W. (1968) 'Report on species trials on mixed wooded savanna sites on hills in Teso, Busoga and Mubende districts', Technical Note No. 146/68, Uganda Forest Department

Kriek, W. (1970) 'Report to the government of Uganda on performance of indigenous and exotic trees in species trials', FAO, Rome

Morton, K. D. (1975) 'A summary of the performance of various species in departmental trial plots', Research Report 5/11/75, Forestry Department, Solomon Islands

Nagel, J., Dobbeler, H., Albert, M. and Schmidt, M. (2003) *BwinPro Programm zur Bestandesanalyse und Prognose*, Niedersächsische Forstliche Versuchsanstalt, Göttingen, Germany

Nutto, L. (1999) 'Neue Perspektiven für die Begründung und Pflege von jungen Eichenbeständen', Schriftenreihe Freiburger Forstliche Forschung, Band 5, Freiburg, Germany

Ohland, C. (1998) 'Entscheidungshilfen zur Bewirtschaftung von *Gmelina arborea* in der Zona Norte Costa-Ricas: Ergebnisse einer Untersuchung zum Dickenwachstum, zur Kronenentwicklung und zur Astreinigung', MSc thesis, Albert-Ludwigs University, Freiburg, Germany

Schmidt, M. and Kändler, G. (2009) 'An analysis of Norway spruce stem quality in Baden-Württemberg: Results from the second German national forest inventory', *European Journal of Forest Research*, vol 128, no 5, pp515–529

Schröder, J., Röhle, H., Gerold, D. and Münder, K. (2007) 'Modeling individual-tree growth in stands under forest conversion in East Germany', *European Journal of Forest Research*, vol 126, no 3, pp459–472

SPGS (2009) 'Timber and forest/plantation situation in Uganda', Sawlog Production Grant Scheme, www.sawlog.ug/gpage3.html, accessed 11 November 2009

Spiecker, H. (2009) 'Controlling the diameter growth and the natural pruning of sessile and pedunculate oaks (*Quercus petraea (Matt.) Liebl.* and *Quercus robur L.*)', www.freidok.uni-freiburg.de/volltexte/6533/, accessed 11 November 2009

Tack, C. H. (1956) 'Uganda timbers', Forest Department Uganda, Ministry of Agriculture and Forestry, Kampala, Uganda, pp56–57

Tennigkeit, T. and Weinreich, A. (2001) 'Tree density dependent yield development of *Araucaria hunsteinii* and *A. cunninghamii* at Kifu Forest Reserve, Mukono District, Uganda', consultant report to the EC Natural Forest Management and Conservation Project, Freiburg, Germany

Uzoh, F. C. C. and Oliver, W. W. (2008) 'Individual tree diameter increment model for managed even-aged stands of ponderosa pine throughout the western United States using a multilevel linear mixed effects model', *Forest Ecology and Management*, vol 256, no 3, pp438–445

Van Laar, A. and Akça, A. (1997) *Forest Mensuration*, Cuvillier Verlag, Göttingen, Germany

Vargas-Larreta, B., Castedo-Dorado, F., Álvarez-González, J. G., Barrio-Anta, M. and Cruz-Cobos, F. (2009) 'A generalized height-diameter model with random coefficients for uneven-aged stands in El Salto, Durango (Mexico)', *Forestry*, vol 82, no 4, pp445–462

Warsopranoto, R. S., Soer Jono, R. and Ardikusuma, R. I. (1966) 'Results of an investigation of *Maesopsis eminii* plantations in the S. Bandung forest management unit', *Rimba Indonesia*, vol 11, no 1/4

# 12

# Silvicultural Management of Community Forests towards Multiple Uses in the Bale Mountains of Ethiopia

*Girma Amente, Jürgen Huss, Timm Tennigkeit and Yonas Yemshaw*

## INTRODUCTION

The livelihoods of the people living in rural Ethiopia are closely linked to the utilization of natural resources, particularly forests. Forest resources contribute directly to poor people's livelihoods by providing subsistence goods (fuel, timber, food, medicine and fodder), goods for sale to generate additional income, and other indirect social and ecological benefits. The contribution of forests to the livelihoods can only be maintained if they are managed sustainably (Angelsen and Wunder, 2003; Warner, 2000).

However, the exclusion of local communities from decisions relating to the use and management of forests has resulted in a situation whereby they utilize the forest resources illegally and feel no responsibility for the condition of the forests. As a result, the remaining natural forests and woodlands are suffering from uncontrolled access. Consequently, the forest resources are diminishing while the demand for forest products and services is increasing. The proportion of area of Ethiopia defined as forest is about 11.9 per cent (FAO, 2005), but in reality some of these areas defined as forest land have very poor forest stock.

This process has negative impacts on the contribution of the forestry sector to the national income. Moreover, the degradation and depletion of the forest resource base has a major impact on other natural resource uses and sectors in the economy, such as agriculture, and water resources, energy and biodiversity.

Protection of forests from people is the conventional mode of forest conservation. However, prohibiting access to forests accelerates degradation and encroachment of remaining forest resources. Poor, forest-dependent people see no need to control their use, as they are not sure whether they will have access to the forests in the future (Mengesha, 2004). Consequently, the degradation of the forest resources is continuing at a time when the Forest Service lacks the capacity to safeguard them. One approach to stop forest degradation is to involve forest-dependent communities in decision-making on how to manage the forest in their locality. This is now a global trend, particularly in developing countries, where governments increasingly turn to local communities to manage forests. In general, it has been difficult to conserve forest resources without the cooperation of those who depend on them for their livelihood (Arnold, 1998; FAO, 2001; Amente, 2006). Besides, the strategy of wise use is more effective for conservation than a purely defensive strategy (De Graaf, 1986).

In Ethiopia, the first attempt to implement a community forestry programme was started in the late 1970s with the objective of promoting soil and water conservation by planting hilltops and hillsides (EFAP, 1994). However, this programme was state-driven and carried out with little or no participation by local communities. There were also no clear objectives and procedures with respect to management and utilization of these forests (Poschen-Eiche, 1987). As a result, the communities viewed most of these types of community forests as state properties. Ultimately, most of these forests were destroyed during the political transition period of 1990–1991.

The second attempt to establish community-based forest management was started in the 1990s. This incorporated participatory forest management initiatives after the state-controlled management system proved to be inefficient in slowing the depletion of the forest resource. There are now a numbers of projects implementing community-based forest management approaches in the country, such as:

- the Integrated Forest Management Project Adaba-Dodola (IFMP);
- the Chilimo Participatory Forest Management Project;
- the Borena Collaborative Forest Management Project; and
- the Participatory Forest Management Project Belete-Gera.

In areas where the approaches have been implemented, forest user groups took responsibility for the conservation of forests. Forest user groups in the Bale Mountains of Ethiopia provide one important example, successfully managing part of the previously state-owned Adaba-Dodola forest priority area. The impact assessment conducted in forest areas managed by the user groups has shown an improvement in the forest condition and the livelihoods of the people (Bekele et al, 2004).

However, most of the participatory forest management approaches are initiated with support from non-governmental organizations and donors. This implies the need for right-sizing and institutionalizing the approaches so that they will be implementable within the capacity of the government institutions.

Forest areas placed under community-based management are mostly degraded and little productive. Transforming these degraded forests into productive forests that meet the needs and demands of the community requires the careful planning of silvicultural activities. Until recently, community forestry research addressed mainly social, legal and institutional aspects. Little progress has been made in the development of appropriate silvicultural techniques and practices which would enable the natural forests to better meet communities' needs for different forest products and services (Miagostovich, 2001; Ojha, 1999). Consequently, foresters lack the range of tools and skills necessary to assist the communities in developing more productive forests.

Conventional forest management practices emphasize the management of forests for the production of timber or other industrial forest products. In community forestry, communities expect a wider range of products and services from their forests, which necessitates the development of management practices reflecting these needs. To this effect, silvicultural research in community forests has great importance to enhance the productivity of those forests and improve the livelihoods of the communities. Therefore, there is a need to develop simple and practical silvicultural management techniques that can be implemented using locally available facilities and capacities.

In this chapter we aim to provide locally adapted silvicultural tools to transform existing degraded forests into managed selection forests and discuss the conditions needed to develop such tools.

# THE ADABA-DODOLA FOREST PRIORITY AREA AND STUDY METHODS

## The study area

The study was conducted in forests managed by forest-dwellers' associations, which are hereafter referred to as user groups, in the Adaba-Dodola forest priority area.

The forest area is located on the northern slopes of the Bale Mountains, ca. 320km from Addis Ababa. The forest formation reveals strong dominance of conifers up to 2850m asl, with *Juniperus excelsa* dominating other species, followed by *Podocarpus falcatus*. At the middle altitude between 2850m and 3000m asl, *Juniperus excelsa* is still dominant but now associated with broadleaf hardwood species. *Podocarpus falcatus* no longer occurs at this altitude as a canopy species, although some individuals may appear in the understorey. Between 3000m and 3400m asl, *Juniperus excelsa*, *Hagenia abyssinica*,

*Hypericum lanceolatum* and *Erica arborea* dominate the forest formation. *Erica arborea* occurs as a shrub at its uppermost distribution range.

In the early 1990s, the three timber species *Juniperus excelsa*, *Podocarpus falcatus* and *Hagenia abyssinica* accounted for 73 per cent of all trees (Trainer, 1996). Before the establishment of the community-based forest management, the forests were seriously affected by uncontrolled wood extraction and grazing. Consequently, the forests are deprived of their valuable species and good quality trees. The standing volume consists mainly of over-mature and over-sized trees, whereas the younger and medium-sized intermediate trees have already disappeared.

## The WAJIB approach

The failure of conventional forest management approaches to safeguard the Adaba-Dodola forests led to an attempt to involve the people living in and around the forests. Broad discussions have been conducted with the stakeholders concerned to try and come up with a feasible and participatory forest conservation approach as an alternative. The discussions resulted in shared objectives and strategies to conserve the forests. It was through such discussions that a new participatory forest management model called the WAJIB approach developed. WAJIB stands for forest-dwellers' associations in the local language, hereafter referred to as 'user groups'. The process of developing the new approach was facilitated by the Integrated Forest Management Project Adaba-Dodola (IFMP). The project is a technical cooperation project between the governments of Ethiopia and Germany, implemented by the German Technical Cooperation (GTZ) and the Oromia Bureau of Agriculture and Rural Development. The central concept of the new approach is to grant exclusive user rights to organized user groups (Kubsa and Tadesse, 2002). This approach empowers the forest-dwellers to get organized and systematically access the forest resource with clearly defined rights and responsibilities.

To establish user groups a forest in a given village is subdivided into forest blocks with areas of 300–500ha. During the implementation, discussions and negotiations with the adjacent communities and villages are held frequently in order to reach consensus on the issues requiring common agreement, such as border demarcation. Such processes involve important persons from the community groups, villages, forest service and district administration. Based on the forest carrying capacity of 12 hectares per household (Unkovsky, 1998), each forest block is managed by a user group of not more than 30 households (Figure 12.1).

Each user group signs a forest block allocation contract with the forest service, which gives them legal entitlement to use and manage the forest. The contract document clearly defines the rights and duties of the user groups and the Forest Service:

**Figure 12.1** *Regular meeting of a forest user group at Sokora,
Adaba-Dodola forest, Bale Mountains*

- The rights of the user groups include settlement in the block, grazing, maintaining the already existing farm plots, and use of forest products both for consumption and sale.
- The duties are to restrict further settlement and agricultural expansion, maintain the initial forest cover and pay forest rent in exchange for the use rights they have been granted.
- The Forest Service is obliged to provide organizational and technical support and carry out annual and periodic forest cover assessments and settlement censuses. Moreover, the Forest Service is expected to respect the rights of the user groups on the one hand and safeguard against free-riders on the other.

There are sanctioning mechanisms in place in case of non-compliance with the terms of the contract.

## Study methods

Information about the composition and structure of the forest are indispensable for the planning and implementation of silvicultural treatments in the user group forests. To this end, diagnostic inventories were conducted in the three selected forests to investigate the abundance and distribution of established young regrowth and mature trees. Special emphasis was given to the young regrowth of the dominant timber species. Advanced regeneration of current and potential commercial species is the most crucial target for silvicultural intervention in degraded natural forests (ITTO, 2002). This is because tending the already established regrowth is easier than inducing new natural regrowth or planting seedlings. According to Lamprecht (1989), the young regrowth of the desirable species with the potential and quality to form the final stand are referred to as 'potential crop trees' (PCTs).

## Selection of the sample user group forests

During the time in which this study was initiated there were three villages in the study area, in which 19 user groups were established. Three user group forests were selected for the purposes of this study. The district forest service and the staff of the Adaba-Dodola Integrated Forest Management Project were involved in the selection process. The selection was made taking into account the representativeness of the sample forest areas in terms of altitudinal variation, time of establishment as a user group forest, forest area per user household, livestock density in tropical livestock units (TLU) and access to local markets. Additionally, the accessibility of the sample forests to be used as demonstration centres for the testing, implementing and monitoring of the proposed silvicultural treatments was considered.

As there is strong inclination on the part of the user groups to follow their respective village lines, the sample forest areas were selected in such a way that they represented the three villages. Accordingly, the forest areas attended by the Jaldo, Gede and Changiti user groups were selected from the villages of Bura-adelle, Deneba and Berisa respectively to serve as case studies for this research.

The features of the selected forest areas are described in Table 12.1. The lower edges of the three selected forest areas are almost at the same altitude. However, the uppermost limit of the forest areas varies slightly from one to another. With respect to time of establishment as a user group forest, the three forest areas vary amongst one another by a period of one year.

**Table 12.1** *Description of the sample forest areas*

| Features | Unit | Jaldo | Forest area Changiti | Gede |
|---|---|---|---|---|
| Altitude | m asl | 2600–3200 | 2600–3300 | 2600–3400 |
| Time of establishment | year | 2002 | 2001 | 2000 |
| Size of the forest area | ha | 364 | 554 | 489 |
| Forest area per household | ha | 18 | 31 | 17 |
| Livestock density | TLU/ha | 0.4 | 0.3 | 0.6 |
| Distance from Dodola (major local market) | km | 20 | 10 | 15 |

*Source:* Schmill (2003) and IFMP database

**Table 12.2** *Local classification of tree species based on diameter at maturity*

| Group | Diameter at maturity cm | Tree species |
|---|---|---|
| 1 | ⩾40 | *Juniperus excelsa, Podocarpus falcatus, Hagenia abyssinica, Ekebergia capensis, Scheffleria abyssinica, Olea europea* ssp. *africana* and *Mytenus* species |
| 2 | ⩾25 | *Hypericum lanceolatum, Rapanea melanphloeos, Nuxia congesta, Buddleya polystachia, Erica arborea, Pittosporum viridiflorum* and others |

## Local classification of tree species

The different tree species occurring in the forests were classified based on the diameter that can be attained at maturity. As there is no adequate information about the rotation age and growth rate of the species, the classification was made based on personal observations, discussion with members of the user groups and local experts who have local knowledge of the species. This classification should be improved through long-term monitoring and research inputs. Based on the local classifications the species were categorized according to two groups:

1   **Group 1 species** attaining a diameter of ⩾40cm at maturity. This category included the major timber species used mainly for the production of lumber and construction materials (Table 12.2). From personal observation it was known that species like *Juniperus excelsa* and *Podocarpus falcatus* tend to form heavy spiral grain above this diameter. From discussions with members of the user groups it was also understood that they prefer medium-sized trees to oversized ones. This is because of difficulties in felling, processing and transporting timber of larger dimensions. The slow-growing nature of the species also discourages users from waiting for trees to attain bigger dimension.

2   **Group 2 species** attaining a diameter of ⩾25cm at maturity. Most of the species in this group are currently not used for lumber production. They are used as firewood, for charcoal, for local construction materials and in the production of farm implements. According to the members of the user groups it made no sense to wait for the species in this group to attain larger sizes as they are not used for lumber production.

## Criteria for the selection of potential crop trees

Potential crop trees were selected from young regrowth in the forests on the basis of quality and vitality. The quality of the young regrowth selected as potential crop trees should be good enough to justify further management. However, in the case of the investigated forests, which are very degraded, it is not reasonable to set high quality standards. At the initial stage, the existing healthy young trees should be accepted as they offer the only chance to renew the forests, unless good quality regrowth is established by planting. Moreover, the quality criteria can be improved in the long run. Therefore, in the context of this study the following selection criteria were used to identify the potential crop trees:

- healthy young trees without damage to the main stem;
- >2m in height for both Group 1 and 2 species, based on the assumption that they are above the reach of the animals and will escape further browsing damage;
- <40cm and <25cm diameter for Group 1 and 2 species respectively, based on the assumption that any management intervention aimed at trees above this size may not have a significant impact on improving the quality of the trees;
- no forking above breast height; if there is forking below breast height the best shoot will be selected;
- in the case of patches of regrowth, the minimum distance between the selected trees should be 4m.

## Defining potential crop tree classes

Potential crop tree classes were defined based on the abundance and distribution of potential crop trees in the forests. The reason for this is that future interventions focus on the tending and promotion of these trees to improve the production potential of the forests. The optimum number of crop trees that can be accommodated in a selection forest type taking into consideration the grazing needs of the user groups was found to be 170–210 per hectare (Amente, 2005). Based on the optimum number of crop trees, the presence of at least 10 per cent (20 potential crop trees per hectare) of the total at the initial stage was regarded as sufficient to start the transformation process. The same number of potential crop trees is of course expected in subsequent decades. For future management interventions three potential crop tree classes were defined:

1  sufficient (areas with ≥20 potential crop trees per hectare);
2  moderate (areas with 5–15 potential crop trees per hectare); and
3  areas with no potential crop trees per hectare).

The per hectare values of the potential crop trees were obtained by multiplying values per plot by five, as a sampling intensity of 20 per cent is used in the forest inventory. As a result, all per hectare values are a multiple of five. This created a gap between potential crop tree classes even though there are no omitted values.

## Inventory design

A systematic sampling technique was applied to assess the abundance and distribution of potential crop trees and mature trees in the three selected forests.

The following procedures were used to establish the sample plots. First a 100 × 100m grid was located across the entire forest area, which implies one plot per hectare. Then, sample plots were established of 2000m² (r = 25.23m) around each intersection on the grid, the equivalent of 20 per cent of a hectare. Finally, GPS (geographic positioning systems) and a handheld compass were used to locate the plot centres.

Within each of the sample plots we determined the number and species of potential crop and mature trees, the diameter at breast height for both potential crop and mature trees measured with a diameter tape, and the quality class of the mature trees of the timber species.

## Data analysis

The data collected were analysed using descriptive statistical tools. To complement the results from the descriptive analysis, the spatial distribution of the potential crop and mature trees was analysed using ArcView GIS software. In particular, the spatial distribution of potential crop tree classes and of the three timber species was analysed, according to altitude, settlement patterns and rivers. The altitudinal classes employed for this analysis followed the prevailing forest formation in the forest areas.

# PRECONDITIONS FOR SUCCESSFUL SILVICULTURE

## Existing legal and policy framework for community forestry

The success of community-based forest management depends on creating appropriate legal and policy framework conditions to secure community interest in forest management. To ensure the sustainable management of the remaining forests of Ethiopia, it is important to make sure that appropriate policies are in place that take into consideration the historical and legitimate rights of local communities. In Ethiopia's land tenure system, land and natural resources are the common property of the state and the people, and shall not be subject to sale or other means of exchange. Given this condition, one option to support community-based forest management is by granting and securing of the user rights.

Ethiopia did not have a formal forest policy till the first forest policy was enacted in 2007. Although forest proclamations existed for a long time, the absence of a formal law resulted in an absence of systematic guidance for forest conservation activities. The overall objective of the new forest policy is to meet public demand in forest products and foster the contribution of forests in enhancing the economy of the country through appropriately conserving and developing forest resources. The current forest proclamation, which was also enacted in 2007, states that sustainable utilization of the country's forest resources is possible through ensuring the participation of, and benefit-sharing by, the concerned communities as well as by harmonizing forest policies and programmes with those of other economic sectors, particularly with agriculture and rural development policy. Moreover, the proclamation recognizes two major types of forest ownership: state and private forests. The state forests are further categorized into federal and regional forests, while the private forest by definition includes a forest area developed by peasant associations or by associations organized by private individuals. According to this definition the forest-dwellers association in Adaba-Dodola forest can be classified under the private type of forest ownership.

Experience gained so far from the implementation of the WAJIB approach in Adaba-Dodola has contributed to the consideration of community forestry as a viable pathway for forest management in the country. The regional forest proclamation enacted in July 2003 in Oromia, the regional state in which the study area is located, bears witness to this fact. Article 4(3) of Forest Proclamation No. 72/2003 of the Oromia Regional State stipulates community-based forest management as a strategy for forest conservation in the region. Article 4(6) elaborates that the concerned regional executive office can conclude agreements with appropriate parties to strengthen forest protection, development and management. Article 20(3) of the land-use proclamation indicates that the delineation, demarcation, development, protection, rehabilitation and management of protected areas shall be done with the participation of the local communities. This proclamation also states that pockets of natural forest lands should be identified, demarcated, protected, managed and sustainably used by the local communities. These supportive legal framework conditions indicate the interest of the government in promoting community-based forest management in the country.

Nevertheless, the existing proclamations lack detailed implementation guidelines for most of their provisions. One aspect that needs clarity is the use of prohibited tree species in the community forests. In both federal and regional proclamations it is stated that tree species such as *Juniperus excelsa*, *Podocarpus falcatus* and *Hagenia abyssinica* are prohibited from use in state and regional forests. This means it is possible to sustainably utilize these species in private and community forests, which is also recognized in the proclamations. Despite this, misunderstandings abound among the implementing institutions and foresters regarding whether or not the ban on these species also concerns community forests. If these species are protected from use in community forests there is no incentive for the user groups to invest their labour and time towards silvicultural management of their forests. Therefore, clear and detailed implementation guidelines are much needed. This will in turn enhance scaling-up of community-based forest management initiatives.

Due to frequent restructuring and limited institutional capacity, the forest service lacks the capability to engage in sustainable forest resource management. Particularly, community-based forest management requires experts to play facilitation and advisory roles rather than controlling ones. Currently, the forest service lacks financial and material means and also expertise to work closely with the people. This gap should be bridged with training and capacity-building.

## Socio-economic situations of the user groups

Silvicultural management of community forests requires a profound understanding of the existing legal framework and socio-economic and forest conditions. Understanding the socio-economic attributes of the user groups is an important precondition for silvicultural management of the user group forests. This is because silviculture is simply a tool to achieve the objectives and expectations of the user groups within the natural capacity of the forests. Hence the socio-economic information about the user groups was reviewed to disclose the social and organizational structure, livelihood strategies, forest use patterns, management objectives and capacity of the user groups. A household-level socio-economic study conducted by Schmitt (2003) provided most of the information for this study.

The society as a whole is very traditional and male-dominated. Hence the male partners decide most of the family issues. There is a very clear division of labour in the investigated households. Normally, it is the duty of the children to care for the livestock unless they are at school. In the absence of children the wives cover this duty. Household activities such as preparing food, caring for children, fetching water and firewood collection are shouldered by the women. When it comes to forestry activities, the men are most involved in patrolling, wood processing and tending activities. In some user groups, women also participate in patrolling the forest during the daytime. Mixed farming is a widely practised strategy in the area, serving as a means to diversify livelihoods. Subsistence agriculture, livestock production and forest utilization are the three main livelihood sources of the user groups (Schmitt, 2003). These three strategies are interlinked and depend on one another.

The user groups depend entirely on their forests for any wood-related needs. Pit-sawing is the traditional method of wood processing widely used by forest-dwellers. Using this method the trees are felled, debranched, cross-cut and converted into different products inside the forest to be transported out of the forest by people or animals. Axes and two-man saws are the most commonly used tools for processing. The recovery rate employing this method is reported to be 30 per cent (Demesa, 2002). An average household needs about six cubic metres of wood for consumption annually (Schmitt, 2003). The most favoured species for the aforementioned uses are *Juniperus excelsa*, *Podocarpus falcatus* and *Hagenia abyssinica* (Regassa, 2003; Schmitt, 2003; Shiferaw, 2003). *Juniperus excelsa* is particularly valued by the user groups for all sorts of construction wood, furniture and firewood.

Grazing in the forest is the only means for the user groups to maintain their livestock. They consider the forests as a source of fodder and shade for their livestock. Studies conducted in the area reported that the two indigenous conifer species, *Juniperus excelsa* and *Podocarpus falcatus*, are less sensitive to browsing (Tesfaye et al, 2002; Regassa, 2003). The studies also revealed that grazing reduced mainly the young regrowth of broadleaf species. *Hagenia abyssinica*, in particular, is very palatable to all types of livestock. The implication of this is that it is necessary to regulate grazing in the broadleaf-dominated parts of the forests. There is strong interest and commitment from the user groups to establish grazing systems that allow the forests to regenerate. Some user groups have already started to set aside some portion of their forests for natural regeneration by barring their animals from entering such areas. Therefore, there is a great demand for a management system that supports the integration of the two production systems.

According to a study conducted by Shiferaw (2003), the forests provide the user groups with a wide range of non-wood forest products. In addition to providing wood products, grazing and ecotourism service, they also provide medicines, food (edible fruits, leaves, roots and mushrooms), detergents, honey, bamboo, spices and dyes.

Forest management requires a long-term commitment by the user groups to undertake forest improvement work and to regulate use. To transform the existing degraded forests into productive forests that generate tangible benefits on a sustainable basis, the user groups have to commit different resources to forestry activities. These resources include human resources, tools and skills, as well as financial resources. All these resources should be put together and used properly to bring about improvement in both forest condition and livelihoods.

The user groups possess only a few traditional tools for forestry activities and the processing of forest products. Moreover, they have few skills with respect to tending activities such as pruning and thinning, and low-impact harvesting techniques are not well practised by the user groups. However, considerable numbers of forest-dwellers have the traditional skills of wood processing using the pit-sawing technique.

## THE CURRENT STATUS OF THE FORESTS

### Species distribution of potential crop trees

The two indigenous conifers, *Juniperus excelsa* and *Podocarpus falcatus*, account for the great majority of the PCTs found at the lower altitude. In addition to its occurrence at lower altitudes, *Juniperus excelsa* is also dominant in middle and upper altitudes, which makes it the most abundant species in the forests. The two species, particularly *Podocarpus falcatus*, are reported to be less susceptible to browsing, unless there is shortage of fodder on the forest floor (Tadesse, 1999; Tesfaye et al, 2002). Consequently, they can be regarded as key species for the rehabilitation of the forests in the context of the user group forests where grazing cannot be avoided.

Unlike the two conifers, the PCTs of *Hagenia abyssinica* were found to be rare in the forests (Figure 12.2). This is consistent with the results of previous studies (Holweg, 1998; Tadesse, 1999; Regassa, 2003). The same authors also described *Hagenia* as the preferred tree species browsed by all types of livestock. The areas where the PCTs of *Hagenia* occurred were mostly charac-

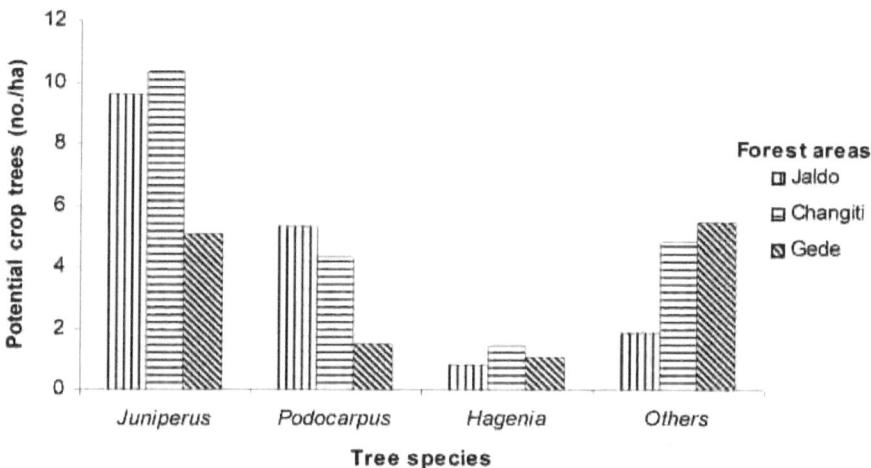

**Figure 12.2** *Species distribution of potential crop trees in the forests of the Jaldo, Changiti and Gede forest user groups in the Adaba-Dodola forest, Bale Mountains*

terized by steep riverbanks and inaccessible and dense shrub-covered areas. Its susceptibility to browsing implies the need for protection and care for the young regrowth until it outgrows the browsing zone.

With respect to the distribution of mature trees, *Juniperus excelsa* contributed 60–85 per cent of the total number of mature trees, equivalent to 70–80 per cent of total basal area (Table 12.3). This is consistent with a study conducted by Regassa (2003). There, *Juniperus* accounted for 70 per cent of total basal area. *Podocarpus falcatus* and *Hagenia abyssinica* made up only a small amount of the mature standing stock. A high demand for the two species contributed to their overextraction. The great majority of the available mature trees are oversized and of poor quality and not suitable for lumber production, implying that the best quality trees had already been extracted. This finding is supported by a stump analysis conducted by Regassa (2003) – 80 per cent of the stumps recorded were within the diameter range of 10–50cm. This could be due, on the one hand, to the better quality of trees in the indicated sizes. On the other hand, the ease of felling, processing and transportation of these inter-mediate dimension trees using locally available tools might have caused their overexploitation. This implies the need for capacity-building in relation to the development of efficient tools and techniques in order to encourage the user groups to utilize the over-mature trees, which will in turn increase the growing space available to the PCTs.

Table 12.3 *Basal area and proportion of tree species*

| | Unit | Jaldo | Forest area Changiti | Gede |
|---|---|---|---|---|
| Basal area | m²/ha | 22 | 29 | 22 |
| *Juniperus* | % | 81 | 82 | 71 |
| *Podocarpus* | % | 3 | 5 | 1 |
| *Hagenia* | % | 16 | 13 | 25 |

## Size distribution of potential crop trees

When all of the recorded species are combined, the diameter distribution of the potential crop trees showed an irregular pattern. It showed a reversed-J distribution only in Changiti forest (Figure 12.3). In this forest area the majority of the potential crop trees were in the lowest diameter class, whereas the intermediate classes were not well represented. In the remaining two forests, on the contrary, there was a good representation of potential crop trees in the intermediate classes. However, the difference between the size classes is quite small. The distribution of potential crop trees is in agreement with studies conducted in Dindin forest (Shibru and Balcha, 2004). Some size classes are over-represented, while others are missing. On the one hand, this can be attributed to the selective cutting of the intermediate size classes during past illegal logging. On the other hand, grazing might have caused the low proportion of the smaller size classes. The localized patchy type of grazing causes more damage in some areas, while other areas are less utilized. However, grazing may not be the only factor that affects the establishment and growth of young regrowth in the forests. Tadesse (1999) found no difference in the abundance of small-sized individuals under closed forest conditions between grazed and

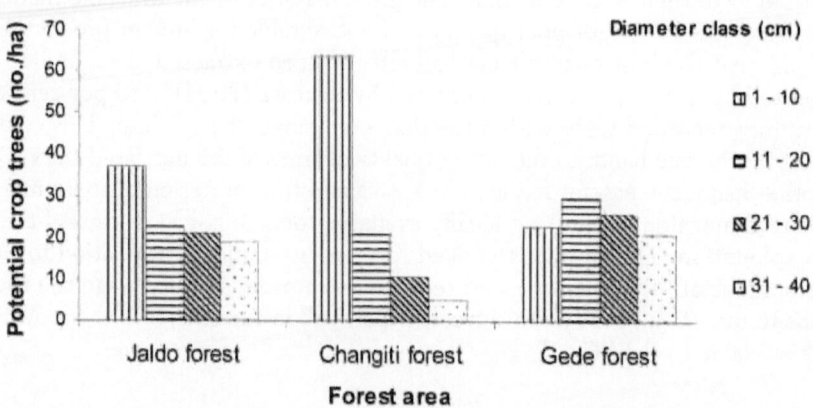

Figure 12.3 *Diameter distribution of the potential crop trees in the forests of the Jaldo, Changiti and Gede forest user groups in the Adaba-Dodola forest, Bale Mountains*

non-grazed forests. *Juniperus* is generally known for its low establishment potential under the closed canopy of mature parent trees (Friis, 1992; Teketay and Bekele, 1995). Bussmann (2002) also reported the sporadic occurrence of *Juniperus* regrowth, a light-demanding species, under a dense vegetation cover. Active opening of the canopy is necessary to support the growth of *Juniperus*. Regassa (2003) recorded higher numbers of regrowth in intensively logged-over forest areas with open canopies compared to forests under user group management, with regulated use.

# SILVICULTURAL MANAGEMENT TOWARDS MULTIPLE USE

Forest management in community forestry usually aims to provide a wider range of products and services compared to the conventional timber-oriented management. In the study area, the following management objectives have been defined based on the current use patterns and future expectations of the user groups as well as the conservation interest of the forest service:

- timber and livestock production;
- watershed protection;
- ecotourism; and
- utilization of non-wood forest products.

## Selection of an appropriate silvicultural system

The purpose of silviculture is to create and maintain the kind of forests that will best fulfil the objectives of the owner/user within the capacity of local site and stand conditions (Lamprecht, 1989; Smith et al, 1997; Nyland, 2002). Hence the selection of a silvicultural system for the user group forests was made based on the forest management objectives and the current status of the forests. The silvicultural system regarded as appropriate for the transformation of the degraded user group forests into managed selection forests is a single-tree selection approach. This is because the single-tree selection system forms uneven-aged stands and involves the selective removal of individual trees at short intervals (Nyland, 2002), in other words harvesting causes few changes to the structure and composition of the forests. The continuous forest cover and the array of age classes in the resulting selection forests makes it ideal for managing forests with multiple use objectives, where protection against soil erosion, aesthetic values and riparian zones are also important (Smith et al, 1997; Nyland, 2002). Therefore, it enables achieving the multiple management objectives of the user groups.

# Do existing potential crop trees suffice to transform the forests?

The establishment of natural regrowth in the user group forests is considered to be difficult due to grazing pressure. However, there is young regrowth >2m in height which escaped the danger of browsing. To determine the number of crop trees that can be maintained under the management objectives set by the user groups, a study on crown development of the dominant species was conducted. The results showed that about 200 PCTs/ha of different age and species can be maintained considering the grazing needs of the user groups (Amente, 2005). It is assumed that this number of trees per hectare will ensure the maintenance of a relatively open forest canopy with adequate pasture on the forest floor. However, the relatively open forest condition arising from integration of grazing will enable crop trees to develop larger and deeper crowns, as the lower branches stay alive longer. In such a situation pruning becomes a compulsory activity to maintain the growing space and the wood quality of the potential crop trees.

Assuming a transformation period of 100 years, the recruitment of 20 PCTs/ha and decade was deemed to be sufficient in the context of the studied forests to achieve the final target of 200 PCTs/ha at the end of the transformation period. Currently, 30–45 per cent of the total areas of the studied forests have at least 20 PCTs/ha, while a considerable proportion of the area hosts a moderate number of PCTs (Amente et al, 2006). Therefore, while tending areas with sufficient PCTs it is possible to regenerate areas with poor and moderate regrowth.

## Improvement treatments

The investigated forests consist mainly of poor quality trees that are not suitable for lumber production; most of the better quality trees have been extracted illegally in the past. This calls for the implementation of improvement treatments to enhance the production potential of the forests, in terms of both quantity and quality of production. The term improvement covers all operations carried out in growing stands intended to improve future yields (Lamprecht, 1989). It includes liberation cutting, selective thinning, pruning and climber cutting, among others. In uneven-aged systems, most of these treatments are conducted simultaneously in a single operation.

The main goal of the improvement treatments in the user group forests is to promote the growth of PCTs by removing competitors. According to Nyland (2002), in forests that were previously not subject to planned management, treatments shouldn't necessarily remove all less-than-perfect trees in a single intervention. Rather he advised a period of adjustment to progressively upgrade stand conditions and gradually develop balanced structures. In the context of the user group forests the first nine decades may be interpreted as a period of adjustment.

Improvement treatments also produce pre-harvest products through the removal of potential crop tree competitors. This may include the cutting of old trees overtopping the PCTs and pole-sized trees in even-aged groups. In the case of the user group forests, improvement treatments contribute to the great majority of the harvest cuttings until a time when the first generation of the PCTs reach harvestable dimensions. This is because most of the existing PCTs are under competition from the mature and over-mature trees.

## Determining sustainable harvest

To date, forest utilization in natural forests is mostly conducted without planning or pre-harvest inventories (Mengistu, 2003; Amente, 2005). The valuable species of good quality are removed through selective cuttings. Only a small number of species like *Juniperus excelsa*, *Podocarpus falcatus* and *Hagenia abyssinica* are in high demand. The potential uses of most other species have not yet been investigated, the outcome being the depletion of these few species. Improvement operations after logging are not even considered, resulting in forests stocked with poor quality over-aged trees and very scarce natural regrowth. After logging, in most cases, there is the chance that forests are converted to agricultural and grazing lands.

From discussions held with members of the user groups, it is understood that in most cases they are utilizing trees that fell or were left behind on the forest floor during illegal logging operations. They are very cautious when it comes to using living trees for fear of causing a reduction of the forest cover, which results in the cancellation of the agreement made with the forest service. Neither the user groups nor the experts of the forest service know how much can be extracted from the forest on a sustainable basis without significantly reducing the forest cover. Consequently, user groups under-utilize their forests. In the near future, when the dead trees and leftovers from the past illegal logging have been exhausted, there will be a possibility of over-utilization. To address this problem, a simple allowable cut model was developed. The allowable cut model is based on the concept of gradual harvesting of the existing mature trees while at the same time promoting the continuous recruitment and growth of the PCTs (Figure 12.4). This figure can only apply to the user group forests in the study area. To identify mature and potential crop trees a local classification of tree species based on diameter at maturity was used (Table 12.2)

The determination of the allowable cut was based on the recruitment rate of PCTs and the amount of mature trees available in the forest. Based on a 100-year planning period, it is assumed that the first generation of PCTs reach harvesting DBH after 90 years. Until then, the existing mature and over-mature trees should be utilized gradually. One option to do this is to evenly distribute the existing mature trees over the nine decades. Accordingly, the currently available mature trees in a given forest area are divided by nine decades to determine the possible allowable cut for one decade.

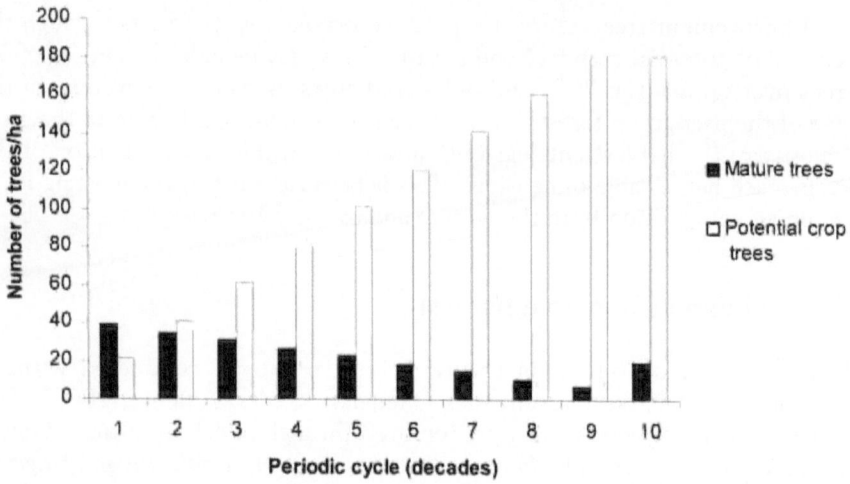

**Figure 12.4** *Predicted trends in the number of mature and potential crop trees after decades following the allowable cut harvesting model and stimulating the growth of potential crop trees in the studied forests of Adaba-Dodola, Bale Mountains*

PCT-focused management is the main principle in the management of the user group forests. Therefore the utilization of the existing mature trees is only possible if there is recruitment of PCTs on the ground to ensure the continuity of the forests. This implies that user groups will be allowed to harvest allowable cuts set for the decade only if the required number of PCTs have been recruited in the preceding decade. To attain the optimum number of PCTs set for the planning period (about 200 PCTs/ha), a minimum recruitment rate of 10 per cent is expected every decade. However, the user groups cannot wait ten years between periodic harvests, as they depend on the forests for their subsistence needs. This calls for the conversion of the periodic allowable cuts into annual cutting levels. Accordingly, an annual allowable cut can be derived from the periodic allowable cut by dividing the periodic allowable cut per timber utilization area by the number of years in the period. This gives the number of mature trees that can be harvested per year in the timber utilization area. The annual allowable cut was then determined in terms of the species distribution of mature trees in the forests.

## Integrating grazing into forest management

Grazing is still practised in traditional forest management systems in many parts of the world. In mountain environments, where agricultural activities are constrained by climate, animal husbandry it is one of the livelihood options available to many farmers. In the Mediterranean region of North Africa, forest grazing is still a popular tradition (Karmouni, 1997). Forest grazing is also widely practised in Bhutan and Himalayan coniferous forests (Norbu, 2000; Roder et al, 2002). According to Mayer (2003), the multipurpose utilization of forest resources including forest grazing has a long tradition in the Swiss Alps. It is also a very common practice in the montane forests of Ethiopia (Teketay and Bekele, 1995).

Opinions among foresters, livestock experts, ecologists and forest user groups on grazing as a forest function are still divided. Nevertheless, the issue of forest grazing cannot be discussed from a solely ecological or legal point of view. According to Wangchuk (2002), the role of livestock in the socio-economic structure of the rural households has to be examined before making any decision. Livestock production is one of the main livelihood sources for the user groups living in the studied forests. The forests are the main source of fodder for their livestock. Consequently, it is difficult to exclude grazing from the forest management system without disrupting livelihoods.

Many foresters believe that cattle grazing and timber production are not compatible. The main argument raised against forest grazing is that animals damage regrowth by trampling and browsing. However, the magnitude of damage varies with tree species and type of animals. *Juniperus excelsa* and *Podocarpus falcatus* are reported to be not very sensitive to browsing, except in situations where there is a shortage of fodder (Tadesse, 1999; Tesfaye et al, 2002). *Hagenia abyssinica*, on the other hand, is the species most preferred by all types of livestock. Studies conducted in the mountain forests of the Swiss Alps also indicated that conifers are less sensitive to browsing than deciduous trees (Mayer et al, 2003). Studies conducted in mixed coniferous forests in central Bhutan concluded that browsing damage due to grazing was negligible for conifer species (Roder et al, 2002). Therefore, grazing can be integrated into forest management at the lower and middle altitudes of the forests where conifers are the dominant species, provided that the stocking rates do not exceed the sustainable limits set for the forest area. At the upper altitude, where *Hagenia* is dominant, portions of the forest areas should be strictly protected from grazing until the young plants are above the reach of the animals.

Meanwhile, forest grazing can also be beneficial provided it is well regulated. It may diversify income-generating sources and decrease dependency on wood extraction, reduce fire hazard by minimizing the ground vegetation, promote seed dispersal, and enhance seed germination by disturbing the organic layer and uncovering the mineral soil.

To ensure the continuous recruitment of PCTs, crucial for the continuity of the forests, the type and intensity of grazing should be regulated by feasible measures. The following are suggested:

- **Monitoring stocking rates.** There is a fear that the livestock population may increase in the future, as the user groups regard livestock as a symbol of prestige. Therefore, the stocking rates should be monitored in the frame-work of the periodic forest inventories. Moreover, the user groups should be encouraged to improve their livestock husbandry system by focusing on the quality of production rather than the number of animals.

- **Rotational grazing.** Although the current stocking rates are not beyond the sustainable limit, the prevailing patchy type of grazing exerts high pressure in some areas while others are not efficiently utilized. Livestock more frequently graze areas close to the settlements and watering points than the remaining parts. Consequently, these areas are overgrazed and devoid of regrowth. To distribute grazing pressure over the entire forest area and efficiently utilize the available pasture in the forests, the implementation of a rotational grazing regime is advisable. This can be based on existing seasonal shifting of animals between different pasture types. Open pasture areas, forest and farmlands are some of the areas upon which traditional grazing regimes are exercised. In order to enhance the recruitment of PCTs, 10 per cent of the forest area should be regenerated every decade. Such areas should remain exempt from grazing until the young regrowth is above the browsing zone. This will increase the chances for establishment of susceptible tree species such as *Hagenia*. Rotational grazing can be practised on the remaining 90 per cent of the forest area. In areas where grazing is permitted, young plants <2m height can be fenced individually or in groups using thorny shrubs, branches of trees or other materials available in the forest in order to prevent damage.
- **Restricting grazing rights for certain livestock types.** The damage caused by grazing varies according to livestock type. Goats are regarded as most dangerous in relation to browsing of young plants; most user groups in the study area already banned goats from the regeneration areas. In countries like Switzerland grazing rights apply only to cattle, which ultimately discouraged farmers from keeping goats (Mayer, 2003). User groups should be encouraged to implement similar regulations. The effect of wild ungulates in browsing young plants should not be ignored either. Mayer et al (2003) reported that the browsing effect of wild ungulates is three times higher than that of cattle. As witnessed by the members of the user groups, the population of wild animals has been increasing since they took over the management of the forests because of reduction in poaching. This in turn increases the threat of browsing. The introduction of controlled hunting will minimize this impact while also generating additional income for the user groups.

## Watershed protection

As described in the management objectives, timber production is not the only objective of managing the user group forests. Maintaining other functions of forests may mean setting aside of some areas for specified purposes. In the study area, there is a tradition of preserving trees along riverbanks. Even during past illegal logging, local communities did not fell trees around water points and riverbanks. The reason for this was their understanding of the protection value of these strips, quite apart from serving as shade for their livestock during watering. With the exception of the watering points, most of the riparian zones lie in deep valleys and in inaccessible areas due to which they are also important wildlife habitats. Moreover, these areas are economically important as they provide water for a hydroelectric dam constructed further downstream. Rivers also provide fish to local communities. Therefore, the protection of riparian zones from logging helps to maintain the functions indicated.

Activities to be permitted or prohibited in riparian zones vary from place to place based on the management objectives. In the case of user group forests where these areas are exempt from logging, it is deemed adequate that their functions be maintained. In return, the forest service has to exempt the user groups from the payment of rent for those areas. This will encourage the user groups to continue protecting these areas that benefit not only the user groups but also the communities downstream. In addition, institutions such as the Ethiopian electric light and power authority, which is benefiting from the protection of the watershed at no cost to itself, should support the user groups in their efforts at sustainably managing the forests. Payments for such off-site benefits from the forests enjoyed at local, national, regional and global levels contribute greatly to the improvement of the livelihood of the user groups (Angelsen and Wunder, 2003).

# IMPLEMENTATION CAPACITY

Successful implementation of silvicultural treatments in selection forests requires trained personnel. The user groups and the local experts are very inter- ested in engaging in the management of forests. However, they lack practical experience with respect to proposed silvicultural treatments. Studies indicated that labour and time are not limiting factors in the management of user group forests (Schmitt, 2003). Therefore, on-site training of the user group members and the local experts is important.

Improving forest management techniques alone does not guarantee increased benefits. It also requires efficient processing techniques. Existing on- site conversion of logs through pit-sawing and transportation of products manually and using animals is well suited to harvesting of individual trees scattered throughout the forests. However, overall efficiency is very low. According to Demesa (2002), the recovery rate of traditional pit-sawing is only 30 per cent.

Most available tools are of low quality and poorly maintained, which hinders user groups and prevents them from enjoying greater benefits from their management efforts. This implies the need for training in relation to tool handling and maintenance and highlights the necessity to upgrade processing capacities of user groups.

# CONCLUSION

The study revealed that user group forests have potential to be transformed into managed selection forests through a PCT approach. To maintain and harmonize the manifold functions of the forest, a multiple-use forest management system is proposed. This can be achieved provided that certain uses are regulated.

Successful grazing management is essential to sustainably manage the forests. When regulated, forest grazing can serve as a management tool in addition to its contribution to livelihoods. If proposed measures are implemented, grazing in user group forests can be regulated.

Currently, the wood consumption rate of the user groups is lower than the sustainable allowable cut. At the same time, the current productivity of the forests is significantly lower than their production potential. To make the conservation efforts of the user groups even more beneficial and to accommodate the growing needs of the user groups it is necessary to improve the productivity of the forests. To achieve this goal, the introduction of a PCT-focused silvicultural management is suggested. The successful implementation of the proposed treatments requires practical training both for members of the user groups and for forest service staff.

Finally, monitoring of silvicultural interventions directed at readjustment of management practices where necessary is highly needed.

# REFERENCES

Amente, G. (2005) 'Rehabilitation and sustainable use of degraded community forests in the Bale Mountains of Ethiopia study on tree quality', PhD thesis, Freiburg University, Freiburg, Germany

Amente, G. (2006) 'Integrated and participatory forest management in the Bale Mountains of Ethiopia', in *Proceedings of International Conference on Sustainable Livelihoods and Ecosystems in Mountain Regions*, 6–9 March, Chiang Mai, Thailand

Amente, G., Huss, J. and Tennigkeit, T. (2006) 'Forest regeneration without planting: The case of community managed forests in the Bale Mountains of Ethiopia', *Journal of the Dry lands*, vol 1, no 1, pp26–34

Angelson, A. and Wunder, S. (2003) 'Exploring the forest-poverty link: Key concepts, issues and research implications', CIFOR Occasional Paper, No. 40, Bogor, Indonesia

Arnold, A. E. M. (1998) 'Managing forests as common properties', FAO Forestry Paper, No. 136, Rome

Bekele, T., Senbeta, F. and Ameha, A. (2004) 'Impact of participatory forest management practices in Adaba-Dodola forest priority area of Oromia, Ethiopia', *Ethiopian Journal of Natural Resources*, vol 6, no 1, pp89–109

Bussmann, R. W. (2002) 'Vegetation ecology and regeneration of tropical mountain forests', in R. S. Ambasht and N. K. Ambasht (eds) *Modern Trends in Applied Terrestrial Ecology*, Kluwer, Dordrecht, The Netherlands, pp195–223

De Graaf, N. R. (1986) *A Silvicultural System for Natural Regeneration of Tropical Rain Forest in Surname*, Agricultural University, Wageningen, The Netherlands

Demesa, G. (2002) 'Forest utilization and comparison between traditional lumber conversion methods and a portable sawmill in Arero', MSc thesis, Swedish University of Agricultural Sciences and Wondo Genet College of Forestry, Wondo Genet, Ethiopia

EFAP (1994) Ethiopian Forestry Action Program, Addis Ababa

FAO (2001) 'How forests can reduce poverty', FAO, Rome

FAO (2005) 'Global forest resources assessment 2000: Main report', FAO, Rome

Friis, I. (1992) 'Forests and forest trees of northeast tropical Africa', *Kew Bulletin Additional Series*, vol XV, HMSO, London

Holweg, C. (1998) 'Impact of silvipasturalism on natural regeneration of main tree species in Afromontane forests in the Bale mountains, Ethiopia', MSc thesis, Albert-Ludwigs-University of Freiburg, Institute of Silviculture, Freiburg, Germany

ITTO (2002) 'ITTO guidelines for the restoration, management and rehabilitation of degraded and secondary tropical forests', ITTO Policy Development Series No. 13, ITTO, Yokohama, Japan

Karmouni, A. (1997) 'Forest grazing: Case study of Maghreb countries', XI World Forestry Congress, Antalya, Turkey, Session 17, Volume 7, FAO, Rome

Kubsa, A. and Tadesse, T. (2002) 'Granting exclusive user rights to the forest dwellers in the state owned forest: The WAJIB approach in Ethiopia', *Proceedings of Second International Workshop on Participatory Forestry in Africa*, Arusha, Tanzania, 18–22 February 2002, FAO, Rome

Lamprecht, H. (1989) *Silviculture in the Tropics. Tropical Forest Ecosystems and their Tree Species – Possibilities and Methods for their Long-Term Utilization*, GTZ, Eschborn, Germany

Mayer, A. C. (2003) 'Range management on wood pastures of an alpine valley', *Australian Journal of Forest Science*, vol 120, no 1, pp19–28

Mayer, A. C., Stockli, V., Huovinen, C., Konold, W., Estermann, B. L. and Kreuzer, M. (2003) 'Herbage selection by cattle on subalpine wood pastures', *Forest Ecology and Management*, vol 181, pp39–50

Mengesha, B. (2004) 'Community-based environmental protection and natural resources conservation: An alternative to the conservation of *Acacia* woodlands of the Rift Valley of Ethiopia', *Kosso Newsletter*, Addis Ababa

Mengistu, K. (2003) 'Ethiopia country paper', in *Proceedings of Workshop on Tropical Secondary Forest Management in Africa: Reality and Perspectives*, 9–13 December, Nairobi, Kenya, FAO, Rome

Miagostovich, M. (2001) 'Building forest users' capacity to develop silvicultural practices', in *International Workshop on Participatory Technology Development and Local Knowledge for Sustainable Land Use in Southeast Asia*, Regional Community Forestry Training Centre for Asia and the Pacific (RECOFTC), Bangkok

Norbu, L. (2000) 'Cattle grazing – An integral part of broadleaf forest management planning in Bhutan', Doctoral dissertation, Swiss Federal Institute of Technology, Zurich, Switzerland

Nyland, R. D. (2002) *Silviculture: Concepts and Applications* (second edition), McGraw-Hill, London, UK

Ojia, H. (1999) 'Silviculture in community forestry: Conceptual and practical issues emerging from the middle hills of Nepal', NepalNet (Electronic Networking for Sustainable Development in Nepal)

Poschen-Eiche, P. (1987) *The Application of Farming Systems Research to Community Forestry. A Case Study in the Hararge Highlands, Eastern Ethiopia*, Tropical Agriculture Series, vol 1, TRIPOS Verlag, Langen

Regassa, A. (2003) 'Impact of WAJIB forest management on structure and regeneration of indigenous tree species – The case of Adaba-Dodola regional forest priority area', MSc thesis, Swedish University of Agricultural Sciences and Wondo Genet College of Forestry, Wondo Genet, Ethiopia

Regional State of Oromia (2002) 'Oromia Rural Land Use and Administration Proclamation, No. 56/2002', Adama, Ethiopia

Regional State of Oromia (2003) 'Regional Forest Proclamation No. 72/2003', Adama, Ethiopia

Roder, W., Gratzer, G. and Wangdi, K. (2002) 'Cattle grazing in the conifer forests of Bhutan', *Mountain Research and Development*, vol 22, no 4, pp368–374

Schmitt, J. (2003) 'The contribution of community forest management to the rural household economy: A case study from Bale mountains, Ethiopia', I-TOO working paper No. 11, Freiburg, Germany

Shibru, S. and Balcha, G. (2004) 'Composition, structure and regeneration status of woody species in Dindin natural forest, Southeast Ethiopia: An implication for conservation', *Ethiopian Journal of Biological sciences*, vol 3, no 1, pp15–35

Shiferaw, E. (2003) 'Identifying income generating potentials with respect to the forest resources of the Adaba-Dodola regional forest priority area: With special attention to non-timber forest products, the case of "forest blocks" under WAJIB ownership', BSc thesis, Wondo Genet College of Forestry, Wondo Genet, Ethiopia

Smith, D. M., Larson, B. C., Kelty, M. J. and Ashton, P. M. S. (1997) *The Practice of Silviculture: Applied Forest Ecology* (ninth edition), John Wiley and Sons, Inc., New York

Tadesse, T. (1999) 'Effects of grazing and fire on tree regeneration in coniferous montane forests of the Dodola area, Ethiopia', MSc thesis, Georg-August University of Gottingen, Germany

Teketay, D. and Bekele, T. (1995) 'Floristic composition of Wof-Washa natural forest, Central Ethiopia: Implications for the conservation of biodiversity', *Feddes Repertorium*, vol 106, pp1–2, 127–147

Tesfaye, G., Teketay, D. and Fetene, M. (2002) 'Regeneration of fourteen tree species in Harenna forest, southern Ethiopia', *Flora*, vol 197, pp461–474

Trainer, J. (1996) *Forest Inventory of Indigenous Forest of Pilot Area Peasant Associations: Deneba, Berisa, Burachele-Adele*, IFMP, Dodola, Ethiopia

Unkovsky, S. (1998) *Model Calculation of Minimum Forest Area per Household*, IFMP, Dodola, Ethiopia

Wangchuk, S. (2002) 'Grazing management in national parks and protected areas: Science, socio-economics and legislation (tenure)', *The Journal of Bhutan Studies*, vol 7, pp61–81

Warner, K. (2000) 'Forestry and sustainable livelihoods', *Unasylva*, vol 51, no 02, pp3–12

# 13

# *Juniperus procera* (Cupressaceae) in Afromontane Forests in Ethiopia: From Tree Growth and Population Dynamics to Sustainable Forest Use

*Frank J. Sterck, Camille Couralet, Grace Nangendo, Alemayehu Wassie, Yishak Sahle, Ute Sass-Klaassen, Lars Markesteijn, Tesfaye Bekele and Frans Bongers*

## INTRODUCTION

*Juniperus procera* Hochst. Ex Endl. forests once covered some 200,000 hectares in the highlands of Ethiopia (Teketay and Bekele, 2002). These forests formed the largest juniper forests on the African continent (Kelecha, 1979). In 1955, juniper forests constituted 3 per cent of the total forest area of Ethiopia (Jansen, 1981; FAO, 1986). Juniper forests are nowadays very much reduced in area, fragmented or very much degraded (Javeed et al, 1980; Hall, 1984; FAO, 1986; Borghesio et al, 2004), due to a combination of intensive deforestation, timber harvests and grazing inhibiting successful regeneration (Teketay and Bekele, 2002; Burgess et al, 2004; Couralet et al, 2005). In this study we compare juniper populations in remaining semi-natural forests, secondary forests and church forests. We use this comparison for sketching the key problems to restore and maintain *Juniperus* populations in these different forest types and consider possible consequences of climate change such as increasing temperature (Bonnefille and Umer, 1994) and increasing frequency of drought events (Verschuren et al, 2000).

*Juniperus procera* is the only indigenous member of the total of 60 *Juniperus* species in Africa, where it mainly occurs in east African tropical highlands (Negussie, 1995). The species is commercially known as African pencil cedar because its wood characteristics closely resemble that of *J. virginiana*, the Eastern red cedar grown in North America and used in pencil-making (Noad and Birnie, 1989). In Ethiopia juniper can reach heights of 40m and diameters of >200cm. It produces valuable durable timber with a wood density of 450–650kg/m$^3$ (Eggeling and Dale, 1951; Pohjonen and Pukkala, 1992). Seeds are mainly dispersed by birds (Young and Young, 1992). In its original distribution area, it is considered a multipurpose species as it provides a lot of services to the people. In this chapter we give a brief overview of those purposes to sketch the potential importance of such a species for local people.

## *JUNIPERUS PROCERA* – A MULTIPURPOSE SPECIES

Juniper trees are known to protect watersheds, nutrient-deficient soils, and dry and exposed sites (Dallimore and Jackson, 1966). Juniper wood seasons well and is used for applications that do not need nailing. Though over-mature trees often show core rot (Gardner, 1926; Chalk et al, 1932), seasoned heartwood is very stable and used for indoor and outdoor constructions (doors, windows, furniture and poles) and also for fuel (Chaffey, 1982). Moreover, the wood contains cedar oil that is used in microscopy, soap, perfumes and medicines and for abortions (Jansen, 1981). Bark strips are used for roofing huts, covering beehives and as pads to carry water jars (Negussie, 1995). Berries have been used in earlier times as an embalming agent (Vivi and Drar, 1941). These berries contain oil that is also beneficial in pharmaceutical preparations and for flavouring alcoholic drinks and other food products (Zaman et al, 1968). It is believed to have relaxing and revitalizing properties.

Leaves and shoots produce oil that is used in medicine (Dallimore and Jackson, 1966), and are also used as incense in India, Iran, Afghanistan and Baluchistan (Dallimore and Jackson, 1966). Chopped and finely ground leaves, mixed with water, are used as a drench for horses and mules suffering from stomach disorders. Dry powdered leaf parts are used to cure wounds. A cold aqueous extract of the leaf is active against *Mycobacterium* tuberculosis. The podophyllotoxin (antibiotic) present in juniper leaves is said to be active against tumours (Jansen, 1981). The smoke of fruiting branches relieves rheumatic pains. Ground twigs and buds serve against intestinal worms and a decoction of dry young branches is used as medicine against itch for camels. The fruits are of medicinal value for example, to treati headaches and skin diseases. The resin is used as a stimulant and for treatment of ulcers and liver diseases. While this list of possible purposes for using *Juniperus* is not complete, it shows the potential for local communities and markets, and maybe even for markets at a larger scale.

## GEOGRAPHIC DISTRIBUTION

*Juniperus procera* ranges from Hungary in the west, Russia in the north, China and Myanmar in the east down to Africa through Arabia in the south (Brandis, 1906; Zaman et al, 1968; Migahid and Hammouda, 1974; Diavanshir, 1974; Hara et al, 1978; Chaudhry and Wali-ur, 1979; Hall, 1984). In Africa, juniper extends from the Nubian Hills in southern Egypt (Vivi and Drar, 1941) all along the east African plateau west of the Nile to Malawi (Kelecha, 1979) and Zimbabwe (Kerfoot, 1966). The species has been reported in Sudan, Ethiopia, Djibouti, Somalia, Kenya, Zaire, Uganda, Tanzania, Malawi and Zimbabwe (Jansen, 1981; Hall, 1984). It occurs abundantly in western Kenya and in the Ethiopian highlands. In Ethiopia, individuals are found at 1500–3500m asl, but mostly at 1750–2500m, where trees occur on rocky, well-drained soils and within a wide range of rainfall zones, from 350 to 2000mm/year. Precipitation in the main juniper forest belt ranges from 1000 to 1400mm/year (Friis, 1992).

## POPULATION DISTRIBUTIONS

The last natural juniper-dominated forests in Ethiopia are now the property of the state (national forest priority areas) for production, protection and biological conservation services (Teketay and Bekele, 2002). The four investigated state forests (Figure 13.1) had reasonable stocks of juniper (112–493 individuals per hectare). Maximum stem diameter varied from 72cm to 205cm between sites. The two secondary state forests (Hugumburda and Chilimo) had similar DBH distributions (Figure 13.2a), but these distributions differed from those of the semi-natural state forests (Adaba-Dodola and Menagesha-Suba). The secondary forests had individuals in a narrower DBH range (1–80cm) than the semi-natural forests (1–155cm). The four state forests all had individuals in the smallest size class (DBH <10cm), but the secondary forests had the highest frequency in this class. The four state forests showed gradually decreasing tree numbers with increasing size class. Relatively low tree numbers of intermediate size (10–40cm DBH) were found in the semi-natural forests. The high number of stumps (left over from cut trees) in these latter forests suggests that recent harvesting contributed to these low numbers.

Tree allometry patterns (Figure 13.3) show a typical DBH–height relation: DBH and height initially increase, but later height ceases to increase while DBH continues to increase. We estimated the maximum height of juniper in the

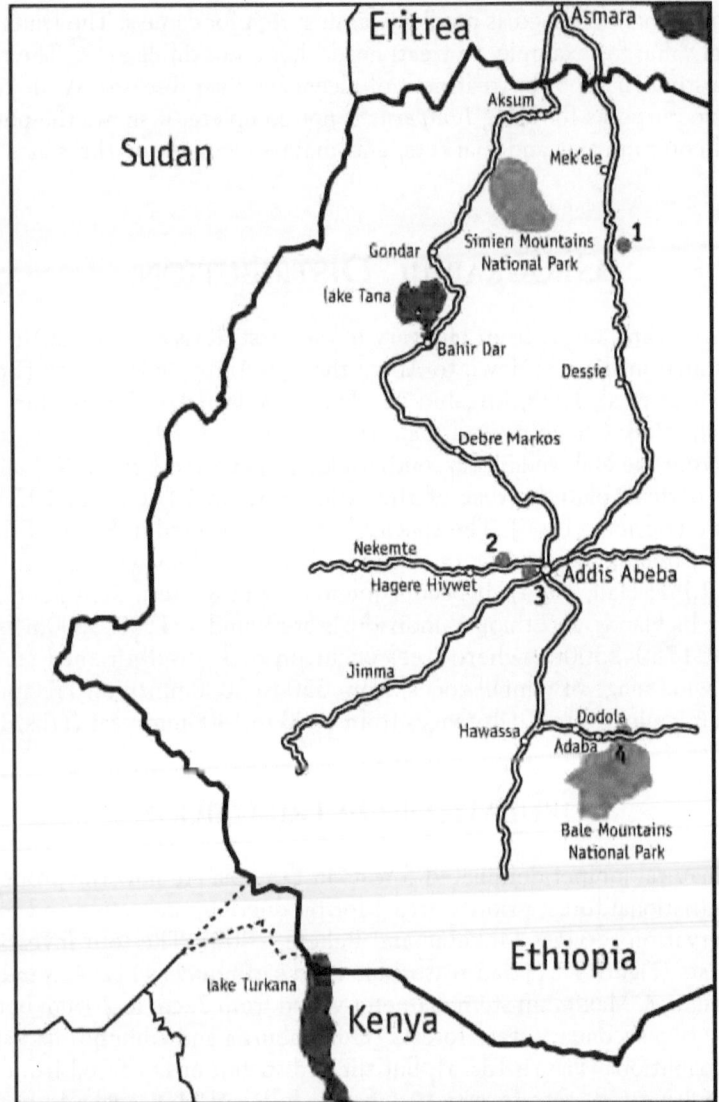

**Figure 13.1** *Location of the four study sites in Ethiopia*

*Notes:* 1 = Hugumburda, 2 = Chilimo, 3 = Menagesha-Suba, 4 = Adaba-Dodola.

Menagesha-Suba and Adaba-Dodola forests at 39m and 33.4m respectively. For the Hugumburda and Chilimo forests, we could not calculate a maximum height because the biggest trees were still growing in size.

Overall, there is representative abundance of juniper in these forests, including considerable regeneration (Figure 13.2). A closer evaluation of the <10cm DBH class, however, shows that there are low numbers for the smallest sizes (<1m tall), especially for Chilimo and Hugumburda. Although juniper is a

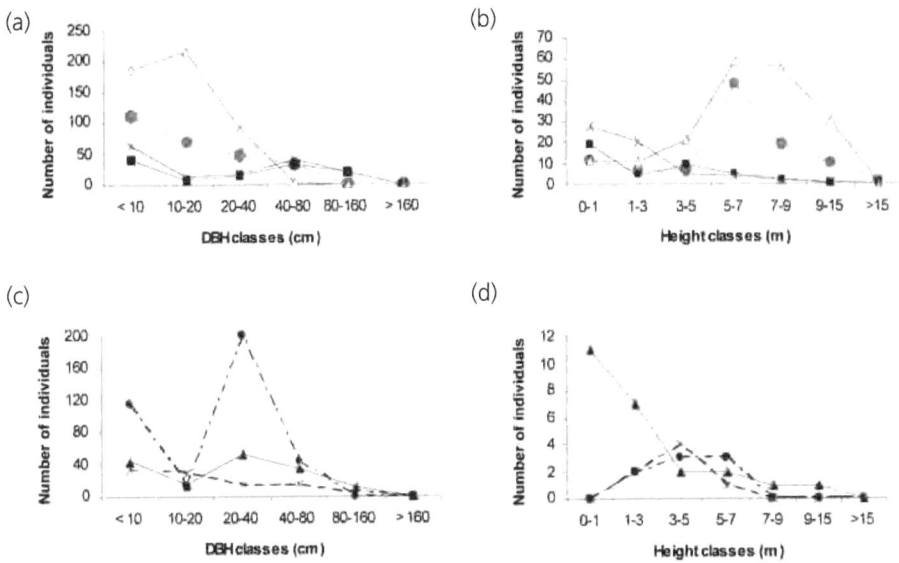

**Figure 13.2** Juniperus procera *abundances for all DBH classes in
(a) and (c) state forests and (b) and (d) church forests*

Notes: For small trees (<10cm DBH), abundances are given per height class in the same (c) state forests
and (d) church forests. For state forests (a and c): circles (Chilimo), squares (Adaba-Dodola), triangles
(Hugumburda) and crosses (Menagesha-Suba). For church forests (b and d): crosses (Gibstawit),
diamonds (Asha) and triangles (Dengolt – Wassie, 2007).

prolific seed-bearer and seeds are fertile (Negussie et al, 1991), seeds may not
germinate due to ample shade and humus under dense secondary forest
(Gardner, 1926; Yirdaw and Leinonen, 2002). In open patches seeds germinate
better (Wimbush, 1937; Friis, 1992), as can be found more frequently in semi-
natural forests (C. Couralet, Adaba-Dodola, 2003, personal communication; Y.
Sahle, Menagesha-Suba, 2004, personal communication). Controlled local
thinning may be considered to create some open patches to encourage natural
regeneration. In the secondary forests, nursery beds and enrichment planting
may be considered in such open patches to promote regeneration and to avoid a
future age-class gap in the tree population.

Apart from state forests, there are many small remnant forests, most of
which are owned by churches (Wassie, 2007; Wassie et al, Chapter 6 of this
volume). Of the church forests studied, the two largest church forests (ca. 9ha
and 25ha respectively) showed a gradually declining tree number with increasing DBH class. The smallest church forest (1.6ha) showed a rather erratic
pattern. The two smallest church forests had no individuals in the smallest size
class (<1m height). Apparently, regeneration was not successful there. For the
other five church forests, more information is needed to ascertain if juniper
once existed in the forests. Two of these five had only a few large juniper trees.
These observations suggest that juniper populations may be too small to be vital.

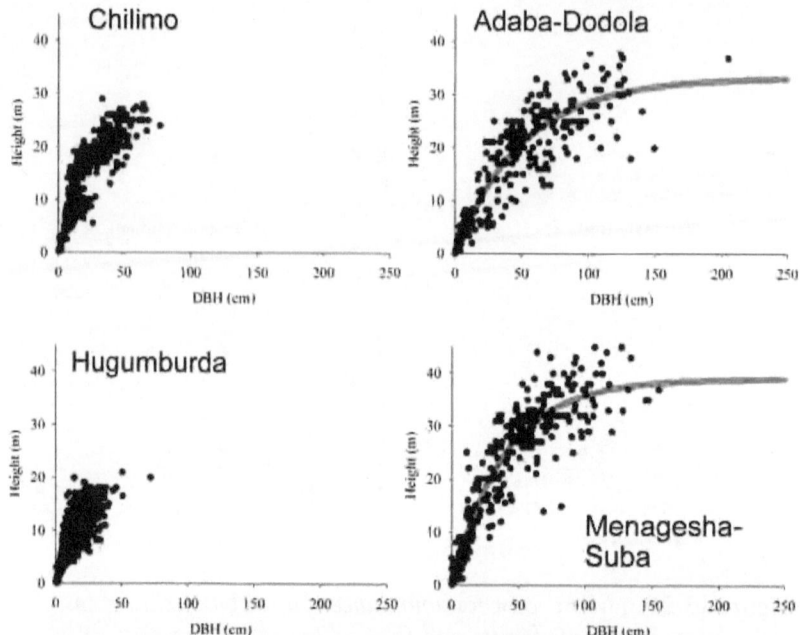

**Figure 13.3** *Height vs. diameter of* Juniperus procera *individuals in the four state forests*

*Notes:* For two semi-natural forests (with individuals over a wider DBH range), curves have been fitted according Thomas (1996): Height = $H_{max} \cdot (1-e^{-\alpha \cdot DBH^{\beta}})$, where $H_{max}$ (maximum height), $\alpha$ and $\beta$ are the estimated parameters.

These populations would benefit from the protection and planting of the regenerating seedling and saplings and, most importantly, from building larger church forest areas to provide larger, more viable populations.

In general, the results suggest that semi-natural state forests, secondary state forests and church forests differ in population structure and regeneration problems, and that such different forests require different actions to guarantee the vital juniper population for the future.

# DENDROCHRONOLOGY: TREE GROWTH IN RESPONSE TO CLIMATE

We were able to calculate climate–growth relationships for *Juniperus procera* from Adaba-Dodola (Couralet et al, 2005) and Menagesha–Suba (Sass-Klaassen et al, 2008) as juniper from southern highland forests appear to form clearly distinguishable growth rings (Conway et al, 1997) in response to variation in precipitation. At other highland locations such as the Northern Rift Valley, distinct ring formation has been proved in response to Blue Nile base flow (Wils et al, 2009), while in other highland regions such as Doda and Denkoro forest,

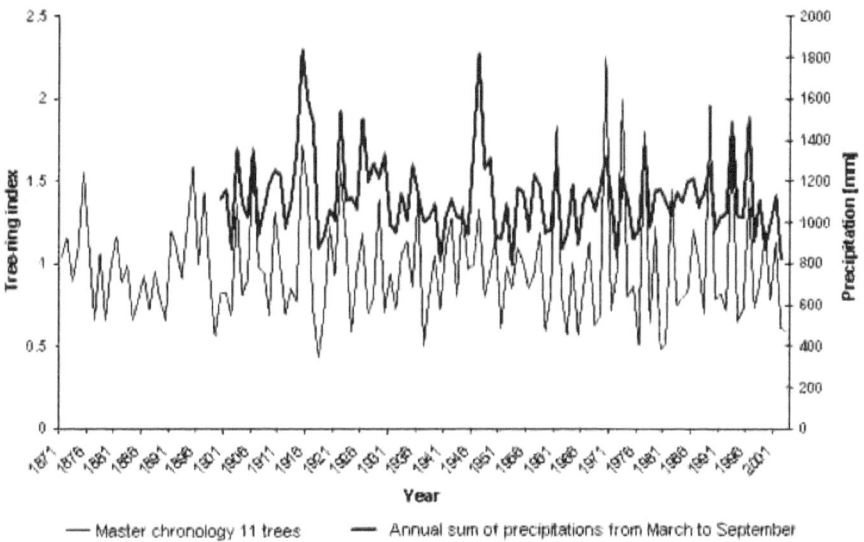

**Figure 13.4** *Comparison of tree-ring chronology from 11* Juniperus procera *trees with record of annual sum of precipitation during the rainy season in the Adaba-Dodola semi-natural forest*

*Note:* Tree-ring index values indicate average radial stem growth of the 11 trees in each year.
*Source:* Couralet et al (2005)

north of Addis Ababa, growth-ring patterns have found to be more erratic (Wils et al, 2009). Growth boundaries in juniper are characterized by an abrupt change in wood density between the radial flattened tracheïds with thick cell walls (produced at the end of the previous growing season) and the large thin-walled tracheïds formed at the beginning of the following growing season. Based on ring counts of plantation trees of known age, it was verified for juniper from Adaba-Dodola that growth rings are annual (Couralet et al, 2005). Dendrochronological techniques were used for age determination and to study growth dynamics of juniper in relation to changing climate conditions. Stem discs were collected of nine trees from Adaba-Dodola forest and two from a nearby plantation. After sanding of the wood surface growth rings were visually detected under the microscope and measured to a precision of 1/100mm.

The age of the 11 study trees ranged from 71 to 135 years (Table 13.1), with trees 1, 2, 3 and 9 being of similar age, which might indicate favourable conditions for tree establishment around 1925. A so-called site chronology, in other words a mean growth ring, was calculated reflecting the average growth pattern of the 11 cross-dated growth-ring series and related to monthly data of precipitation and temperature. The growth was found to be positively corre-lated with the amount of precipitation in the growing season from March to September (Figure 13.4). This indicates that changes towards a drier climate will have considerable influence on growth rates of juniper populations in Afromontane forests of Ethiopia. The occurrence of so-called double rings,

**Table 13.1** *Tree age and the growing period of* Juniperus procera

| Sample | Diameter (cm) | Number of rings | Estimated age (years) | Mean growth rate per year (cm) | Growth period |
|--------|---------------|-----------------|-----------------------|--------------------------------|---------------|
| 1 | 32 | 64 | 77 | 0.42 | 1924–2003 |
| 2 | 27 | 64 | 77 | 0.35 | 1926–2003 |
| 3 | 18 | 69 | 71 | 0.25 | 1932–2003 |
| 4 | 32 | 133 | 147 | 0.22 | 1856–2003 |
| 5 | 40 | 154 | 168 | 0.24 | 1835–2003 |
| 6 | 23 | 97 | 109 | 0.21 | 1893–2002 |
| 7 | 25 | 81 | 93 | 0.27 | 1910–2003 |
| 8 | 24 | 120 | 135 | 0.18 | 1883–2003 |
| 9 | 26 | 77 | 82 | 0.32 | 1924–2001 |
| 10 | 31 | 68 | 78 | 0.40 | 1926–2003 |
| 11 | 38 | 69 | 78 | 0.49 | 1926–2003 |

*Note:* Estimated from growth-ring measurements for Adaba-Dodola forest; trees 10 and 11 are plantation trees.
*Source:* Couralet et al (2005)

manifested as density variation in the growth ring, points to a growth response on the shorter peak of precipitation during early spring (Sass-Klaassen et al, 2008) and can frustrate growth-ring detection (Wils et al, 2009).

At Menagesha-Suba state forest the 24 studied juniper trees followed a linear age–DBH increment pattern (Figure 13.5). With a mean growth rate of 0.26cm/yr (range: 0.13 to 0.48cm/yr) the junipers at Menagesha-Suba are on average slower growing than those at Adaba-Dodola, which on average reach 0.40cm/yr (range: 0.29 to 0.51cm/yr). However, Figure 13.5 also illustrates that growth levels differ considerably between trees at Menagesha-Suba. A minority of individuals on both sites showed a curved age–DBH pattern. The majority of trees show a linear age–DBH pattern with rapidly growing trees remaining rapidly growing and the slow trees remaining slow over their whole lifespan. It remains unclear whether such strong autocorrelation in growth resulted from variation in site conditions, disturbances, or genetic information.

# POPULATION DYNAMICS

A pilot study was carried out to assess the population dynamics of junipers in Adaba-Dodola forest, and similar studies are currently being undertaken in various other sites (Bekele, 2005, unpublished results). Trees of the inventoried populations were assigned to different DBH classes. By using the age–DBH patterns (Figure 13.5), these DBH classes were transformed to age distributions for the whole forest (Couralet et al, 2005). Using matrix-model techniques, the stable population structure (Figure 13.6) and the population growth rates were calculated. A population growth rate of 1 indicates a stable population structure. Both the observed and predicted stable populations are characterized by a high frequency of small individuals, fewer individuals in the next intermediate size classes, a higher frequency in the subsequent size class

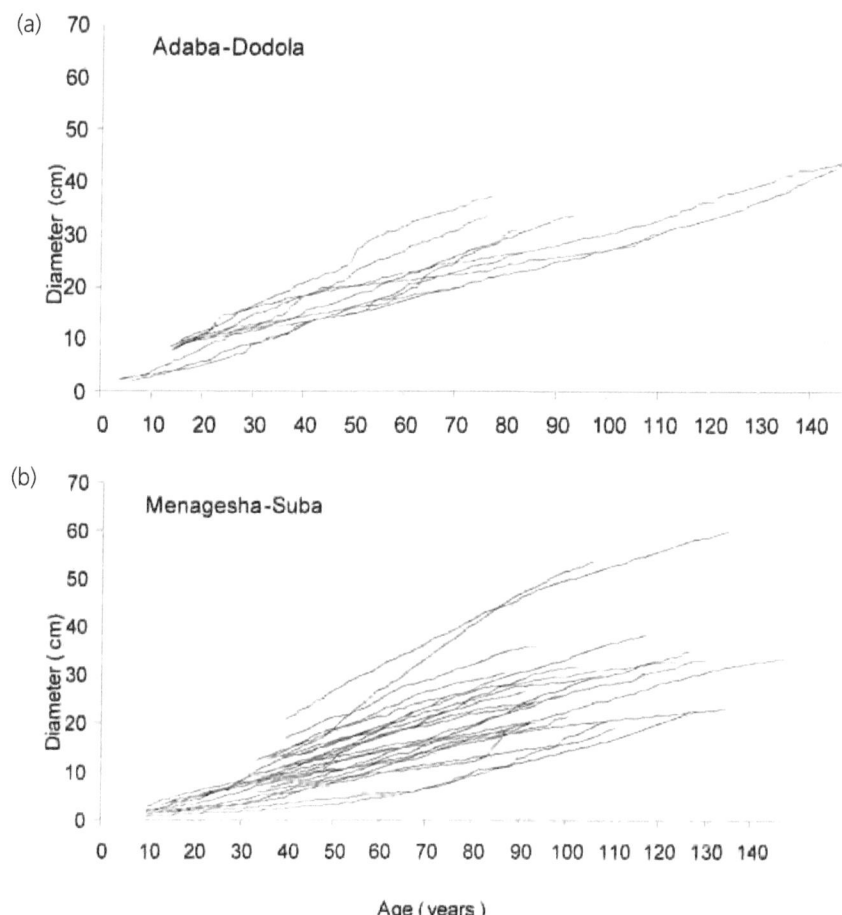

**Figure 13.5** *Cumulative diameter increment of Juniperus procera trees*

*Notes:* (a) 9 trees from Adaba-Dodola; (b) 24 trees from Menagesha-Suba.
*Source:* Couralet et al (2005) and Sahle et al (unpublished results)

and then gradually the frequency decreases with increasing size class. The observed population had lower frequencies in the 5 to 40cm DBH classes (Figure 13.6) than the predicted stable population structure. This suggests that in the actual situation these intermediate size classes are preferentially harvested, for example for stems serving as poles for construction.

An elasticity analysis (De Kroon et al, 1986; Caswell, 1989) was performed to indicate the proportional change in $\lambda$ (elasticity of the population) as a result of the proportional change in each of the matrix elements. Elasticities suggest that the population growth rate of the juniper at the Adaba-Dodola site is most sensitive to changes in survival of trees in the intermediate size classes, between 5 and 40cm diameter (Couralet et al, 2005). As already suggested, this coincides with the size classes that are actually preferably harvested. To avoid a

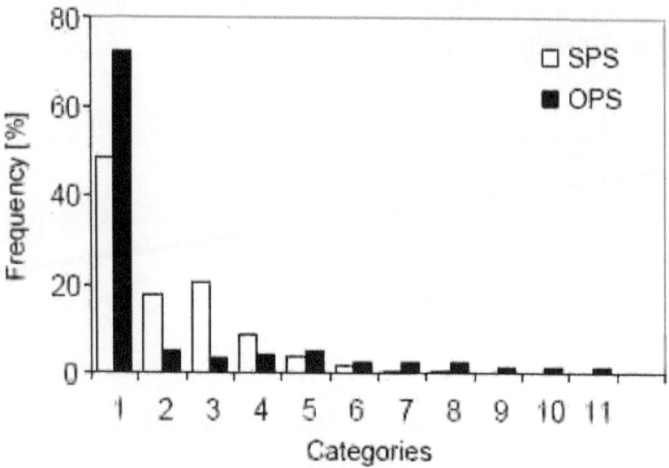

**Figure 13.6** *Results from matrix modelling for* Juniperus procera *in the Adaba-Dodola forest*

*Notes:* Comparison of observed (OPS) with stable (SPS) population structures. Categories are based on diameter values: 1: 0–5cm, 2: 5–10cm, 3: 10–25cm, 4: 25–40cm, 5: 40–55cm, 6: 55–70cm, 7: 70–85cm, 8: 85–100cm, 9: 100–115cm, 10: 115–130cm, 11: >130cm.
*Source:* Couralet et al (2005)

negative influence of these harvesting activities on the population in the long run, strong control is recommended. The population growth was, surprisingly, hardly sensitive to changes in the smallest size classes. Apparently, the population of the Adaba-Dodola state forest is at present not much affected by grazing. Of course, this is no guarantee for stable populations when grazing levels further increase in the future. Moreover, this situation may differ for secondary state forests and church forests that largely differed in structure from the semi-natural state forests.

## CONCLUSIONS

Although populations of *Juniperus procera* have been greatly reduced in their natural range, compensation by establishing plantations has not been considered an option mainly because of the slow growth of the trees, low wood quality due to irregular and fluted stems (Yirdaw, 2001), susceptibility of young trees to fire (FAO, 1958), and uncertainty of future landownership (Teketay and Bekele, 2002). Consequently, the semi-natural state forest, secondary state forests and church forests remain the most important sources of juniper in Ethiopia and initiation points for forest restoration. We showed how juniper populations differed between semi-natural state forests, secondary state forests and church forests. Research in semi-natural forests suggests stable population structures under the current levels of tree harvesting and (low) grazing, but the tree populations are sensitive to tree harvesting in intermediate, widely preferred, size classes. Therefore we suggest that tree harvesting in these size classes should be controlled to keep these populations stable, particularly at the forest borders where high harvest intensities reduce forest areas and transfer them into open pastures (F. J. Sterck, Adaba-Dodola and Menagesha-Suba, 2002 and 2003, personal communication). The secondary state forests were not yet fully grown and probably lack frequent regeneration due to shading and humus in the understorey. Thinning might be considered to create open spots and promote natural regeneration locally, and thus create a more natural population structure. The church forests showed the most irregular population distributions. The lack of regeneration in the smallest church forests suggests that these small forests need intensive management, for example protection of regeneration from grazing, possibly after (enrichment) planting or seeding. Moreover, expansion of the remaining church forest fragments should be considered to create conditions for viable tree populations in the long run.

We observed a strong relationship between radial growth rates and the amount of precipitation during the rainy season. This means that the current production levels, but also tree vitality in general, will be reduced in Afromontane forests when droughts become more frequent (Sass-Klaassen et al, 2008). As climate–growth responses can strongly differ depending on the study area and climate constellations, for example from generally dry conditions such as in the Dessa state forest in Tigray to more unimodal (Gondar) or bimodal (Kombolcha) rainfall patterns, we recommend conducting studies on climate–growth relationships and population dynamics in different regions across the Ethiopian highlands. We also recommend that management focuses on 'edge-effects' at the borders of the forests where the greatest disturbances occur and concentrates on promoting enlargement of the smaller forest fragments, such as the church forests, to create conditions for viable populations. Of course, such management advice needs to be properly tuned to local community and political conditions.

# REFERENCES

Bonnefille, R. and Umer, M. (1994) 'Pollen-inferred climatic fluctuations in Ethiopia during the last 3000 years', *Palaeogeography, Palaeoclimatology, Palaeoecology*, vol 109, pp331–343

Borghesio, L., Giannetti, F., Ndang'ang'a, K. and Shimmelis, A. (2004) 'The present conservation status of *Juniperus* forests in the south Ethiopian Endemic Bird Area', *African Journal of Ecology*, vol 42, pp137–143

Brandis, D. (1906) *Indian Trees*, Archibald Constable and Co. Ltd., London

Burgess, N., Hales, D. A. J., Underwood, E., Dinerstein, E., Oslon, D., Itoua, I., Schipper, J., Ricketts, T. and Newman, K. (2004) *Terrestrial Ecoregions of Africa and Madagascar: A Conservation Assessment*, Island Press, Washington, DC

Caswell, H. (1989) *Matrix Population Models*, Sinauer Associates, Sunderland, MA

Chaffey, D. R. (1982) *South-West Ethiopia Forest Inventory Project. A Reconnaissance Inventory of Forests in South-West Ethiopia*, Forestry and Wildlife Conservation and Development Authority, Addis Ababa

Chalk, L., Burtt Davy, J. and Desch, H. E. (1932) *Some East African Coniferae and Leguminosae*, Clarendon Press, Oxford, UK

Chaudhry, M. I. and Wali-ur, R. (1979) 'Insect pests of juniper, their parasites and predators', *Pakistan Journal of Forestry*, vol 29, no 1, pp21–24

Conway, D., Brooks, N., Briffa, K. R., Desta, S., Merrin, P. D. and Jones, P. D. (1997) *Exploring the Potential for Dendroclimatic Analysis in Northern Ethiopia*, University of East Anglia, Climate Research Unit, Norwich, UK

Couralet, C., Sass-Klaassen, U., Sterck, F. J., Bekele, T. and Zuidema, P. A. (2005) 'Combining dendrochronology and matrix modelling in rapid and data limited demographic studies: An evaluation of *Juniperus procera* from the Ethiopian highlands', *Forest Ecology and Management*, vol 216, pp317–330

Dallimore, W. and Jackson, B. (1966) *A Handbook of Coniferae and Ginkgoaceae*, Edward Arnold Ltd., London

De Kroon, H., Plaisier, A., Van Groenendael, J. and Caswell, H. (1986) 'Elasticity: The relative contribution of demographic parameters to population growth rate', *Ecology*, vol 67, pp1427–1431

Diavanshir, K. (1974) ''Problems of regeneration of *Juniperus polycarpos* C. Koch in the forests of Iran', *Silvicultural Genetics*, vol 23, no 4, pp106–108

Eggeling, W. J. and Dale, I. R. (1951) 'Indigenous trees of the Uganda Protectorate', The Government Printer, Entebbe, Uganda

FAO (1958) 'Choice of tree species', FAO, Rome

FAO (1986) 'Databook on endangered tree and shrub species and provenances', FAO, Rome

Friis, I. (1992) *Forests and Forest Trees of Northeast Tropical Africa. Their Natural Habitats and Distribution Patterns in Ethiopia, Djibouti and Somalia*, HMSO, London

Gardner, B. A. (1926) 'East African pencil cedar', *Empire Forestry Journal*, vol 5, pp39–53

Hall, J. B. (1984) '*Juniperus excelsa* in Africa: A biogeographical study of an Afromontane tree', *Journal of Biogeography*, vol 11, pp47–61

Hara, H., Stearn, W. T. and Williams, H. J. (1978) *An Enumeration of the Flowering Plants of Nepal*, Trustees of British Museum (Natural History), London

Jansen, P. C. M. (1981) *Condiments and Medicinal Plants in Ethiopia, Their Taxonomy and Agricultural Significance*, Pudoc, Wageningen, The Netherlands

Javeed, Q. N., Perveen, R., Imtiazul-Haq and Ilahi, I. (1980) 'Propagation of *Juniperus polycarpos* C. Koch through tissue culture 1. Induction of callus', *Pakistan Journal of Forestry*, vol 30, pp72–77

Kelecha, W. M. (1979) *Reports on Ethiopian Forests*, Forestry and Wildlife Conservation and Development Authority, Addis Ababa

Kerfoot, O. (1966) 'Distribution of Coniferae: The Cupressaceae in Africa', *Nature*, vol 212, p961

Migahid, A. M. and Hammouda, M. A. (1974) *Flora of Saudi Arabia*, National Enterprises, Riyadh

Negussie, A. (1995) *A Monographic Review on Juniperus Excelsa*, Faculty of Forestry, Alemaya University of Agriculture, Alemaya, Ethiopia

Negussie, A., Mayhead, G. J. and Good, J. E. (1991) 'The effect of pre-treatments and diurnal temperature variations on the germination of *Juniperus excelsa*', *International Tree Crops Journal*, vol 7, pp57–66

Noad, T. and Birnie, A. (1989) *Trees of Kenya*, NHBS, Totnes, UK

Pohjonen, V. and Pukkala, T. (1992) *'Juniperus procera* Hocht. ex Endl. in Ethiopian forestry', *Forest Ecology and Management*, vol 49, pp75–85

Sass-Klaassen, U., Couralet, C., Sahle, Y. and Sterck, F. J. (2008) 'Juniper from Ethiopia contains a large-scale precipitation signal', *International Journal of Plant Sciences*, vol 169, no 8, pp1057–1065

Teketay, D. and Bekele, T. (2002) *Indicators and Tools for Restoration and Sustainable Management of Closed-Deciduous Forests in East Africa. Prevailing Forest Types and their Management History: Country Report for Ethiopia*, I-TOO project, Addis Ababa

Thomas, S. C. (1996) 'Asymptotic height as a predictor of growth and allometric characteristics in Malaysian rain forest trees', *American Journal of Botany*, vol 83, pp556–566

Verschuren, D., Laird, K. R. and Cumming, B. F. (2000) 'Rainfall and drought in equatorial east Africa during the past 1100 years', *Nature*, vol 403, pp410–414

Vivi, G. T. and Drar, M. (1941) *Flora of Egypt*, Fouad I University, Cairo

Wils, T. H. G., Robertson, I., Eshetu, Z., Sass-Klaassen, U. G. W. and Koprowski, M. (2009) 'Periodicity of growth rings in *Juniperus procera* from Ethiopia inferred from crossdating and radiocarbon dating', *Dendrochronologia*, vol 27, pp45–58

Wassie, A. (2007) 'Ethiopian church forests: Opportunities and challenges for restoration', PhD dissertation, Wageningen University, Wageningen, The Netherlands

Wimbush, S. H. (1937) 'Natural succession of the Pencil Cedar Forest of Kenya Colony', *Empire Forestry Journal*, vol 16, pp49–53

Yirdaw, E. (2001) 'Diversity of naturally-regenerated native woody species in forest plantations in the Ethiopian highlands', *New Forests*, vol 22, pp159–177

Yirdaw, E. and Leinonen, K. (2002) 'Seed germination responses of four Afromontane tree species to red/far-red ratio and temperature', *Forest Ecology and Management*, vol 168, pp53–61

Young, J. A. and Young, C. G. (1992) *Seeds of Woody Plants in North America* (revised and enlarged edition), Dioscorides Press, Portland, OR

Zaman, M. B., Khan, A. A. and Khan, M. S. (1968) 'Survey of juniper berries in Baluchistan forests and prospects for their exploitation', *Pakistan Journal of Forestry*, vol 17, pp503–509

## 14

# Degradation, Species Invasion and Management Responsibility in Tanzanian Dry Forests: What Do Local People Say?

*John A. F. Obiri, John B. Hall and John R. Healey*

## INTRODUCTION

Forest policymakers and managers are increasingly being asked to develop forest land-use plans and policies that incorporate the general public's views and more importantly public preferences (Bass, 2001). To do this, forest managers and policymakers need to understand the public's attitudes towards alternative situations and options for the forest (Obiri and Lawes, 2002; Tindall, 2003; Elands et al, 2004; Flinkman, 2004). Here we have investigated attitudes of local forest stakeholders towards degradation, plant species invasion, regulation of tree harvesting and alternative structures for management responsibility in dry forests of east-central Tanzania. The current policies for reducing forest degradation are compared with the views of local communities to ascertain whether or not the two are in agreement.

In Tanzania, forests represent a key natural capital in the livelihoods of rural communities (Nduwamungu, 1996; Luoga et al, 2000) and they cover 50 per cent of the country. However, these forests have, in the past and up to the present, faced clearance and heavy exploitation, which is evident through declining land area under forest cover and degradation of the remaining forest

(Kikula, 1986; Schwartz et al, 2002; Schwartz and Caro, 2003). This has resulted in major changes in the condition of large areas of forest, including a decline in species diversity and a marked increase in the abundance of 'weedy' plant species that benefit from disturbance (including trees, shrubs and herbs). While most of these (but not all) are native to Tanzania, their populations have expanded so greatly into forest habitats where they were previously absent or rare that local perceptions of them as 'invasive' are independent of their biogeographical status.

Forest degradation has been linked to weaknesses in the policy of landownership (Luoga et al, 2005). Landownership and forest management can be classified into four tenure types where forests are under:

1 state control;
2 private ownership;
3 community ownership; or
4 left as open-access areas referred to as 'general lands' (Ministry of Natural Resources and Tourism, 1998).

Since the mid-1990s Tanzanian national forest policy has promoted two forms of participatory forest management: wholly community-based forest management and joint forest management between communities and the state (Blomley and Ramadhani, 2006); a recent survey showed that participatory forest management was established (or in the process of being established) in more than 1800 villages and over more than 3.6 million hectares of forest (Blomley et al, 2008).

Several studies have shown that the intensities of forest disturbance and degradation differ under these alternative tenure systems. Early research showed both the community-owned forest and open-access general lands to be the most degraded, hold the least species diversity and require close monitoring if their resources are to be conserved (Nduwamungu, 1996; Zahabu, 2001; Luoga et al, 2005). The situation is clearly complex, however, in terms of variation both among locations and over time. In a more recent study Blomley et al (2008) found an increase in tree basal area and volume in miombo forest under community-based and joint management in contrast to a decline under state or open-access management. They also found evidence that joint forest management reduced rates of tree cutting and other forms of forest disturbance. However, in a separate study of montane forest, Persha and Blomley (2009) found that rates of tree cutting had been greater under joint forest management than under state or community-based management.

It is notable that this research has tended to be based on inventories of tree abundance and biomass and has hardly considered the views of local communities. Where information has been obtained from local people, this has generally been about their use of forest resources and not on forest degradation, the regulation of forest use and forest tenure/management responsibility. However, in a notable recent development, in the West Usambara Mountains of Tanzania, Persha and Blomley (2009) did use semi-structured interviews and group discussions to obtain information on how alternative forms of forest management operate in terms of local governance system structure and functioning (such as tenure rights, rule formation and enforcement, and roles and interaction among actors). To date, the study of Vyamana (2009) stands out as the best example of where the *opinions* of local communities about the success of forest management have been sought in Tanzania, in this case based on a comparison, with respondents stratified into wellbeing categories, among nine communities that experienced different forms of forest management responsibility.

Many studies have indicated that local communities, especially those closest to and dependent on forests, hold the key to sustainable management and conservation of forest resources (for example Gibson et al, 2000; Bass, 2001). A lack of positive local support because of neglect of people's opinions and preferences by government institutions can lead to an escalation of forest exploitation and degradation (Nhira et al, 1998; Wily, 2000; Weis, 2000). Reducing degradation and instituting practical restorative management of forests can be better achieved when informed by local people's attitudes towards resource use and conservation (Ylhaisi, 2003). Because the role of households and individuals in forest use and management is closely linked to their position within the socio-economic structure of communities, it is important that variation in opinion among these actors is understood. In Tanzania there is still a lack of information on the opinions in local communities about the threats to forests, the problems caused to local livelihoods by the resulting increased abundance of invasive species, preferences for different forms of regulation of tree harvesting and people's anticipation of which alternative forms of responsibility for forest management are most likely to improve forest condition. Given the high level of degradation in large areas of Tanzanian forest, and the importance of local stakeholders' views for the adoption of participatory forest management and its potential success to reduce this problem, it is this information gap that the present study sought to address.

# SOCIAL CONTEXT AND STUDY PROCEDURE

## Area and stakeholder community

The study was undertaken in the Kitulanghalo area of Morogoro District in the Morogoro Region of Tanzania. This is a dry forest area that is largely settled by two main ethnic groups: the Kwere- and Zigua-speaking people. The settlements are typical of rural Tanzania, characterized by poverty, lack of formal employment opportunities, inability to meet household subsistence needs solely from agriculture and heavy reliance on natural resources for income (Shemwetta et al, 2004). Monthly household incomes range from US$20 to US$100.

The study was carried out in two large villages (Maseyu and Lubungo) adjacent to a large area of state-controlled forest and to community-controlled forest and general lands. These villages are administered under well-structured village governments with democratically elected leaders and councils of assistants who are largely elders (Luoga et al, 2005). The village communities include several forest stakeholder groups, such as charcoal makers, middlemen in the charcoal and timber trades, animal hunters, medicine and fuelwood gatherers, and leaders from traditional groups, religions and local schools. Stakeholders were taken as social actors who were keenly interested in the area's natural resources and their management. They were generally willing to offer specific contributions towards the resource management for which they possessed skills or knowledge or had a comparative advantage (for example by living in close proximity or being mandated).

## Participatory rural appraisal

To win the confidence of the villagers and gain initial rapport we first conducted a participatory rural appraisal (PRA) and consulted stakeholders on their awareness of invasive species in and around Kitulanghalo Forest before soliciting views on a series of statements. The PRA process brings together the development project/programme planning skills of external practitioners and the knowledge and awareness of a local community. These are combined to identify rural problems handicapping development progress and to devise and apply measures feasible in the local context for alleviating them. In the present case, the PRA involved timeline analysis, resource mapping, transect walks, wealth ranking and visual methods via use of pictures. For the last, interviewee comments were elicited on pictures of the forest under different conditions of degradation and invasion by pioneer/invasive species.

## Statements offered for respondent views

75 households, sampled randomly from a total of 892 for the two villages together (a sample of 8.4 per cent), were consulted about their views on forest change from degradation, species invasions and management responsibility alternatives. The themes of forest degradation and options for management were chosen following concerns expressed in Tanzanian forest policy documents (Ministry of Lands, Natural Resources and Tourism, 1989; Ministry of Natural Resources and Tourism, 1998; Kajembe et al, 2004b) and, independently, by social and environmental scientists working in Tanzanian dryland forest (for example Ylhaisi, 2003; Luoga et al, 2005). Equally, they are issues relevant to the contemporary global debate about forest degradation and conservation (see, for example, Campbell and Vainio-Mattila, 2003).

Six statements were made to respondents, each respondent being requested to indicate her/his attitude towards each statement by choosing one of five levels of agreement (strongly agree; agree; neutral; disagree; strongly disagree):

1   Harvesting of trees should be controlled.
2   Harvesting of trees should be banned.
3   Invasive plant species are not a problem to farming/gardening and other similar activities.
4   Forest degradation and species invasion would be high under community-based management.
5   Forest degradation and species invasion would be high under state management.
6   Forest degradation and species invasion would be high under participatory (joint) management.

## Data processing

To investigate how views varied across various socio-demographic characteristics, the responses offered were related to categories of respondent age (young, <35; middle-aged, 35–55; elderly, >55), residency period (new settler, <15 years; mid-term settler, 15–30 years; long-term resident, >30 years), and gender – increasingly emerging as a source of diverging opinion on forest policies, governance and management in southern Africa (Chitinga and Nemarundwe, 2003). Comparisons across income levels, a characteristic commonly considered in such studies (Tindall, 2003), were not considered as almost all residents were poor (90 per cent having a household income of less than US$100 per month).

The attitudes of respondents with respect to each statement were summarized overall and by the various stakeholder categories as frequency distributions of the agreement levels. The most commonly offered response (modal) was noted. $\chi^2$ goodness-of-fit tests were used to determine if there was a significant departure from equal representation of every agreement level. For each statement, a $\chi^2$ test of independence was used to determine if there was a gender difference in the views expressed.

# RESULTS

## Tree harvesting

A majority of respondents agreed that tree harvesting should be controlled (55 per cent strongly agreed or agreed) but strongly disagreed (69 per cent) with the idea of banning it (Tables 14.1 and 14.2). Opinion was not notably affected by age or residency period. There was evidence that males were more strongly in favour of control than females, but this difference was not significant.

Table 14.1 *Summaries by agreement level, in total and by gender, of views expressed by 75 respondents on six statements relating to regulation of the harvesting of trees, the impact of invasive plants and responsibility for forest management, in Kitulanghalo Forest, Tanzania*

| Statements | Levels of agreement with statements | | | | | | | | | | | | | | |
| --- | --- | --- | --- | --- | --- | --- | --- | --- | --- | --- | --- | --- | --- | --- | --- |
| | All responses (75) | | | | | Female responses (30) | | | | | Male responses (45) | | | | |
| | SA[1] | A | N | D | SD | SA | A | N | D | SD | SA | A | N | D | SD |
| Harvesting of trees should be controlled. | **23** | 17 | 18 | 4 | 13 | 6 | **12** | 6 | 1 | 5 | **17** | 5 | 12 | 3 | 8 |
| Harvesting of trees should be banned. | 6 | 3 | 4 | 10 | **52** | 2 | 1 | 2 | 0 | **25** | 4 | 2 | 2 | 10 | **27** |
| Invasive plant species are not a problem to farming/ gardening and other similar activities. | 6 | 10 | 7 | 19 | **33** | 0 | 4 | 5 | 7 | **14** | 6 | 6 | 2 | 12 | **19** |
| Forest degradation and species invasion would be high under community-based management. | **27** | 15 | 4 | 23 | 6 | 5 | 7 | 3 | **11** | 4 | **22** | 8 | 1 | 12 | 2 |
| Forest degradation and species invasion would be high under state management. | 7 | 7 | 10 | **28** | 23 | 0 | 4 | 3 | **13** | 10 | 7 | 3 | 7 | **15** | 13 |
| Forest degradation and species invasion would be high under joint (participatory) management. | 7 | 10 | 12 | **25** | 21 | 3 | 6 | **9** | 7 | 5 | 4 | 4 | 3 | **18** | 16 |

*Notes:* [1] SA, strongly agree; A, agree; N, neutral; D, disagree; SD, strongly disagree; modal responses in bold.

## Invasive species

Two aspects of invasive species were considered: awareness of their occurrence and the likelihood that they would prove problematic. Awareness was high: 81 per cent of respondents were aware of invasive plants and familiar with certain species, including trees such as *Acacia nigrescens* Oliv. and *A. polyacantha* Willd., and particularly those which are shrubs or herbs (for example *Lantana camara* L. (an exotic species) and *Solanum incanum* L.). Within the villages, awareness of invasives did not vary with gender or residency period but a much higher proportion of young respondents (39 per cent) were ignorant of these plants than those who were middle-aged (12 per cent) or elderly (0 per cent), ($\chi^2 = 11.316$, 2 d.f., P<0.01). A majority of respondents (of each gender and age group) believed that these invasive species are a problem: overall 44 per cent strongly disagreed (and another 25 per cent disagreed) with the statement that invasive plants were not a problem for farming/gardening or similar activities, with no significant difference between genders or amongst age or residency period groups (Tables 14.1 and 14.2).

Table 14.2 *Interpretation of respondent views expressed on six statements relating to regulation of the harvesting of trees, the impact of invasive plants and responsibility for forest management, in Kitulanghalo Forest, Tanzania*

| Statements | $\chi^2$ test of goodness-of-fit (for uniform agreement level) | | | $\chi^2$ test of independence (female vs. male) |
|---|---|---|---|---|
| | All respondents $\chi^2$ | Female respondents $\chi^2$ | Male respondents $\chi^2$ | $\chi^2$ |
| Harvesting of trees should be controlled. | $13.468^{**}$ | $10.334^{*}$ | $14.00^{**}$ | $9.203^{ns}$ |
| Harvesting of trees should be banned. | $116.001^{**}$ | $75.664^{**}$ | $49.777^{**}$ | $8.413^{ns}$ |
| Invasive plant species are not a problem to farming/gardening and other similar activities. | $32.651^{**}$ | $17.668^{**}$ | $19.555^{**}$ | $7.011^{ns}$ |
| Forest degradation and species invasion would be high under community-based management. | $27.934^{**}$ | $6.668^{ns}$ | $32.444^{**}$ | $9.877^{*}$ |
| Forest degradation and species invasion would be high under state management. | $25.735^{**}$ | $19.001^{**}$ | $10.666^{*}$ | $6.539^{ns}$ |
| Forest degradation and species invasion would be high under joint (participatory) management. | $15.601^{**}$ | $3.334^{ns}$ | $24.000^{**}$ | $11.588^{*}$ |

*Notes:* In all tests there were 4 degrees of freedom (d.f.); $*P<0.05$, $**P<0.01$; ns = not significant.

## Responsibility for forest management

Views expressed on the likelihood of forest degradation and species invasion varied amongst the management responsibility alternatives. The majority view (56 per cent) was that with community-based responsibility there was a likelihood of degradation and invasion (Table 14.1). However, the profile of responses was clearly bimodal – only 5 per cent were neutral, whereas a large minority disagreed (39 per cent). In contrast, the majority felt that degradation and invasion would not be a likely outcome of state professionals taking full (state management, 68 per cent) or part (joint participatory management, 61 per cent) responsibility. No differences in view were associated with respondent age or residency period. However, a gender effect was apparent, with no clear opinion offered by women on the likely consequences of community-based or joint (participatory) management, whereas 49 per cent of men strongly agreed, and another 18 per cent agreed, that community-based management would lead to high levels of forest degradation and species invasion (Table 14.2). In contrast 76 per cent of men disagreed or strongly disagreed that this would result from joint (participatory) forest management. While a majority of both genders felt that state management would not result in these problems, this opinion was particularly prevalent amongst women (77 per cent).

# DISCUSSION

## Communities and forest resource use

Until the new Forestry Act of 1998 was enacted, previous Tanzanian govern-ment policies had weakened local institutional systems with respect to the control and management of land and forests (Ministry of Natural Resources and Tourism, 1998; Luoga et al, 2005). The previous land and forest policies had not specifically considered communities but were strongly orientated towards centralized state control of resources, with land being a public resource (Kikula, 1986; Ylhaisi, 2003). The consequence was a substantial destruction and decline in Tanzanian forest cover (Kiunsi, 1994; National Environment Management Council, 1994; Kaoneka and Solberg, 1994; Yanda and Mung'ong'o, 1997); some of the effects continued after the 1998 act (Schwartz et al, 2002; Luoga et al, 2005).

In the present study, the local community respondents disagreed with the banning of tree harvesting from the forest. This is not surprising given their high dependence on forest resources. However, it was surprising that although there was overall majority agreement with some form of control of forest resource use, this agreement was expressed less strongly among the female respondents. This gender difference may be because of women's greater reliance on forest products for their livelihoods compared with men (Chitinga and Nemarundwe, 2003).

## Knowledge of invasive plant species

Our study showed that knowledge about invasive plants varied within the communities. Although invasive plants appear to be a widespread problem for the villages, not all of the respondents were aware of these species. Contrary to our expectation, knowledge about invasive plant species was not related to the time respondents had resided in the area, nor to their gender, but it was related to their age – ignorance of invasive species was more prevalent among the young. This may be attributed to elderly respondents having memory of the condition of land and its vegetation before the more recent major increase in the abundance of invasive species, and to barriers to the transfer of indigenous knowledge between age groups through local initiation systems.

A large majority of community members (especially those with the longest memories) perceived invasive species to be a problem and they recognized this to be the case for species that are biogeographically both exotic and native to the country. Shepherd (1989) found that farmers on the lower slopes of Mount Kenya were far more aware of whether a tree species was already familiar within the local landscape than whether it was native or exotic to the country in its origin. Therefore, for participatory forest management, and specifically the problems created by the abundance of weedy invasive species, there seems to be no value in distinguishing between those that are exotic or native to the country as a whole (which in the case of Tanzania covers a large area with a wide range of habitats and plant communities). Whether exotic or native, these species fall into one category in terms of the problems they create for land management, and they share traits such as prolific seeding, early age of first reproduction and easy establishment in degraded environments (Rejmanek, 1995). Therefore, a common approach to their management should be adopted, including the targeted reduction in their populations and measures to increase the resistance of the forest habitat to invasion.

## Responsibility for forest management

Despite the widespread international trend towards formal implementation of community-based, joint and other forms of participatory forest management, in a systematic review Bowler et al (2010) found that overall evidence of its benefit for forest condition is only weak and there is insufficient evidence to draw any conclusions about its impact on local livelihoods. For Tanzania, the strongest evidence has been provided by Blomley et al (2008), who used forest inventory plot data to show that in forests with participatory management involving local communities, forest condition (tree basal area, density, low rates of tree cutting and so on) was better than in forests under state or open-access management. Vyamana (2009) found that 63–93 per cent of the respondents in seven villages that had experienced joint forest management and community-based forest management perceived that forest condition had improved, whereas 63 per cent and 75 per cent of the respondents in the two control villages perceived that forest condition had worsened. Any inference of causality is, however, weakened by inadequate baseline information and lack of clarity about the criteria originally used to select villages for intervention by projects to promote joint/community-based forest management (Bowler et al, 2010).

Many studies in sub-Saharan Africa indicate that forest stakeholders, particularly those living close to forests, do give more support to community-based and joint (participatory) forms of forest management than to state forest management (Dème, 1998; Hilhorst and Coulibaly, 1998; Wily, 2000; Bass, 2001). Our results from two communities which had direct experience of both state and community-based forest management differed from these studies. When asked which type of responsibility for management would be most effective at combating forest degradation and plant invasion, the respondents gave little support to their own (community-based) management system and supported forests being managed by the state; women were uncertain about joint (participatory) management, though it was the system most strongly supported by men. Similar views, which contradict the idea of the supremacy of common property regimes, have also been found in other recent studies in sub-Saharan Africa (Campbell et al, 2001; Malleson, 2002; Obiri and Lawes, 2002). Some studies (for example Shackleton and Shackleton, 2000) have challenged the notion of communities being the best custodians of forests and

313

argued that it depends on whether or not the communities have a sense of ownership of the forests, and have strong rights to resources supported by laws. Luoga et al (2005) noted that such rights should include the entitlement to impose taxes on commercial wood harvesters and use the proceeds to facilitate community management of the forest (for example through policing). In the two communities of the present study, leaders had the right to distribute community forest land to local people, but the right to harvest certain tree species, such as the state-protected ones like *Pterocarpus angolensis* DC., remained with the state forestry authority. This division of power may have weakened the overall authority over forest resource use of community leaders in the eyes of local people. Persha and Blomley (2009) also highlighted the importance of well-designed institutional and governance arrangements for the success of community-based forest management; they emphasized the importance of tenure security and institutional autonomy.

## Community-based forest management

It is not surprising that the communities showed little confidence in their own collective forest management. In the first instance, the community-controlled forest has become severely degraded over a long period of time (Nduwamungu, 1996; Luoga, 2000; Zahabu, 2001), a situation that continued to the time of the present study (Luoga et al, 2005). This degradation of resources has reduced their value and lowered local people's support for community-based management (Dubois, 1999). There are, however, Tanzanian studies that provide evidence for the success, by some criteria (especially forest condition), of community-based forest management (Kajembe et al, 2004a and c; Mialla et al, 2004; Blomley et al, 2008; Persha and Blomley, 2009; Vyamana, 2009). Success was associated with those communities that had a much greater stake in the forests because they had unique user groups such as traditional ritual healers and leaders whose sacred activities were intimately connected with the forests. In such forests, conservation was found to be motivated by fear among the local population that tampering with trees could lead to bad omens befalling them. Such user groups, absent from the communities of the present study, strongly reinforce the socio-cultural ethics of forest use. However, in some cases in Tanzania, even with the existence of these ritual leaders and healers, communal management is still unsuccessful because people no longer carry out the traditional ways of nurturing the land with their accompanying customs following the villagization (resettlement) policies of the early 1960s to mid-1980s (Tanganyika African National Union, 1967) that disrupted the traditional land management systems (Kikula, 1986). Current policy favours communities moving on from the truly traditional management systems to a situation where there is cooperation between people of different ethnic backgrounds who jointly develop and follow agreed management procedures. However, the capacity of communities to enforce such new control mechanisms, especially over a prolonged period, is questionable (Ylhaisi, 2003; Luoga et al, 2005).

Also contributing to the lack of support for community-based forest management is the fact that community management does not guarantee an equal sharing of forest resources even if it does halt forest degradation. Such inequality has arisen under long-standing programmes of community-based forest management, where resources are unfairly distributed on the basis of clan, tribe or a household's influence: poor households often lose out to the wealthier, such that community-based forest management may even increase inequality (Vyamana, 2009). In the present study the relationship between distribution of benefits from forest resources and the economic levels of households could not be tested as most households were poor and inseparable economically. Nonetheless, unfairness in resource access rights could still exist because of other social differences, such as between clans or religions, prevalent in the region.

Although many successes in forest conservation have been reported under community-based management, we question how general and sustainable this is under the current rapidly changing demographic, social and economic conditions facing most communities, and the lack of evidence for benefits to the poor. We believe that this perception by local stakeholders strongly influenced the responses obtained in the present study, and our view is that successful forest conservation through community-based management depends on subtle issues such as equity in rights to access resources, in support of the conclusions of Vyamana (2009).

## State forest management

It is ironic that the local communities perceived the state to be the best custodian of the forest whilst the government is putting very little investment into safeguarding the state-managed forests in the study area. For instance, Luoga et al (2005), working in the same area, noted that a government-controlled forest covering over 2452 hectares had only two guards, who were ill-equipped and patrolled on foot. Two factors could be associated with the respondents' preference for state forest management. First, there is a prevalent view in Tanzania, particularly in rural areas, that the state has sufficient manpower and tools to handle a wide range of societal and environment issues, including forest management. Second, under existing conditions, local people harvest forest products from everywhere including within the 'guarded' state forests, and respondents will be uncertain if this situation would continue if responsibility for the management of forests changed hands. While the state may not be efficient in guarding against illegal tree harvesting, this situation might well change if the community joined in the forest management. Thus, it is in the short-term interests of those local people who do harvest illegally from the state forest for the status quo (of poor forest guarding by the state authority) to continue in preference to the uncertain future represented by a change in forest management responsibility where access to the resource might become more restricted for most community members.

## Joint (participatory) forest management

The lack of a common view amongst the respondents about joint (participatory) forest management points to the community's uncertainty towards this form of participation. This uncertainty is largely linked to a lack of clarity about models of participation among local people (Ylhaisi, 2003). There is a plethora of literature on participatory forest management whose definitions and style of implementation are in many cases dissimilar and contradictory (Ballabh et al, 2002; Campbell and Vainio-Mattila, 2003; Polansky, 2003). However, in broad terms participatory initiatives vary along a wide spectrum of power-sharing options where at one end there is a tendency towards de facto state control while at the other there is a tendency towards community control, with power more balanced in the centre (Campbell and Vainio-Mattila, 2003). While a balance of power is often advocated, it seems hardly ever to be achieved to the satisfaction of both categories of stakeholder. In fact, documentation describing many different participatory forest management programmes does not make it clear where on this continuum each lies. For instance, Hackel (1998) warned of the incoherence of the concept of participatory management largely because different people have such divergent views regarding any balance point; participatory forest management has often been advocated without people appreciating the difficulties involved in reconciling forest conservation (often driven by the state) and community development needs. It is this uncertainty that could have contributed to the doubts and thus neutral responses on joint (participatory) management expressed by the female respondents in the present study (only a minority of them thought it would be beneficial for the forest, whereas men were much more supportive). Vyamana (2009) found that benefits from joint forest management were captured by village elites.

## Gender and responsibility for forest management

In a number of southern African countries, including Tanzania, it can be argued that government rural development policies have neglected women, driving them into more dependency on harvesting forest resources. Land allocation and resettlement policies, for example, mainly support male-headed households and give less preference to female-headed ones (Chitinga and Nemarundwe, 2003). Further, studies in Tanzania and Zimbabwe (Chisanga, 1990; Tichagwa, 1998; Chitinga and Nemarundwe, 2003) have shown that government agricultural policies which support only farmers who own land (for example with loans for crops) favour men over women. This may be a major factor in women becoming more dependent on harvesting forest resources (such as fuelwood) for their basic, unchanging subsistence needs, and their greater responsibility for this within the household. Men are more likely to be able to reduce their exploitation of forest resources (for example harvesting charcoal or honey to sell) because they have more employment and business opportunities elsewhere. This potentially makes women the major agents of forest degradation after the initiation of forest protection measures (Chitinga and Nemarundwe, 2003).

The contrasting views of the female and male respondents on different types of forest management responsibility may be a manifestation of these differences in their forest resource dependency. The current position of loose state forest management over most of the forest area in their neighbourhood enables women to harvest resources at a level sufficient to meet their needs. If a new regime increased restrictions on forest resource harvesting, women would be more adversely affected. This may have been what motivated a large majority of female respondents to favour the status quo of state management over joint (participatory) or community-based management, even though the question asked referred to forest degradation and species invasion rather than their access to forest resources. Further support for this interpretation is provided by female respondents appearing to be less strong in their agreement with the control of tree harvesting than were men. However, the largest difference in responses between the genders was that men were much more confident that joint (participatory) forest management would be beneficial for forest condition than were women; as discussed in the section above this probably reflected greater female distrust that the institutions governing access to forest resources under that complex institutional management arrangement would act in their favour.

## CONCLUSIONS

The results of this study indicate that invasive weedy plant species, whose increase in abundance is associated with forest degradation, are of major concern to local communities, and merit much greater attention in ecologically informed forest management plans. Invasive species are an important neglected issue even when there are no 'headline grabbing' rapid expansions in the population of well-known exotic species. Weeding is a very time-consuming activity for most farmers (with major consequences for the availability of labour for other activities), yet the need for innovations to reduce this problem is often neglected in rural development projects. Forest management that results in a lower level of degradation offers the potential to reduce the source populations of these weeds, with major potential benefits for the livelihoods of those farming within forest landscapes.

Both women and men had major concerns about potential changes in the current forest management regime, being strongly against the banning of tree harvesting, while accepting some form of control of this activity. Men were more certain about the form of forest management responsibility that they considered best for the forest (participatory joint forest management) and which they least favoured (community-based management). The majority of women believed that state forest management was best, and disfavoured the two participatory forest management alternatives (with no clear preference between them), which we have interpreted as a result of women's greater dependency on harvesting forest products for their livelihoods. These results (in parallel with those of a number of other recent studies in sub-Saharan Africa) challenge the hegemony of the view that communities as a whole inherently support community-based management of forests over state management. This community opinion corresponds with the independent research of Vyamana (2009) in other Tanzanian villages showing that the evidence that participatory forest management reduces poverty and social exclusion is weak.

There is, therefore, a strong need for the opinions of crucial local stakeholders to be elicited, rather than assumed, in the process of developing the most appropriate form of participatory forest management to meet local needs and preferences. Negotiations over changes in forest governance should engage all sections of the community (with particular attention to gender issues) and be informed by their current dependence on forest resources and how access will be affected by any change. In the communities researched by this study, there is sufficient support for negotiation to be carried out with the state on joint (participatory) forest management, incorporating some form of control of forest resource use, but this process must be particularly sensitive to women's views and needs. Wider-scale policy development, and the potential to inform communities of what forms of participatory forest management have worked best elsewhere, have a strong need for a much improved evidence base. To achieve this it will be important for robust research to be integrated into participatory forest management project activities.

# REFERENCES

Ballabh, V., Balooni, K. and Dave, S. (2002) 'Why local resource management institutions decline: A comparative analysis of Van (forest) Panchayats and Forest Protection Committees in India', World Development, vol 30, pp2153–2167

Bass, S. (2001) 'Working with forest stakeholders', in J. Evans (ed) The Forest Handbook: Applying Forest Science for Sustainable Management, Blackwell Science, Oxford, UK, pp221–231

Blomley, T. and Ramadhani, H. (2006) 'Going to scale with Participatory Forest Management: Early lessons from Tanzania', International Forestry Review, vol 8, pp93–100

Blomley, T., Pfliegner, K., Isango, J., Zahabu, E., Ahrends, A. and Burgess, N. (2008) 'Seeing the wood for the trees: An assessment of the impact of participatory forest management on forest condition in Tanzania', Oryx, vol 42, pp380–391

Bowler, D., Buyung-Ali, L., Healey, J. R., Jones, J. P. G., Knight, T. and Pullin, A. S. (2010) 'The evidence base for community forest management as a mechanism for supplying global environmental benefits and improving local welfare', Collaboration for Environmental Evidence www.environmentalevidence.org/SR48.html, accessed 21 May 2010

Campbell, B., Mandondo, A., Nemarundwe, N., Sithole, B., Dejong, W., Luckert, M. and Matose F. (2001) 'Challenges to proponents of common property resource systems: Despairing voice from the social forests of Zimbabwe', World Development, vol 29, pp589–600

Campbell, L. M. and Vainio-Mattila, A. (2003) 'Participatory development and community based conservation: Opportunities missed and lessons learnt?', *Human Ecology*, vol 31, pp417–437

Chisanga, M. (1990) *The Social Cost of Structural Adjustment: Some Evidence from Africa*, WID Structural Adjustment Paper 3a

Chitinga, M. and Nemarundwe, N. (2003) 'Policies and gender relationships and roles in the miombo woodland region of southern Africa', in G. Kowero, B. M. Campbell and U. R. Sumaila (eds) *Policies and Governance Structures in Woodlands of Southern Africa*, Center for International Forestry Research (CIFOR), Bogor, Indonesia, pp187–212

Dème, Y. (1998) 'Natural resources management by local associations in the Kelka Region of Mali', Issue Paper No. 74, International Institute for Environment and Development (IIED), London

Dubois, O. (1999) 'Assessing local resilience and getting roles right in collaborative forest management: Some current issues and potential tools, with special reference to sub-Saharan Africa', in *Proceedings of an International Workshop on Pluralism and Sustainable Forestry and Rural Development*, FAO, Rome, pp49–83

Elands, B. H. M., O'Leary, T. N. O., Boerwinkel, H. W. J. and Wiersum, K. F. (2004) 'Forests as a mirror of rural conditions; local views on the role of forests across Europe', *Forest Policy and Economics*, vol 6, pp469–482

Flinkman, M. (2004) 'Institutions and wood supply to construction wood markets in Dar es Salaam and Mwanza: Implications for policy and enforcement', *Scandinavian Journal of Forest Research*, vol 19, pp25–35

Gibson, C. C., McKean, M. A. and Ostrom, E. (2000) *People and Forests: Communities, Institutions, and Governance*, MIT Press, Cambridge, MA

Hackel, J. D. (1998) 'Community conservation and the future of Africa's wildlife', *Conservation Biology*, vol 13, pp726–733

Hilhorst, T. and Coulibaly, A. (1998) 'Elaborating a local convention for managing village woodlands in southern Mali', Issue Paper No. 78, International Institute for Environment and Development, London

Kajembe, G. C., Namubiru, E. L., Shemwetta, D. T. K., Luoga, E. J. and Mwaipopo, C. (2004a) 'The impact of rules in forest conservation in Tanzania: The case of Kwizu Forest Reserve, Same District, Kilimanjaro', in D. T. K. Shemwetta, E. J. Luoga, G. C. Kajembe and S. S. Madoffe (eds) *Institutions, Incentives and Conflicts in Forest Management: A Perspective*, Mzumbe University Press, Morogoro, Tanzania, pp92–107

Kajembe, G. C., Shemwetta, D. T. K. and Luoga, E. J. (2004b) 'IFRI country report', in D. T. K. Shemwetta, E. J. Luoga, G. C. Kajembe and S. S. Madoffe (eds) *Institutions, Incentives and Conflicts in Forest Management: A Perspective*, Mzumbe University Press, Morogoro, Tanzania, pp1–8

Kajembe, G. C., Shemwetta, D. T. K., Luoga, E. J. and Nduwamungu, J. (2004c) 'Incentives for sustainable forest management in Tanzania', in D. T. K. Shemwetta, E. J. Luoga, G. C. Kajembe and S. S. Madoffe (eds) *Institutions, Incentives and Conflicts in Forest Management: A Perspective*, Mzumbe University Press, Morogoro, Tanzania, pp80–91

Kaoneka, A. R. S. and Solberg, B. (1994) 'Forestry related land-use in the west Usambara Mountains, Tanzania', *Agriculture, Ecosystem and Environment*, vol 49, pp207–215

Kikula, I. S. (1986) 'Environmental effects of Tanzania's villagisation programme', PhD thesis, School of Australian Environmental Studies, Griffin University, Australia

Kiunsi, R. B. (1994) *Vegetation Loss in Rural Settlements – Causes and Remedies with Examples from Tanzania*, Dar es Salaam University Press, Dar es Salaam

Luoga, E. J. (2000) 'The effects of human disturbance on diversity and dynamics of eastern Tanzania miombo woodlands', PhD thesis, University of Witwatersrand, Johannesburg, South Africa

Luoga, E. L., Witkowski, E. T. F. and Balkwill, K. (2000) 'Differential utilisation and ethnobotany of trees in Kitulangalo Forest Reserves and the surrounding community lands, eastern Tanzania', *Economic Botany*, vol 54, pp328–343

Luoga, E. L., Witkowski, E. T. F. and Balkwill, K. (2005) 'Land cover and use changes in relation to the institutional framework and tenure of land and resources in eastern Tanzania miombo woodlands', *Environment, Development and Sustainability*, vol 7, pp71–93

Malleson, R. (2002) 'Changing perspectives on forests, people and "development": Reflections on the case of the Korup forest', *IDS Bulletin*, vol 33, pp94–101

Mialla, Y. S., Kajembe, G. C. Malimbwi, R. E. and Nduwamungu, J. (2004) 'Participatory and conventional forest assessments: A case study of Monduli-Mlimani catchment forest, Monduli District, Tanzania', in D. T. K. Shemwetta, E. J. Luoga, G. C. Kajembe and S. S. Madoffe (eds) *Institutions, Incentives and Conflicts in Forest Management: A Perspective*, Mzumbe University Press, Morogoro, Tanzania, pp69–79

Ministry of Lands, Natural Resources and Tourism (1989) *Tanzania Forestry Action Plan 1990/91–2007/08*, The United Republic of Tanzania, Government Printer, Dar es Salaam

Ministry of Natural Resources and Tourism (1998) *Natural Forest Policy*, The United Republic of Tanzania, Government Printer, Dar es Salaam

National Environment Management Council (1994) *The Proposal for the National Conservation Strategy for Sustainable Development*, National Environment Council, Dar es Salaam

Nduwamungu, J. (1996) 'Tree and shrub diversity in miombo woodlands: A case study of SUA Kitulanghalo Forest Reserve, Morogoro, Tanzania', MSc dissertation, Faculty of Forestry, Sokoine University of Agriculture, Morogoro, Tanzania

Nhira, C., Baker, S., Gondo, P., Mangono, J. J. and Marunda, C. (1998) *Contesting Inequality in Access to Forests*, Policy that Works for Forests and People Series No. 5, Centre for Applied Social Services and Forestry Commission, Harare, and International Institute for Environment and Development, London

Obiri, J. A. F. and Lawes, M. J. (2002) 'Challenges facing new forest policies in South Africa: Attitudes of forest users towards management of the coastal forests of Eastern Cape Province', *Environmental Conservation*, vol 29, pp519–529

Persha, L. and Blomley, T. (2009) 'Management decentralization and montane forest conditions in Tanzania', *Conservation Biology*, vol 23, pp1485–1496

Polansky, C. (2003) 'Participatory forest management in Africa: Lessons not learned', *International Journal of Sustainable Development and World Ecology*, vol 10, pp109–112

Rejmanek, M. (1995) 'What makes a species invasive?', in P. Pysek, K. Prack, M. Rejmanek and M. Wade (eds) *Plant Invasions – General Aspects and Special Problems*, SPB Academic Publishers, Amsterdam, pp3–13

Schwartz, M. W. and Caro, T. M. (2003) 'Effect of selective logging on tree and under-story regeneration in miombo woodland in western Tanzania', *African Journal of Ecology*, vol 41, pp75–82

Schwartz, M. W., Caro, T. M. and Banda-Sakala, T. (2002) 'Assessing the sustainability of harvest of *Pterocarpus angolensis* in Rukwa Region, Tanzania', *Forest Ecology and Management*, vol 170, pp259–269

Shackleton, C. M. and Shackleton, S. E. (2000) 'Direct use values of secondary resources from communal savannas in Bushbuckridge lowveld, South Africa', *Journal of Tropical Forest Products*, vol 6, pp28–57

Shemwetta, D. T. K., Kajembe, G. C., Malimbwi, R. E. and Nduwamungu, J. (2004) 'The role of diverse institution in the rule of governance and generation of conflicts over forest products in Tanzania', in D. T. K. Shemwetta, E. J. Luoga, G. C. Kajembe and S. S. Madoffe (eds) *Institutions, Incentives and Conflicts in Forest Management: A Perspective*, Mzumbe University Press, Morogoro, Tanzania, pp29–39

Shepherd, G. (1989) 'Putting trees into the farming system: Land adjudication and agroforestry on the lower slopes of Mount Kenya', Social Forestry Network Paper 8a, Overseas Development Institute, London

Tanganyika African National Union, (1967) *The Arusha Declaration and TANU's Policy on Socialism and Self-Reliance*, Government Printer, Dar es Salaam

Tichagwa, W. (1998) *Beyond Inequalities: Women in Zimbabwe*, Zimbabwe Women's Resource Centre and Network and Southern Africa Research and Documentation Centre, Harare

Tindall, D. B. (2003) 'Social values and the contingent nature of public opinion and attitudes about forests', *The Forestry Chronicle*, vol 79, pp692–705

Vyamana, V. G. (2009) 'Participatory forest management in the Eastern Arc Mountains of Tanzania: Who benefits?', *International Forestry Review*, vol 1, pp239–253

Weis, T. (2000) 'Beyond peasant deforestation: Environment and development in rural Jamaica', *Global Environmental Change: Human and Policy Dimensions*, vol 10, pp299–305

Wily, L. A. (2000) 'Forest laws in eastern and southern Africa: Moving towards a community based forest future?', *Unasylva*, vol 203, pp19–26

Yanda, P. Z. and Mung'ong'o, C. G. (1997) 'The farming systems of Kasulu District, western Tanzania: A case study of Buhoro, Ruhita and Titye Villages', consultancy report to ENRECA's joint research committee, Institute of Resource Assessment, University of Dar es Salaam, Dar es Salaam

Ylhaisi, J. (2003) 'Forest privatisation and the role of community in forests and nature protection in Tanzania', *Environmental Science and Policy*, vol 6, pp279–290

Zahabu, E. (2001) 'Impact of charcoal extraction on the miombo woodlands: The case of Kitulangalo Area, Tanzania', thesis, Sokoine University of Agriculture, Morogoro, Tanzania

15

# Forest Dynamics in Southwest Ethiopia: Interfaces between Ecological Degradation and Resource Enrichment

*K. Freerk Wiersum*

## INTRODUCTION

Human impact on forests and tree resources may take two different forms. From an ecological conservation perspective attention is mostly focused on negative human impacts resulting in degradation in the form of loss of biodiversity and ecosystem services. This negative human impact is related to human overexploitation and to decreased production potential. Alternatively, from a socio-economic development perspective, attention is also given to human agency in enriching natural resources through a process of domestication resulting in resource enrichment. For understanding the relations between these different types of human impacts it is useful to consider the process of domestication in more detail. In a narrow sense, domestication is considered as a biological process involving changes in morphological and genetic make-up of selected crop, tree or animal species in order to increase its productivity (Simons and Leakey, 2004). In a wider sense, it is considered as also involving an ecological process in the form of homogenization of production conditions: ecosystems are adjusted to the human domain by increasing the production of human-valued natural resources. And in an even more inclusive interpretation,

323

domestication is conceived of as also including an acculturalization process involving the adaptation of species to human-controlled production conditions (Wiersum, 1997a; Simons and Leakey, 2004), including the implementation of conscious human practices for the conservation and stimulation of the provisioning of valuable natural resources. As a result of the multidimensional nature of domestication, there are different opinions about the relation between domestication and degradation in forest ecosystems. It can be argued that domestication as a process of tree improvement involves a loss in genetic diversity and that it results in monospecific tree plantations of genetically selected trees (Simons and Leakey, 2004). From this point of view, domestication of forest resources is often considered as reflecting negative human impacts on the forest ecological structure and functioning, and hence as a degradation process. However, increasingly attention is given to domestication as a process of increasing production in biodiverse agroforestry systems (Simons and Leakey, 2004), and it is argued that domestication involves a process of human agency in enriching forest resources and improving livelihood conditions. Several authors (Michon and De Foresta, 1997; Wiersum, 1997a) have indicated that the process of forest domestication does not involve a simple process of creating monocultural plantations, but rather a multi-trajectory pathway involving various stages ranging from forest conservation to forest modification and forest transformation. From this perspective, domestication of forest resources is considered as reflecting human agency in adapting forests to human environments.

Several authors have indicated that for better understanding of the multi-faceted nature of forest dynamics, and for understanding forest dynamics, the interactions between the processes of degradation due to human exploitation and resource enrichment through human creativity in stimulating the production capacity of valuable forest resources need to be assessed (McNeely, 2004). A better understanding is needed of the multiple dimensions of biocultural relations as a means towards identifying tradeoffs between ecological conservation and human development (Lamb et al, 2005). Several authors have argued that in order to better understand these tradeoffs, more attention should be given to the intermediate phases in the process of domesticating forests (Michon and De Foresta, 1997; Wiersum, 2004; Perfecto and Vandermeer, 2008). This is not only of relevance for better understanding of the multifaceted impact of humans on forests, but may also provide valuable lessons for forest restoration programmes. Such programmes will be most effective if they, apart from restoring forest ecological conditions, consider restoration as a co-evolutionary biocultural process including the stimulation of human agency in adjusting forest conditions to multiple human values in relation to the conservation and use of forest resources.

To date, few systematic studies have analysed domestication as a co-evolutionary process involving changes in forest ecosystem properties and production and management conditions (Homma, 1992; Hladik et al, 1993; Den Hertog and Wiersum, 2000; Paudel and Wiersum, 2004; Michon, 2005). The dynamics in coffee (*Coffea arabica*) production systems in the southwestern highlands of

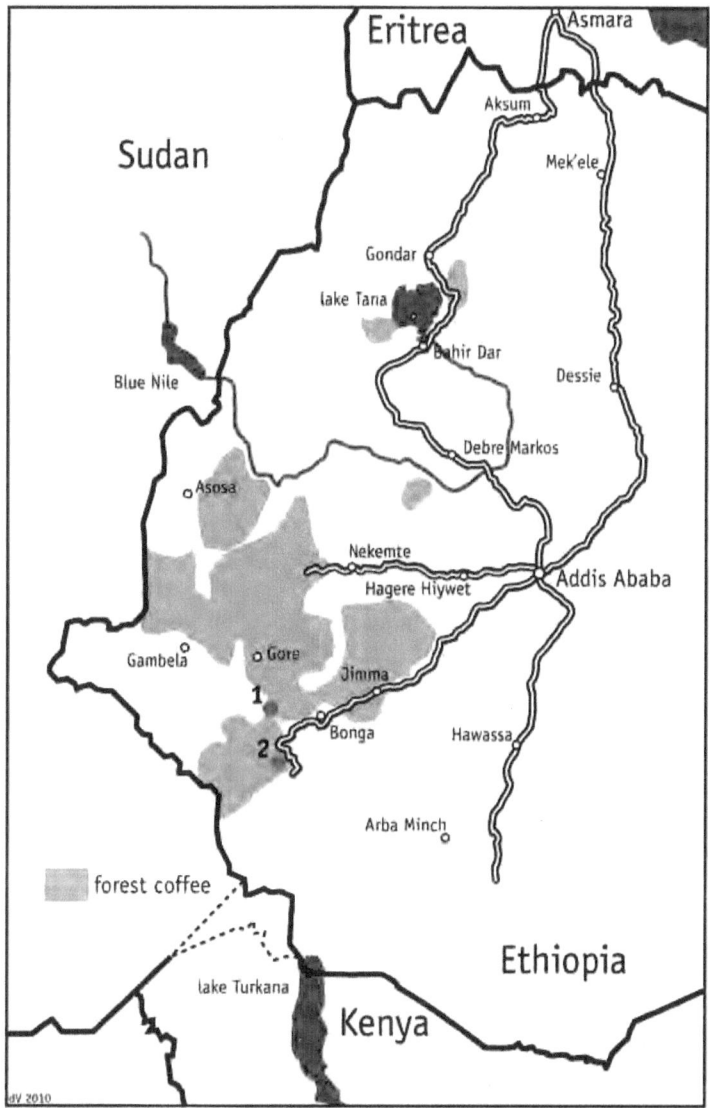

**Figure 15.1** *The main forest coffee-producing areas in Ethiopia,*
*and study locations*

*Notes:* 1 = Sheka forest, 2 = Bench Maji forest.

Ethiopia provide an interesting case. Coffee grows naturally in Afromontane
forest at an altitude of about 1000m to 2000m (Gole et al, 2008). At the
altitude of 1050–1500m coffee is a gregarious sub-canopy species. According to
legend, coffee as a crop originated in the Kaffa region, from where it has spread
over the world (Gole et al, 2000). In several regions in southwest Ethiopia
coffee is still being collected in the natural forests as wild forest coffee (Figure

15.1; Gole et al, 2000; Senbeta and Denich, 2006; Wiersum et al, 2008). Several forests have been modified to stimulate coffee production; this has resulted in the creation of semi-forest coffee systems in which coffee production is stimulated by different management practices. In Ethiopia about 10 per cent of all forest production is obtained from forest coffee systems and 35 per cent from semi-coffee forest systems (Wiersum et al, 2008). Moreover, coffee production was further stimulated by the development of transformed production systems in the form of forest gardens (35 per cent of national coffee production) and plantations (20 per cent).

This chapter describes the multifaceted nature of the biocultural process of domestication of the coffee forests in the southwest Ethiopian highlands. It assesses the tradeoffs between ecological degradation and resource enrichment by analysing the major dimensions of the domestication process. The following questions are addressed:

1   What role do forests play in the local livelihood conditions in southwest Ethiopia?
2   What are the ecological and management characteristics of the coffee production systems present in the region?
3   What were the main processes of acculturalization influencing the production systems?

The discussion includes the consequences of such biocultural assessment for planning of forest restoration programmes.

## FOREST AND LIVELIHOOD CONDITIONS IN SOUTHWEST ETHIOPIA

### Forest conditions

The southwest highlands of Ethiopia currently include the largest of the two remaining continuous blocks of Afromontane forest vegetation in the country. The highlands cover an altitudinal range from 900m to 2700m asl and form the upper catchments of several important rivers, such as the Baro and Akobo (tributaries of the Nile) and the Omo. The forests in this region do not only play a major role in water regulation of these rivers, but this region is also a Biodiversity Hotspot of global interest with *Coffea arabica* as a flagship species (Gole et al, 2000). Historically, the region was rather remote from the main populated regions in Ethiopia. Consequently, the local communities are highly dependent on the forest resources for their livelihoods, with coffee forming the most important non-timber forest product. However, for a variety of reasons including immigration and opening up of lands for commercial estates, at present the conservation of these forests is not assured and they are already considerably fragmented (Bognetteau et al, 2007).

**Table 15.1** *Main forest and land-use characteristics in selected upland and mid hill regions in SW Ethiopia*

|  | Upland region | Mid hill region |
|---|---|---|
| Administrative zone | Sheka | Bench Maji |
| Altitude | 1800–2600m asl | 900–1800m asl |
| Natural vegetation | Mixed deciduous forest; Bamboo forests | Mixed deciduous forests with coffee as a characteristic under-storey species |
| Forest cover | 50–60% | ca. 15% |
| Land use | Forest use | Various types of coffee exploitation: <br> – coffee extraction from natural and semi-natural forests <br> – garden coffee cultivation <br> – coffee plantations |
|  | Small-scale subsistence-oriented agriculture | Small-scale agriculture, with some locally marketable products |
| Average size cropland/ household |  |  |
| Rich households | 3.1ha | 9ha, mainly coffee land |
| Medium rich households | 2.2ha | 4.2ha, mainly coffee land |
| Poor households | 0.8ha | 0.7ha |

*Source:* Bognetteau et al (2007)

The degree of this forest degradation varies between the two main geographic zones within the region (Bognetteau et al, 2007). These are characterized by different forest and land-use types as well as different degrees of forest cover (Table 15.1). In the mid hill area mixed deciduous forests occur with coffee as a characteristic understorey crop. The production of forest coffee forms a major household activity, supplying a significant contribution to household income; this livelihood activity together with collection of other forest products supplements subsistence agriculture. The economic importance of the forests for the local communities has contributed to their conservation. However, due to the need for smallholder agricultural lands for food production, and the gradual establishment of specialized coffee cultivation systems, as well as the opening up of lands for commercial estates, the forest cover is fragmented and ranges in different areas from 28 to 80 per cent with an average of 59 per cent (Bognetteau et al, 2007). At higher altitudes coffee is absent and deciduous forests are complemented by bamboo forests. In these uplands the forest cover is much higher than in the mid hills and most agriculture is subsistence-oriented. In this area traditionally wild honey is the main non-timber forest product.

## Main livelihoods strategies

The importance of forested landscapes for contributing towards human liveli-hoods can be demonstrated by the results of a comparative case study on the livelihood activities and income earning of villagers in two administrative zones in the mid hill region and the upland region respectively (Chilalo, 2007). In general, local people were engaged in multiple livelihood activities. The largest number of respondents (48 per cent) indicated agriculture in the form of food crop and garden coffee production as being their main occupation, while 25 per cent were engaged in both agricultural production and forest product extrac-tion and 27 per cent were engaged in the sole collection of forest coffee (22 per cent) or honey (5 per cent).

The importance of the different household activities is reflected by the incomes derived from the separate livelihood activities (Table 15.2). The income from farm production is relatively low: these activities contributed only 48 per cent to the household cash income. This indicates its predominant subsistence nature, with most households growing crops primarily for home consumption. On average about 73 per cent of crop produced was consumed within the households. In the mid hills, the food crop production is supple-mented by coffee production in garden systems, but these amounts are relatively small among the respondents. The major sources of cash income for the households are forest products, providing 49 per cent (albeit only slightly higher than farming income). Although not well developed, households were also deriving some limited income (on average 2 per cent) from off-farm activi-ties and remittances.

Upland and mid hill zones clearly differ in the contribution of farming production and forest products to household incomes (Table 15.2). The average annual household cash income is almost twice as high in the mid hills as in the uplands. Notably the contribution from forest products is much higher, whereas the difference in average income from farming is much smaller. Overall forest products contribute 52 per cent to average household income in the mid hills against 41 per cent in the uplands. In the mid hills 69 per cent and 24 per cent of the household are engaged in forest coffee and honey collection respectively; the average income of households engaged in these activities is 1249 Ethiopian

**Table 15.2** *Average annual household cash income (Birr/year and relative contribution) in Bench Maji and Sheka administrative zones in SW Ethiopia*

| Income source | Sheka zone (N = 55) | Bench Maji zone (N = 90) | Combined zones (N = 145) |
|---|---|---|---|
| Geographic position | Uplands | Mid hills | |
| Total cash income | 1038 (100%) | 1878 (100%) | 1560 (100%) |
| Forest product income | 428 (41%) | 980 (52%) | 771 (49%) |
| Farming income | 559 (54%) | 858 (46%) | 745 (48%) |
| Off-farm income and remittances | 51 (5%) | 45 (2%) | 44 (2%) |

*Source:* Chilalo (2007)

Birr (ETB) per year and ETB405 per year respectively. In the uplands 13 per cent and 73 per cent of the households are engaged in these two activities, earning annually ETB653 and ETB464 respectively. These data reflect the lower commercial value of the main upland product, honey, in comparison with the main mid hill product, coffee, as well as the less favourable marketing infrastructure in the uplands.

The role of forests in the local livelihoods does not only differ between the mid hill and upland zones, but also between different household categories. For some households, the presence of different forest and cultivation systems offers the opportunity to be involved in a diversity of livelihood activities. These multi-enterprise households engage in both commercial crop production and coffee and honey collection. This diversified livelihood strategy with high-value forest production supplementing agricultural production provides an annual income of over ETB3500. Other households specialize in either forest coffee collection or honey collection, often with additional commercial or subsistence-oriented crop production. The specialization strategy may include activities for further processing and manufacturing of the commodities, thus adding value to their production. For instance, farmers specialized in honey production often manufacture the honey into the local honey beer. These more specialized livelihood strategies provide fair livelihood options with incomes between ETB2000 and ETB2700 per year. A third category of households is engaged mostly in farming and supplements this activity with collection of honey or low-value non-timber forest products such as spices and bamboo. These households earn less than ETB1500 per year. They are engaged in a coping strategy with forests providing the opportunity to supplement their marginal farming activities.

Similar results of forest-derived incomes being as important as incomes from farming were found in other studies. Bognetteau et al (2007) reported for the same region as Chilalo's study contributions of honey and forest coffee to local livelihoods of 18 per cent and 6 per cent respectively in the uplands and 8 per cent and 22 per cent respectively in the mid hills. As these estimates relate to net household incomes rather than gross household incomes as in Chilalo's study, they cannot be compared directly. Nonetheless, the trends are similar, with a greater livelihood impact of forest products in the mid hills as compared to the uplands, and a greater importance of coffee in the mid hills against honey in the uplands. In an adjacent region Mamo et al (2007) found that 40 per cent of household income was derived from farming practices, 39 per cent from forest products and 21 per cent from other sources.

## Conclusion

In SW Ethiopia the area of natural forest cover is decreasing due to conversion of forests to other forms of land use. This conversion does not only involve food crop production systems, but also mixed tree crop production systems. The presence of different types of modified and transformed forest system demonstrates that local people have not just decreased the forest area, but they have also created a mosaic of forest and forest-analogue and resource-enriched land-use systems under different management intensity. This diversified production base of forest products reflects global patterns in the use and management of commercial forest products (Belcher et al, 2005b). This process is most advanced in the more accessible mid hill regions, where marketing conditions are more favourable than in the uplands. Notwithstanding the higher loss of original forest cover in this area as compared to the upland areas, the incomes derived from the forests are higher in the mid hill region than in the upland region. This indicates the important livelihood role of the remaining forests in this region. The landscape transformation can be considered from an ecological point of view as involving a process of forest degradation characterized by the decrease in the original forest cover. However, from a human perspective the creation of a forested mosaic landscape has contributed towards an improvement in livelihood options and increased incomes from both cultivated and forest lands.

# ECOLOGICAL AND MANAGEMENT CHARACTERISTICS OF COFFEE PRODUCTION SYSTEMS

In order to better understand the nature of the forested mosaic lands, it is useful to consider the ecological and management characteristics of the different coffee production systems in more detail. Coffee is not just collected in the forests from wild plants, but also in various types of forest-analogue production systems, in which the coffee production is stimulated by different management practices such as thinning of overstorey trees, removal of ground vegetation, stimulating coffee coppice shoots, and/or transplanting naturally regenerated or artificially raised coffee seedlings (Schmitt, 2006; Senbeta and Denich, 2006). Depending on the vegetation structure, species composition and intensity of management, four main forest-based coffee production systems can be distinguished (Schmitt, 2006; Senbeta and Denich, 2006; Wiersum et al, 2008):

1  Forest coffee system: extraction of coffee from wild trees in little disturbed natural forests.
2  Semi-forest coffee system: in these modified forests limited management practices are carried out, such as removal of competing trees and stimulation of natural regeneration of coffee plants, complemented with planting of wild coffee seedlings from the forest.
3  Garden coffee systems with more intensive management: coffee plants are mostly regenerated from selected seedlings or with nursery-raised cultivars; These forest gardens can be characterized as forest-analogue systems with a mixed species character consisting of both consciously conserved trees from the natural forests and a variety of planted trees.
4  Smallholder and estate coffee plantations of monocultures with scattered shade trees.

**Table 15.3** *Ecological and management characteristics of different coffee production systems in SW Ethiopia*

|  | Forest coffee system | Semi-forest coffee system | Garden coffee system | Coffee plantation system |
| --- | --- | --- | --- | --- |
| (Agro)ecosystem characteristic | Natural vegetation; Original forest structure and composition | Adapted natural vegetation; Modified forest structure; Stimulation of coffee plants | Transformed forest structure; Polyculture of planted crops with maintained wild species | Monoculture (with shade trees) |
| Dominant management practices in respect to harvesting control and stimulation of production | Social control on coffee extraction | Manipulation of tree vegetation for shade management and decreased competition; Transplanting coffee wildlings | Artificial regeneration of valuable species; Agronomic cultivation practices | Intensive cultivation; Use of external inputs (fertilizers, pesticides) |
| Methods of coffee regeneration | Unassisted natural regeneration | Protection of natural regeneration; Purposeful spacing of wild seedlings | Protection of natural regeneration; Transplanting of wild seedlings or nursery-raised seedlings | Use of nursery-raised seedlings of selected cultivars |
| Genetic composition | Wild genotypes | Wild genotypes; Landraces | Landraces; Cultivars | Cultivars |
| Average Annual yields | About 50kg/ha | 100–300kg/ha | 400–500kg/ha | Over 750kg/ha |

*Source:* Wiersum et al (2008)

The different coffee production systems are characterized by a specific combination of ecological characteristics, management practices and productivity (Table 15.3). The forest and semi-forest systems concern a multi-storeyed and mixed-species forest ecosystem (Schmitt, 2006; Senbeta and Denich, 2006). Due to the limited interference in the forest canopy and low-intensity coffee management, the productivity of the coffee in the forest systems is low, with an average annual yield of about 50kg/ha. As a result of the practices to enrich the coffee resources in the semi-forest coffee systems the forest structure becomes less complex (Figure 15.2). Senbeta and Denich (2006) reported that species richness in the semi-forest coffee system is around 30–40 per cent lower than in the less disturbed forest coffee systems (Table 15.4). As a result of the management practices the coffee yield has increased to around 100–300kg/ha (Senbeta and Denich, 2006; Schmitt, 2006; Wiersum et al, 2008). The coffee gardens mimic the forest structure by having a mixture of coffee trees and other useful species. The original forest species are mostly limited to shade trees and other crops, but also additional crops such as fruit trees, tubers, spices and false banana (*Enset ventricosum*) are grown (Teketay and Tegineh, 1991; Abebe et al, 2006). However, the species diversity has further decreased; Hylander and

331

**Table 15.4** *Species diversity in forest coffee systems and semi-coffee forest systems at two locations in SW Ethiopia*

|  |  | Forest coffee systems | Semi-coffee forest systems |
|---|---|---|---|
| Species diversity | Mean number of species/ha | 30–32 | 17–27 |
|  | Total no. of species | 194–275 | 178–199 |
| Number of species per growth form | Canopy trees | 5–11 | 5–8 |
|  | Medium trees | 27–38 | 22–26 |
|  | Understorey and shrubs | 55–60 | 25–50 |
|  | Lianas | 45–70 | 40–41 |
|  | Herbs | 48–61 | 46–75 |
|  | Epiphytes | 25–35 | 22–38 |
| Shannon diversity index (all vascular species) |  | 2.60–2.82 | 1.23–0.90 |

*Source:* Senbeta and Denich (2006)

Nemomissa (2008) found that 42 per cent of the tree species found in the forests were also present in the coffee gardens. Due to the further intensified management practices the coffee productivity in these gardens has increased to around 400–500kg/ha. Only the coffee plantations can be characterized as essentially being a mono-specific cultivation system, even though in this system scattered shade trees are usually present; the average coffee yields in these coffee plantations are around 750kg/ha (Schmitt, 2006; Wiersum et al, 2008).

The gradual intensification of coffee production systems does not only affect the structure and species-richness of the production systems, but also the genetic diversity of coffee (Table 15.3). Most wild coffee regions possess their own genotypes with high levels of within-region genetic variety. Coffee management practices involve a gradual genetic adaptation of the coffee plants, and with increasing management intensity a gradual decrease in the original genetic diversity occurs. Whereas in wild forest coffee systems mainly wild genotypes are present, in semi-forest coffee and garden systems landraces predominate, and in plantations cultivars (Geletu, 2006; Wiersum et al, 2008).

## ACCULTURALIZATION OF COFFEE PRODUCTION SYSTEMS

The process of domestication of the coffee forests does not only involve a process of biological and ecological change as a result of gradual intensification of management practices, but also a process of acculturalization in the sense of adaptation of species to human control systems. Two main processes of acculturalization may be distinguished: incremental formulation of access rights to forest resources and adaptation of production systems to market conditions.

**Figure 15.2** *Forest coffee transects in Bonga region, southwest Ethiopia*

*Notes:* Upper graph: a forest coffee system with undisturbed forest structure. Both upper and lower canopies are closed. In shrub layer coffee is patchily distributed. No large coffee trees. Lower graph: a semi-forest coffee system with disturbed forest structure and increased coffee cover. The upper canopy here is strongly dominated by *Milletia ferruginea*. The lower canopy and shrub layer are dominated by large coffee trees. In both graphs the shaded plants are coffee plants.
*Source:* After Schmitt (2006)

## Dynamics in regulatory arrangements

Management of forest resources does not only imply the implementation of different management practices for influencing production, but also the organization and control of these management practices. Forest domestication often involves changes in the way management is organized, for example in respect to access rights to the resources and management responsibility (Wiersum, 1997b). The importance of this social dimension of domestication is demonstrated by the different forms of management organization for the various coffee production systems. The access rights to coffee in the various production systems are governed by a combination of agrarian and forest tenure regulations (Table 15.5). The agricultural land tenure regulations have changed considerably during the last decades from a feudal system to a socialist model with semi-collectivist villages, to the present smallholder system based on private (freehold) ownership (Omiti et al, 1999). In contrast, the forest tenure system has historically been vested in the state. The forest coffee and semi-forest coffee systems are partly located in state-owned forest reserves and partly on private lands. In the state forest areas local people may claim collection rights in specific forest blocks, but the local people have no rights to own or inherit trees. Nonetheless, the claimants may carry out some management practices. In the privately owned forest areas the access rights to coffee and the management responsibility are held by the landowner; under customary arrangements the coffee berry collection rights are sometimes given to shareholders. The garden coffee and plantation coffee systems are usually privately owned with access rights and management responsibility being held by the landowner. This situation has gradually emerged under influence of the changing policy and tenure conditions (Table 15.6). During the ancient feudal regime, the coffee forests were controlled by local landlords. Peasants were forced to manage these lands, but could not produce coffee privately. During the subsequent Derg regime land reform and village formation was stimulated and peasants gave priority to (re)developing their farmlands, which resulted in less attention to forest coffee production. During the initial period of the present regime emphasizing smallholder development and state forest demarcation, forest coffee production was given less prominence in contrast to garden coffee production.

These data illustrate that the management of the forest resources is based on two principles. It is considered that the government is responsible for the maintenance of the forests as a public good and the conservation of environmental services. This has resulted in statutory rights on government ownership and control over forests. These formal regulations may be supplemented by more informal customary arrangements in respect to collecting forest products, but the lack of formal access rights limits the management intensity. More intensive management activities depend on more secure access rights, and consequently the more intensive management practices are only carried out on privately controlled lands falling under the agrarian regulations. As demonstrated by the variety of local practices for managing the coffee production

**Table 15.5** *Access rights in different coffee production systems in SW Ethiopia*

| | Landownership | Legal access rights to coffee | Customary access rights to coffee | Management responsibility |
|---|---|---|---|---|
| Forest coffee system | State | Collection right for first claimant; No rights to own or inherit trees | Private rights to claimed trees | Local claimant |
| | Individually owned | Exclusive rights for landowner | Share-cropping rights for collection | Landowner |
| Semi-forest coffee system | State | Collection right for first claimant; No rights to own or inherit trees | Private rights to claimed trees | Local claimant |
| | Individually owned | Exclusive rights for landowner | Share-cropping rights for collection | Landowner |
| Garden coffee system | Individually owned | Exclusive rights for landowner | – | Landowner |
| Coffee plantation system | Individually owned | Exclusive rights for landowner | – | Landowner |

*Source:* Philippe (2003)

systems, depending on the legal land-use conditions, local people have developed a variety of customary practices for maintaining the forest resources that are essential for their livelihoods (Philippe, 2003). In contrast to the statutory government arrangements focused on the forests as an essential ecosystem, these customary rights are primarily focused on valuable forest resources.

With the growing concern on the need to conserve the SW Ethiopian forests, it has recently been identified that such a dichotomized system of resource regulation (government bureaucratic structure vs. customary local community structure) often creates conflicts in authority and a decrease in the efficiency of the separate regulatory systems (Negassa and Wiersum, 2006). In order to overcome such conflicts, recently several initiatives have been undertaken to develop an integrated regulatory system for forest conservation and management in the form of participatory forest management schemes with formally registered forest user groups (Zewdie, 2005; Bognetteau et al, 2007; Gobeze et al, 2009). These approaches are based on the understanding that sustainable forest management is best assured by effectively incorporating the forests in the local livelihood systems and by stimulating local responsibility in maintaining not only forest provisioning services, but also essential forest regulatory services, for example in respect to maintaining coffee genetic resources for use in further breeding programmes and preventing soil and water erosion (Gatzweiler, 2005).

## Adaptation to market conditions

The acculturalization process does not only involve adaptation of management systems to access rights and management responsibility, but also the balancing

**Table 15.6** *Factors influencing management in coffee production systems*

| Factors inducing changes | Political period | Nature of the changing conditions | Impacts on coffee production systems | |
|---|---|---|---|---|
| | | | Positive | Negative |
| Political factors | Ancient feudal regime | Compulsory work; Peasants were forced to manage coffee systems | Intensive management in forest lands owned by lords | Disincentive to develop coffee gardens |
| | Derg regime | Land redistribution (land reform) creating tenure uncertainty | Opportunistic development of semi-forest coffee production | Disincentive to manage all coffee systems (priority for farmlands) |
| | Derg regime | Villagization, involving moving people from forest lands | Incentive to develop coffee gardens and plantations closer to settlement | Disincentive to manage forest coffee |
| | Initial period of present regime | State land demarcation creating tenure insecurity on forest lands | Incentive to develop coffee gardens and plantations on private lands | Disincentive to manage coffee in state forest |
| | Recent developments | Development of participatory forest management systems and forest coffee producer groups | Conservation of original coffee gene pool and increased production of forest coffee | |
| Socio-economic factors | Initiated in the ancient regime | Improvement of road infrastructure | Incentive to manage all coffee systems due to better marketing options | |
| | Beginning of Derg regime | Increase of world coffee price | Incentive to manage all coffee systems | |
| | Since Derg regime | Land fragmentation and diminishing farm size | Incentive to maintain forest coffee production | Disincentive to develop plantations |
| | Present regime | Fall of world coffee price | | Disincentive to manage all systems |

*Source:* Adapted from Philippe (2003)

of supply and demand for forest resources. Supply includes ecological factors such as natural occurrence and ease of cultivation of the species concerned, as well as socio-economic factors such as land and tree tenure conditions and access to labour and capital. The demand factors include marketing conditions, prices and degree of substitutability (Paudel and Wiersum, 2002). The changes in socio-economic factors, for example in respect to accessibility and coffee prices, have played a significant role in the evolution of the coffee production systems (Table 15.6). This adaptive process is still continuing. In tandem with the development of new regulatory and organizational arrangements for forest management as a cooperative effort, also improved arrangements for marketing of coffee and honey are being developed, including the stimulation of marketing cooperatives (Bognetteau et al, 2007) and of certification of the products as being produced in an environmentally sustainable and socially responsible manner (Wiersum et al, 2008).

## Conclusion

There are various social processes involved in the gradual adaptation of forest ecosystems to the human environment. The impact of social conditions such as access regulations, organization of management and marketing conditions illustrates the relevance of considering domestication as a biocultural process involving co-evolutionary changes in forest ecosystem properties and production and management conditions. The presence of an important forest commodity in the form of coffee did not only stimulate the development of different arrangements for controlling the extraction of this product, but also stimulated farmers to develop coffee cultivation garden or plantation systems. Conservationists often fear that such cultivation will replace the natural forests, but this was only partially the case for forest coffee. Notwithstanding the domestication of the coffee resources, forests still plays an important production role in the local livelihoods. Recently forest coffee production is being stimulated by introducing new social arrangements for forest and coffee biodiversity conservation. These activities are directed at incorporating the forest systems more strongly into the local livelihood systems and strengthening the social foundation of the forest mosaic landscape. They combine measures for improving forest conservation as well as coffee genetic resources, and measures for improving local profits from forest use through better marketing of forest products and services. This development signifies a new phase in the process of acculturalization by linking local processes of biocultural dynamics to newly emerging national and international concerns on forest and biodiversity conservation.

# DISCUSSION

## Honey production as another example of co-evolution in biocultural relations

The example of the co-evolutionary process of domestication of the coffee forests is not unique. Within the SW Ethiopian forests a similar biocultural process in respect to honey production also took place (Endalawah, 2005). Occasionally, honey is still collected from wild colonies present in the forests. But most honey is produced in traditional beehives that are hung in forests trees. This form of honey production is controlled by the presence of local regulations regarding access and use rights (locally called kobo rights) to either forest plots or specific trees for hanging of the beehives (Wakjira and Gole, 2007). In addition honey is also produced by hanging beehives in trees in privately owned home gardens or coffee gardens. Recently also techniques for honey production using modern beehives were introduced. These modern beehives are mainly located in farmyards, but the bees still depend on the forests for most of their foraging. In other countries similar trends in co-evolution in forest ecological conditions and socio-economic conditions have been reported (Den Hertog and Wiersum, 2000; Paudel and Wiersum, 2002).

## Dynamics in forest garden systems

The contribution of the intermediate stages of forest domestication to biodiversity conservation has repeatedly been acknowledged (McNeely, 2004; Schroth et al, 2004; Perfecto and Vandermeer, 2008) and several studies comparing the biodiversity of forest gardens and adjacent forests have been carried out. In a recent review of such studies it was found that of the 43 studies comparing species-richness, 34 reported lower richness in the forest gardens than in forests, though this decrease varied considerably depending on the ecological groups considered (Scales and Marsden, 2008). The comparative biodiversity data from the coffee forests and coffee gardens (Table 15.4) confirm this finding. As indicated by Scales and Marsden (2008), several factors affect the biodiversity of forest gardens. Generally, it is positively related to subsistence crop use and large geographic variation. It is negatively related to various processes of modernization such as increasing market proximity, management intensification and loss of tradition. These findings correspond with the list of key drivers in the emergence, intensification and decline of forest garden systems identified by Belcher et al (2005a). Factors such as market access and price, infrastructure, and resource tenure heavily affect the dynamics of these systems. The role of these factors in gradually changing the biodiversity of tree-gardening systems is demonstrated in Ethiopia by the recent trends of change in the mixed coffee garden systems (Abebe et al, 2006). These data seem to indicate that the 'intermediate' forest systems are indeed an intermediate phase in the domestication process: the uniformization and homogenization stimulate the forest provisioning services.

However, at present some new developments are influencing the scope of the intermediate forest systems. When domestication is considered as a biocultural process of adapting forests towards human needs, it can be argued that the human needs in respect to forests increasingly are not focused only on provisioning services, but also on regulatory, supporting and cultural services. In this respect, the ecological service functions of the intermediate forests are increasingly recognized. For instance, regarding regulatory services, it has been noted in Tanzania that erosion in mixed coffee gardens is equally low as in natural forests (Lundgren, 1980). Also the role of forest gardens in carbon sequestration is increasingly being recognized (Van Noordwijk et al, 2008). The intermediate phases in the process of coffee domestication also have important supporting functions. The natural coffee genetic resources in the coffee forests are an essential support for modern coffee production, providing essential breeding material in respect to disease resistance and low-caffeine coffee production (Wiersum et al, 2008). They also provide ecological corridors for forest-dwelling species (Schroth et al, 2004). In view of these diverse ecological services of forest gardens, Michon and De Foresta (1997) suggested that the intermediate forest types should not be considered as temporary intermediate stages of forest domestication, but rather as fully domesticated forests that are well adjusted to newly emerging concerns on balancing ecological conservation and human development (see also Belcher et al, 2005a; Perfecto and Vandermeer, 2008).

As indicated by the land-use situation in southwest Ethiopia, it is important to recognize that the process of forest domestication does not necessarily result in the development of a landscape dominated by one uniform land-use system, but rather in a forested landscape mosaic including a combination of both natural, modified and transformed forests as well as cultivated fields. This diversified forest landscape offers good options for the incorporation of forests in local livelihoods. This spatial aspect in the process of forest domestication and its ecological significance needs wider attention (Scales and Marsden, 2008).

## Forest restoration as biocultural process

The different interpretations of forest domestication as referring to either a process of ecological change operating at the level of forest ecosystems, a process of biocultural change operating at the level of forest production systems or a process of biocultural dynamics operating at the landscape level have their consequences for forest restoration. First, they illustrate that forest restoration should not be primarily considered as an ecological process aimed at restoring natural ecosystems, but rather as a process of stimulating new biocultural relations by readjusting forest ecological conditions to newly evolving socio-economic conditions and livelihood requirements. For effective forest restoration thus not only ecological practices for ecosystem rehabilitation should be considered, but also innovative practices for acculturalization of the new forest systems. In the restoration process attention needs to be given both to the provisioning, supporting, regulatory and cultural services that the restored forests should have for local livelihoods and to the question of how local people can actually benefit from those services as a result of locally adjusted regulatory and marketing conditions.

Second, rather than focusing on the restoration of forests in the form of specific ecosystems, restoration efforts should focus on creating forested landscape mosaics. The creation of a mosaic of forest and tree-based land-use systems interspersed with agricultural fields offers the best options for optimizing the tradeoffs between ecological conservation and human development and for providing an array of ecological services needed for supporting a varied set of specialized, diversified or coping livelihood strategies. Additionally, it provides options for both subsistence and commercial production and offers scope for involving different categories of households engaged in either coping, diversification or specialization strategies in forest resource conservation.

## ACKNOWLEDGEMENTS

The author thanks the staff of the Non-Timber Forest Products Research and Development Project in southwest Ethiopia for their information and study cooperation. The MSc students Laurence Philippe, Tefera Belay Endalawah and Mohamed Chilalo are sincerely acknowledged for their contributions to the research; their studies contributed major information for this chapter.

## REFERENCES

Abebe, T., Wiersum, K. F., Bongers, F. and Sterck, F. (2006) 'Diversity and dynamics in homegardens of southern Ethiopia', in B. M. Kumar and P. K. R. Nair (eds) *Tropical Homegardens: A Time-Tested Example of Sustainable Agroforestry*, Springer, Dordrecht, The Netherlands, pp123–142

Belcher, B., Michon, G., Angelsen, A., Ruiz Perez, M. and Asbjornsen, H. (2005a) 'The socioeconomic conditions determining the development, persistence, and decline of forest garden systems', *Economic Botany*, vol 59, no 3, pp245–253

Belcher, B., Ruiz Perez, M. and Achiawan, R. (2005b) 'Global patterns and trends in the use and management of commercial NTFPs: Implications for livelihoods and conservation', *World Development*, vol 33, no 9, pp1435–1452

Bognetteau, E., Haile, B. and Wiersum, K. F. (2007) 'Linking forests and people, a potential for sustainable development of the Southeast Ethiopian highlands', in E. Kelbessa and C. de Stoop (eds) *Participatory Forest Management, Biodiversity and Livelihoods in Africa*, proceedings of international conference, Addis Ababa, Ethiopia, 19–21 March 2007, FARM Africa/SOS Sahel, Addis Ababa, pp36–53

Chilalo, M. A. (2007) 'The contribution of non-timber forest products to rural livelihood in southwest Ethiopia', MSc thesis, Forest and Nature Conservation Policy Group, Wageningen University, Wageningen, The Netherlands

Den Hertog, W. H. and Wiersum, K. F. (2000) 'Timur (*Zanthoxylum armatum*) production in Nepal: Dynamics in non-timber forest resource management', *Mountain Research and Development*, vol 20, no 2, pp136–145

Endalawah, T. B. (2005) 'Dynamics in the management of honey production in the forest environment of Southwest Ethiopia: Interactions between forests and bee management', MSc thesis, Forest and Nature Conservation Policy Group, Wageningen University, Wageningen, The Netherlands

Gatzweiler, F. W. (2005) 'Institutionalising biodiversity conservation – The case of Ethiopian coffee forests', *Conservation and Society*, vol 3, no 1, pp201–223

Geletu, K. T. (2006) *Genetic Diversity of Wild* Coffea arabica *Populations in Ethiopia as a Contribution to Conservation and Use Planning*, Ecology and Development Series No. 44, Cuvillier Verlag, Göttingen, Germany

Gobeze, T., Bekele, M., Lemenih, M. and Kassa, H., (2009) 'Participatory forest management and its impact on livelihoods and forest status: The case of Bonga forest in Ethiopia', *International Forestry Review*, vol 11, no 3, pp346–358

Gole, T. W. M., Teketay, D., Denich, M. and Vlek, P. G. L. (2000) 'Human impact on *Coffea arabica* genetic pool in Ethiopia and the need for its in situ conservation', in J. Engels, V. Ramanatha Rao, A. H. D. Brown and M. Jackson (eds) *Managing Plant Genetic Diversity*, CAB International, Wallingford, UK, pp237–247

Gole, T. W., Borsch, T., Denich, M. and Teketay, D. (2008) 'Floristic composition and environmental factors characterizing coffee forest in southwest Ethiopia', *Forest Ecology and Management*, vol 255, pp2138–2150

Hladik, C. M., Hladik, A., Linares, O. F., Pagezy, H., Semple, A. and Hadley, M. (1993) *Tropical Forests, People and Food. Biocultural Interactions and Applications in Development*, Man and Biosphere Series No. 13, UNESCO and Parthenon Publishers, New York

Homma, A. K. O. (1992) 'The dynamics of extraction in Amazonia: A historic perspective', *Advances in Economic Botany*, vol 9, pp23–32

Hylander, K. and Nemomissa, S. (2008) 'Complementary roles of homegardens and exotic tree plantations as alternative habitats for plants of the Ethiopian montane rainforests', *Conservation Biology*, vol 23, no 2, pp400–409

Lamb, D., Erskine, P. D. and Parrotta, J. A. (2005) 'Restoring of degraded tropical forest landscapes', *Science*, vol 310, no 5754, pp1628–1632

Lundgren, B. (1980) 'Comparison of surface runoff and soil loss from runoff plots in forest and small-scale agriculture in the Usambara Mountains, Tanzania', *Geografiska Annaler*, vol 62A, nos 3–4, pp113–148

Mamo, G., Sjaastad, E. and Vedeld, P. (2007) 'Economic dependence on forest resources: A case from Dendie District, Ethiopia', *Forest Policy and Economics*, vol 9, pp916–927

McNeely, J. A. (2004) 'Nature vs. nurture: Managing relationships between forests, agroforestry and wild biodiversity', *Agroforestry Systems*, vol 61, pp155–165

Michon, G. (2005) *Domesticating Forests: How Farmers Manage Forest Resources*, Center for International Forestry Research, Bogor, Indonesia, and World Agroforestry Centre, Nairobi

Michon, G. and De Foresta, H. (1997) 'Agroforests: Pre-domestication of forest trees or true domestication of forest ecosystems?', *Netherlands Journal of Agricultural Science*, vol 45, pp451–462

Negassa, A. and Wiersum, K. F. (2006) 'Community perspectives on participatory forest management. The case of Chilimo participatory forest management scheme in Ethiopia', *Ethiopian Journal of Natural Resources*, vol 8, no 1, pp57–75

Omiti, J. M., Parton, K. A., Sinden, J. A. and Ehui, S. K. (1999) 'Monitoring changes in land-use practices following agrarian de-collectivisation in Ethiopia', *Agriculture, Ecosystems and Environment*, vol 72, pp111–118

Paudel, S. and Wiersum, K. F. (2002) 'Tenure arrangements and management intensity of butter tree (*Diploknema butyracea*) in Makawanpur district, Nepal', *International Forestry Review*, vol 4, no 3, pp223–230

Perfecto, I. and Vandermeer, J. (2008) 'Biodiversity conservation in tropical agroecosystems. A new conservation paradigm', *Annals of the New York Academy of Science*, vol 1134, pp173–200

Phillipe, L. (2003) 'Dynamics of coffee production systems in Kaffa. A case study in two villages in Kaffa province of Ethiopia', MSc thesis, Wageningen University, Wageningen, The Netherlands

Scales, B. R. and Marsden, S. J. (2008) Biodiversity in small-scale tropical agroforests: A review of species richness and abundance shifts and the factors influencing them, *Environmental Conservation*, vol 35, no 2, pp160–172

Schmitt, C. B. (2006) *Montane Rainforest with Wild Coffea arabica in the Bonga Region (SW Ethiopia): Plant Diversity, Wild Coffee Management and Implications for Conservation*, Ecology and Development Series No. 47, Cuvilier Verlag, Göttingen, Germany

Schroth, G., Harvey, G. A. and Vincent, G. (2004) 'Complex agroforests: Their structure, diversity, and potential role in landscape conservation', in G. Schroth, G. A. B. da Fonseca, C. A. Harvey, C. Gascon, H. L. Vasconcelos and A. M. N. Izac (eds) *Agroforestry and Biodiversity Conservation in Tropical Landscapes*, Island Press, Washington, DC, pp227–260

Senbeta, F. and Denich, M. (2006) 'Effects of wild coffee management on species diversity in the Afromontane rainforest of Ethiopia', *Forest Ecology and Management*, vol 232, no 1, pp68–74

Simons, A. J. and Leakey, R. R. B. (2004) 'Tree domestication in tropical agroforestry', *Agroforestry Systems*, vol 61, pp167–181

Teketay, D. and Tegineh, A. (1991) 'Traditional tree crop based agroforestry in coffee producing areas of Harerge, Eastern Ethiopia', *Agroforestry Systems*, vol 16, pp257–267

Van Noordwijk, M., Suyamto, D. A., Lusiana, B., Ekadinata, A. and Hariah, K. (2008) 'Facilitating agroforestation of landscapes for sustainable benefits: Tradeoffs between carbon stocks and local development benefits in Indonesia according to the FALLOW model', *Agriculture, Ecosystems and Environment*, vol 126, pp98–112

Wakjira, D. T. and Gole, T. W. (2007) 'Customary forest tenure in southwest Ethiopia', *Forests, Trees and Livelihoods*, vol 17, pp325–338

Wiersum, K. F. (1997a) 'From natural forest to tree crops, co-domestication of forests and tree species, an overview', *Netherlands Journal Agricultural Science*, vol 45, pp425–438

Wiersum, K. F. (1997b) 'Indigenous exploitation and management of tropical forest resources: An evolutionary continuum in forest-people interactions', *Agriculture, Ecosystems and Environment*, vol 63, pp1–16

Wiersum, K. F. (2004) 'Forest gardens as an "intermediate" land use system in the nature–culture continuum: Characteristics and future potential', *Agroforestry Systems*, vol 61, pp123–134

Wiersum, K. F., Gole, T. W., Gatzweiler, F., Volkmann, J., Bognetteau, E. and Wirtu, O. (2008) 'Certification of wild coffee in Ethiopia: Experiences and challenges', *Forests, Trees and Livelihoods*, vol 18, pp9–21

Zewdie, Y. (2005) 'Forest access and rural livelihoods in Southwest Ethiopia: An analysis of the record of forest management partnership', in M. A. F. Ros-Tonen and T. Dietz (eds) *African Forests Between Nature and Livelihood Resource. Interdisciplinary Studies in Conservation and Forest Management*, Edwin Mellen Press, Lewiston, NY, and Lampeter, UK, African Studies No. 81, pp95–111

# 16

# Conditions for Sustainable Forest Management in Community Forestry: A Case Study from the Bale Mountains, Ethiopia

*Julia Schmitt*

## INTRODUCTION

Ethiopia, one of the poorest nations in the world (World Bank, 2010), alarms with a high deforestation rate and a remaining forest cover of less than 3 per cent (Bekele, 2001). To stop this development and to preserve one of the last resources of natural forest in the country, the Integrated Forest Management Project (IFMP) operates in the Bale Mountains of Ethiopia. By providing exclusive forest user rights for forest-dweller families (Figure 16.1a), this project aims to generate income from sustainable forest management of the remaining Afromountain forest (Figure 16.1b), which is scattered with pastures, agricultural plots and isolated settlements and threatened by the conversion into agricultural land (Kubsa, 2002). This study assessed whether the conditions for sustainable forest management, as anticipated by the IFMP, are favourable and whether the participating forest-dwellers are able to implement it.

Created in 1995 by the Ethiopian Forest Administration and the German Technical Cooperation (GTZ), the IFMP implements the so-called WAJIB approach, 'WAJIB' being an abbreviation for forest-dweller association in the local language. Local forest-dwellers living illegally in the state forest of the

**Figure 16.1** *Forest-dwellers in Bale Mountains and forest-related activities*

*Notes:* 1a: A forest-dweller family.
1b: Remaining natural high forest, generally dominated by *Juniperus procera*, *Podocarpus falcatus* and *Hagenia abyssinica*.
1c: Pit-sawing platform.
1d: Transport of Juniper poles to a local market.
1e: Construction of a local hut with Juniper splits.

Bale Mountains were supported to establish associations. They signed a contract with the Forest Administration which enabled them to stay in the forest and provided them with exclusive forest user rights. The rights of the WAJIB members include settlement in the forest area and the utilization of forest products for consumption and commercial purposes. The duties comprise maintaining at least the initial tree cover, restricting further settlement and paying an annual forest rent. The forest rent is about one euro per year per hectare and is calculated on base of a tree cover assessment, conducted by the IFMP (IFMP, 2000). Hence recognized forest-dweller associations that have defined their boundaries and selected their members autonomously protect their user rights against outsiders. Regularly, association members patrol their forest area, expel intruders and confiscate illegally collected forest products.

A typical forest-dweller association, a WAJIB, consists of 20 to 30 families that communally manage a natural forest area of 300 to 500 hectares. The number of families on a given forest area results from a model calculation which identifies a certain minimum forest area per household that endows the forest-dwellers with fuel and construction wood and with a cash income from forest products that is similar to what an average farmer can achieve through agriculture (Unkovsky, 1998).

The principles of sustainable forest management include environmentally responsible, economically viable and socially beneficial management of forests (FSC, 2008). Accordingly, in this study the following three dimensions have been assessed:

1  the forest production and its silvicultural sustainability;
2  the forest production with its actual contribution to the rural household economy, the market value of forest products, and the ability of the forest-dwellers to use the full potential of the forest resource with the available indigenous knowledge, processing skills and management capacity; and
3  the socio-economic stability of the forest-dweller approach within the associations, based on the distribution of resources and on decision-making and benefit-sharing arrangements.

# STUDY SITE

## Location and natural conditions

The study site, the state-owned Forest Priority Area of Adaba-Dodola, is located in the Bale Mountains, in the southeastern part of Ethiopia. It is one of the remaining natural high forests in the country, with a size of 53,000ha. The Forest Priority Area serves as a buffer zone for Bale Mountains National Park and as the water catchment of a hydroelectric dam. The Bale Mountains in the Dodola area reach elevations up to 3700m asl and are surrounded by a nearly treeless agricultural plain at an altitude of 2400m asl. This plain extends over more than 2000km$^2$ and is nourished by several perennial watercourses which originate from the Bale Mountains.

The annual rainfall pattern is bimodal. The minor rainy season begins January–February and ends April–May; the main rainy season starts June–July and ends September–October. According to GTZ weather records, the annual average rainfall lies between 850mm in the plain and 1300mm in the mountains. Average temperature ranges from 7 to 24°C (Amente et al, 2006).

## Forest vegetation

The natural high forest of the Bale Mountains is defined as Afromontane forest (Friis, 1992). It can be divided into moist mountain forests and dry mountain forests. The forest, dominated by hard-leaved and evergreen species like *Podocarpus falcatus* (synonym: *Afrocarpus gracilior*), *Juniperus procera* and *Olea europea* ssp. *subsidata* (synonym: ssp. *africana*) in the northwestern part of the Bale Mountains is classified as a dry mountain forest. At higher altitude (above 2700m asl) and on the southern slopes of the mountain massif, moist forest with broadleaved species like *Hagenia abyssinica* and *Hypericum lanceolatum* can be found (Friis, 1992). A forest inventory by Trainer (1996) has shown a current tree cover of 50 per cent and an average standing volume of 117m$^3$/ha. The annual increment of wood is estimated to be 1m$^3$/ha.

According to the International tropical Timber Organization (ITTO) definition (2002), the forest can be defined as a degraded primary forest. The forest is over-utilized, despite the fact that *Hagenia abyssinica*, *Podocarpus falcatus* and *Juniperus procera* are banned from utilization (Negarit Gazeta, 1994, cited in Demesa, 2002). In addition to that, overgrazing is restricting the natural forest restoration capacity, which leads to poor understorey development and absence of regeneration of *Olea europaea*, *Podocarpus falcatus* and *Hagenia abyssinica* (Hohlweg, 1998).

# METHODOLOGY

A combination of qualitative and quantitative research methods in the framework of a qualitative research approach was applied (Lamnek, 1993; Atteslander, 2000). Data collection took place between October 2002 and January 2003.

## Participatory rural appraisal

Participatory rural appraisal (PRA) is a set of tools to collect information from the local population in research and development projects. According to the PRA scheme, the data collection took place in three different phases (Schönhuth and Kievelitz, 1993):

1 introduction of the research idea and the researcher at WAJIB meetings;
2 field trips to conduct household interviews with all recognized WAJIB members, observations, informal discussions and expert interviews; and
3 final WAJIB meeting with participatory mapping, presentation and validation of results.

During these phases the following PRA tools were used:

- **Group discussions.** An official WAJIB meeting at the beginning of the study was held in each community. All WAJIB members were introduced to the research team and the purpose of the study to request participation. After the community's consent, several group discussions and expert group discussions took place during and at the end of the field trips.
- **Household interviews.** A total enumeration of all households in the selected WAJIBs was conducted with semi-structured interviews. In total, 65 families were questioned, including 54 male and 11 female WAJIB members. The questionnaire covered (a) general information about gender, age, family structure and property, (b) forest use, including products, amounts, production techniques and tools, (c) indigenous knowledge of technical and botanical skills regarding forest management: prevailing forest types and site condition, tree species name and age, tree growth, fructification and regeneration patterns, and impact of grazing on forest structure, (d) organization of work with a seasonal work calendar, division of work, selling of products, benefit-sharing and access to credits, (e) sources of income and a ranking of their importance, and (f) future production goals and commitment to forest management.
- **Expert interviews.** People with special knowledge about the forest and the production system were considered experts and after the personal interviews encouraged to demonstrate their work and to provide more details. The expert interviews aimed to gather data describing the forest production system and disclosed the level of indigenous knowledge available. Elders, and association and village leaders were considered experts on socio-economic aspects of the WAJIB communities and were therefore interviewed in more detail.
- **Participatory mapping.** On maps of the area, the borders and key points of the forest areas were defined. After the interviews, the WAJIB members discussed the following issues and marked the respective areas on the map: natural conditions (topography, land use) of the forest area, forest types and their state of degradation, intensive and extensive grazing areas, and location of natural resources use.
- **Observations.** During the field trips several formal and informal observations of the forest production system, wood consumption and the livelihood system took place.
- **Market survey.** Local wood products are produced and sold in wood shops and local markets. A survey was conducted to reveal prices and dimensions of common products.

Table 16.1 *Information on the selected WAJIBs, Bale Mountains*

| Name | Area (ha) | WAJIB members | Area per household (ha) | Logistics | Dominant tree species |
|---|---|---|---|---|---|
| Gede | 489 | 28 | 17.5 | Relatively inaccessible (only by horse) | *Juniperus procera/ Podocarpus falcatus* and *Hagenia abyssinica/ Hypericum lanceolatum* |
| Changity | 554 | 18 | 30.7 | Accessible (by car) | *Juniperus procera/ Podocarpus falcatus* |
| Jaldo | 364 | 20 | 18.2 | Relatively inaccessible (only by horse) | *Juniperus procera/ Podocarpus falcatus* and *Hagenia abyssinica/ Hypericum lanceolatum* |

*Source:* IFMP database

- **The principle of triangulation.** At the institutional level, the forest-dweller associations and the Forest Administration are key stakeholders of the WAJIB approach. Collecting information from different sources and with different methods, as well as cross-checking of answers, is part of the triangulation principle of PRA (Schönhuth and Kievelitz, 1993). For the triangulation, the data was presented to all project stakeholders, WAJIB members and IFMP staff, and cross-checked with own observations.

## Wood products measurement and gross margin analysis

The volume of poles was calculated as $V = 0.25 \times \pi \times d^2 \times h$, where V = volume of pole, d = mean diameter of pole and h = length of pole. Plank volume was calculated as $V = \text{length} \times \text{width} \times \text{thickness}$. To convert local fuelwood sales unit into cubic metres, a calculation model was used that already existed for project purposes (Unkovsky, 1998).

A gross margin calculation reveals the profit of income-generation activities (Varian, 2003). The profit of different harvesting operations and alternative sources of income was calculated as Gross Margin = Revenue minus Costs, where Revenue is the anticipated product retail price derived directly from the extracted wood products, and Costs are the expenses for tools and transport. Expert interviews provided information on the time demand for harvesting and wood processing operations and on product prices.

## Selection of the case study WAJIBs

Three forest-dweller associations, representing different conditions, were selected from the existing 22 communities (Table 16.1). The selected WAJIBs have also been used to develop and demonstrate appropriate silvicultural stand improvement treatments (Amente et al, 2006). The following criteria were taken into account: differences in access to market, different forest types and degradation status, different motivation levels of the members to participate, different forest area per household and dependence on the forest, different land-use priority of the WAJIB members, and different times of project implementation.

# RESULTS

## Livelihood system of the forest-dwellers

The livelihood system of forest-dweller families is based on three major income generation activities – agriculture, livestock production and forest utilization – which are closely interlinked. The forest provides shade and fodder for the livestock. Animal power is used for crop production and transportation of forest products to local markets. The remains on the fields serve as food for the livestock, while dung fertilizes the agricultural land and is used as building material for local houses. Of these three income-generating activities, agriculture is economically the most important one for the majority of forest-dweller families. In Gede and Jaldo it represents the main source of income for about 70 per cent of WAJIB members (Figure 16.1a). In contrast, Changity relies mainly on livestock as the main source of income. However, between 7 and 24 per cent of the community members in all tree WAJIBs earn their livelihood principally from forestry (Figure 16.1c).

## Forest production

Sustainable forest production is the main endeavour of the forest-dweller approach. At present, forest production is conducted for own consumption by forest-dwellers or for sale at local markets. The range of local products comprises timber, poles, splits and fuelwood.

Timber is processed from *Hagenia abyssinica*, *Juniperus procera* and *Podocarpus gracilior*. Trees are felled, debranched, cross-cut and the log smoothened with an axe. Logs are then processed with a two-man saw on a pit-sawing platform inside the forest and transported by horse (Figure 16.1d). The average wood recovery rate for this processing technique is 10–30 per cent (Demesa, 2002).

Poles, used for constructing houses and fences in towns, are processed from dry, mature juniper trunks, due to the low number of poles (in other words regeneration) in the forest. The logs are cut into 3–4m sections and then split with wedge and hammer. These splits are smoothened with axes and processed into round poles with a diameter of 10–15cm (Figure 16.1d).

Splits are used for frame construction of traditional houses, which are made of a round wooden frame with mud walls and a grass-covered roof (Figure 16.1e). To process splits, dry juniper wood is cut and split. The untreated splits are sold in bundles on the market.

**Table 16.2** *Comparing harvestable volume with wood consumption per household in three WAJIBS, Bale Mountains*

| WAJIB | Annual stand increment = harvestable volume (m³) | Consumption wood (m³) | Additional wood available for Harvesting (m³) |
|---|---|---|---|
| Gede | 17.5 | 5.7 | 11.8 |
| Changity | 30.7 | 5.7 | 25.0 |
| Jaldo | 18.2 | 5.7 | 12.5 |

All forest-dwellers are completely dependent on fuelwood as the main source of energy for cooking and heating. Fuelwood for home consumption is mainly collected by female WAJIB members and children. Besides that, female forest-dwellers living within a reachable distance of a market place (a four-hour walk one way) sell a donkey-load of fuelwood on market days.

In addition to wood products, the Bale Mountains provide a wide range of non-wood forest products (NWFPs). Frequently, leaves and roots are used as medicines, ropes are made from lianas, and traditional perfumes and dyes are produced from plant exudates. Honey, fruits, oils, mushrooms, bamboo and spices are collected. Among these, honey is the only NWFP with an already established market; the remaining NWFP products are only used for subsistence.

## Sustainability of wood consumption

**Construction and fuelwood.** To construct the local round houses between 200 and 600 pieces of juniper splits are needed, amounting to 6–18m³. The houses last between five and ten years. For furniture (beds, cupboards and doors), planks from *Hagenia abyssinica* and *Juniperus procera* are used, which last at least ten years. Annual use for planks is 0.2m³, for juniper poles 0.4m³ and for juniper splits 1.7m³. Women in charge of fuelwood collection explained that consumption depends on the outside temperature (the altitude and the season) and the number of people in the household. On average, 3.4m³ of fuelwood is consumed per household per year (converted from local units with Unkovsky, 1998). The total basic consumption of construction and fuelwood thus amounts to 5.7m³ per household per year.

**Comparing wood consumption and stand increment.** To achieve silvicultural sustainability, the annual harvestable volume may not exceed the annual stand increment of a given forest area. Therefore, the wood consumption of WAJIB members was compared to the annual stand increment in each WAJIB (Table 16.2). The annual harvestable volume was based on an estimated stand increment of 1m³/ha/year (Trainer, 1996), and calculated for the average forest area per household (Table 16.1). In none of the three WAJIBs is the increment totally utilized. It can be tripled (Gede, Jaldo) or even increased fivefold

**Table 16.3** *Gross margin of different forest production activities, compared to daily labour, in Bale Mountains*

| Position | | Income-generating activity | | | |
| --- | --- | --- | --- | --- | --- |
| | | Timber | Poles | Fuelwood | Daily labour |
| Revenue | | | | | |
|   Productivity | m³/day | 0.055 | 0.167 | 0.176 | – |
|   Price | EUR/m³ | 100.00 | 12.00 | 3.40 | – |
|   Gross revenue | EUR/day | 5.50 | 2.00 | 0.60 | 0.80 |
| Costs | | | | | |
|   Tools | EUR/day | 0.20 | 0.10 | 0.10 | 0.00 |
|   Transport to market | EUR/daily production | 0.10 | 0.10 | 0.10 | 0.00 |
|   Total direct costs | EUR/day | 0.30 | 0.20 | 0.20 | 0.00 |
| Gross margin | EUR/day | 5.20 | 1.80 | 0.40 | 0.80 |

(Changity). We conclude that it is possible for WAJIB members to gain additional cash income from timber utilization and still conduct sustainable forestry.

## Economic analysis of income generating activities

To assess the economic potential of forestry in the Bale Mountains, a gross margin calculation was applied (Table 16.3). Of interest was the possible income generation expressed by the gross margin per working day compared to the income alternative as daily labourer.

The gross margin calculation shows that processing timber is economically highly attractive. With this activity a forest-dweller is able to earn about six times more than as a daily labourer. Processing poles still results in a better gross margin than daily labour. Only fuelwood production does not achieve this rate. However, due to the lack of alternatives, it is still an attractive source of income generation for women.

## Indigenous knowledge, processing skill and management capacity

The ability of the forest-dwellers to manage their forest resource on their own, with only technical support from the forest administration, was evaluated.

**Tree species specific and silvicultural knowledge.** All WAJIB members knew the main tree species by name and the respective site conditions where they prevail. Tree flowering and seed production periods could be explained. Furthermore, 81 per cent of the male members divided their forest into three different forest types based on altitude. Women, not involved in timber processing, were less knowledgeable than male WAJIB members. The silvicultural knowledge showed a more diverse picture. Most WAJIB members understood

351

**Table 16.4** *Availability of knowledge, processing skills and management capacity factors of forest-dwellers in Bale Mountains*

| Production Factors | Availability | | |
|---|---|---|---|
| Knowledge/skills | Silviculture ↓ | Botany ↑ | Processing → |
| Means of production | Tools ↑ | Transport ↑ | Markets ↑ |
| Investment | | Funds ↓ | Time → |

*Notes:* Upward-pointing arrows represent a favourable management capacity for forest production, horizontal-pointing arrows express forest production at least not limited by the available management capacity and downward-pointing arrows indicate the serious limitations for forest activities.

the connection between grazing pressure and the lack of regeneration. Additionally, there seemed to be rudimentary ideas of certain silvicultural terms such as pruning. However, their purpose and impact was explained in very different ways. Furthermore, tree age was unknown. The production period to achieve timber of desired dimension was not considered to be important. Therefore, the concept of silviculture – to manage trees in order to achieve certain tree products – was unfamiliar and annual increment as a key parameter to determine the annual harvestable volume was not known.

**Processing skills.** The skill to process forest products is the key factor that endows the WAJIB families with income. The interviews showed that not all male WAJIB members have the knowledge to process forest products. Of the 54 interviewed male WAJIB members only 28 per cent were able to process lumber. Juniper poles can be produced by 63 per cent of the male WAJIB members, splits for houses by all males.

**Management capacity.** It was investigated if WAJIB households can provide the necessary means to sustain forest production, focusing on three production factors: time management, availability of tools and investment in forestry operations. Table 16.4 presents an overview of the available management capacity with regard to forest activities. Production means are generally positive, knowledge skills vary from positive to negative, and investments are generally limited.

**Time management.** The time management for forestry activities is influenced by the time input in agricultural production, which is predetermined by the seasonal calendar. The agricultural season and the necessary work input depend on the location of the farmland. WAJIB members with agricultural land on the plain (21 per cent of WAJIB members) start ploughing and sowing in April. Harvesting time is in October and November. These WAJIB members can devote four months per year (December to March) for forestry activities. WAJIB members with agriculture in the mountains (52 per cent of WAJIB members) plough, sow and weed from May to August. Harvesting time is

January and February. This leaves only two months for forest-related activities (March and April). Between September and January the grains in the field have to be protection from wild animals. Protection activities are often conducted by people that do not have agricultural land in a work- and crop-sharing system. 37 per cent of the WAJIB members that do not have any agricultural land are available for forest activities throughout the year.

**Availability of tools and means of transport.** For production of lumber, poles and splits, a two-man saw, axes, slasher, wedge and hammer are needed. All families, including the female-headed households, have at least one axe, mainly used for fuelwood production. The majority of WAJIB members also have other handtools. The handtools are either manufactured by the WAJIB members or bought on the local market. Saws are provided for rent at the Peasant Association administration. Means of transport are horses and donkeys, which are also available to all WAJIB members.

**Investment in forestry operations.** There are public and private credit systems accessible to the WAJIB members, primarily meant to finance fertilizers and equipment to enhance agricultural production and only available for landholders on the plain. Only two WAJIB members mentioned that they have accessed credit to sustain agricultural production; 83 per cent of the WAJIB members knew about the possibility but never used it. 11 per cent claimed that they never heard about the possibility.

## Socio-economic characteristics of the WAJIB approach

As indicators for the social sustainability of the forest-dweller approach, the distribution of resources, the practice of decision-making and sharing the forest-related benefits were analysed.

**Family background.** 17 per cent of the interviewed, registered WAJIB members were in fact female household heads, due to existing polygamy. In the study area, a man can have several wives depending on his wealth status. In the selected WAJIBs, it was varying between one and three wives. Each wife has her own household. Accordingly, the husband travels from household to household, but prefers in general to live with the youngest wife. During the WAJIB household survey the average family had 4.4 children.

**Distribution of agricultural land and livestock.** The distribution of agricultural land varies considerably between the WAJIBs. Households in Gede (1.1ha per household) and Jaldo (1ha) have about the same average size of agricultural land, while in Changity (0.4ha) this is much lower. This explains the low importance of agriculture as a source of income in Changity. Furthermore, the distribution of resources between WAJIB members is not equal either. In Gede, Changity and Jaldo 14, 58 and 40 per cent respectively of the households have no agricultural land. In contrast to the different distribution of agricultural land in each WAJIB community, on average there is a similar number of livestock per WAJIB member. In Gede, Changity and Jaldo 15.7, 14.6 and 12.2 pieces of livestock per household are owned respectively. However, the distribution of livestock of each household varies.

**Wealth categories.** The access to agricultural land and the number of cattle determine the income of a WAJIB family. In the Oromian culture and due to the Ethiopian land tenure system and its political volatility (the former possibility of land redistribution), cattle are considered the main asset. Subsequently, the number of cattle defines the wealth status of a WAJIB member. In a study conducted in the project area, Fufa (2002) defined three different wealth classes, based on the number of cattle and the amount of agricultural land. According to this, the interviewed WAJIB members were classified into poor (<1ha land, 0–5 cattle), less poor (1–2ha land, >5 cattle) and rich (>2ha land, >10 cattle). The results show that 32 male-headed WAJIB households (59 per cent of males) and 9 female-headed WAJIB households (82 per cent of females) have no or little agricultural land and only a small number of livestock and therefore can be considered as poor. These are mainly the male WAJIB members involved in forest utilization. People considered rich are all male elders (eight) and have more than 2ha of agricultural land and more than 10 cattle. This is only the case for 15 per cent of the male WAJIB members. 14 males (26 per cent) and 2 females (18 per cent) were intermediately poor.

**Decision-making and benefit-sharing.** The survey on decision-making mechanisms showed that each forest-dweller association had one elected leader and one deputy. These leaders have the right to introduce suggestions, but decisions are supposed to be made by all members together. By design, this should also apply to female association members. In practice, the traditional decision-making process, which engages intense discussion and culminates in democratic consensus, involves only male community members. While women attend these discussions on behalf of the project, they do not actively participate. Nonetheless, all interviewed women pointed out that, when the discussion turns to women's responsibilities, they become involved and can address questions and problems.

The analysis of benefit-sharing confirmed that all association members consider the forest as common property, implying that resources are shared equitably. This applies to some of the forest-generated incomes, for example the sale of temporary grazing rights to outsiders or the confiscated fuelwood from illegal collectors. However, wood processing and marketing is done individually, which effectively excludes female association members, as profitable forest utilization is only practised by males. Furthermore, the arrangement of forest rent payment to the forest administration varies between the different WAJIBs. In Changity and Jaldo the payment of forest rent is done by an equal share from each member, regardless of their wealth. In Gede, the forest rent is paid by different share according to the member's wealth. Furthermore, different by-laws have been adopted in Gede to administer the forest utilization. Forest-dwellers who are considered poor are granted permission to cut trees. The permission is normally granted when WAJIB members run out of agricultural products to maintain their livelihood, in other words before the new harvesting season. In fact, most forest-dweller households do not have enough agricultural land to sustain their livelihood throughout the year.

## CONCLUSIONS AND RECOMMENDATIONS

### Conditions for sustainable forestry within the WAJIB approach

Forest-dwellers do not rely solely on forest production, but combine it with the two other sources of income – subsistence agriculture and livestock production. Forest products have an essential function as assets during hardship, for example before the new harvest. This may buffer agricultural production risks and minimize the exposure to poverty and famines. Subsequently, the forest serves as kind of safety net for the poorest of the poor, which was emphasized by the FAO (2003) as a strategy to use forest resources to alleviate poverty.

In all three WAJIBs additional tree increment was available for commercial utilization. Furthermore, the positive gross margins achievable through forest management showed the economic potential for controlled utilization. Yet the limited silvicultural knowledge and management capacity prevented the forest-dwellers from using the full potential, such as the annual harvestable volume. Besides this, the above-described forest utilization techniques and the available tools generally focus on mid-aged trees with a diameter range between 15cm and 30cm. This practice, in combination with high grazing pressure, resulted in an over-aged forest with very low productivity (Hohlweg, 1998). The estimated annual increment of 1m$^3$ per year per hectare (Trainer, 1996) is very low, compared to 3–4m$^3$ per year per hectare as average of a heavily degraded natural forest in Ethiopia (EFAP, 1994).

The question whether people without professional education can manage a forest with their own knowledge and experience is an ongoing discussion in the literature and development projects (Wiersum, 1999). Kessy (1998) pointed out that it is difficult to judge whether communities really have management strategies towards the forest, or if forest sustainability is caused by low human population pressure. Especially in times of resource shortages, 'scientific' forest management knowledge is needed to produce sustainability, in other words balancing increment and utilization. However, community forestry approaches have mainly focused on non-timber forest products; therefore experiences with timber production are rare (de Blas et al, 2008).

A second, related issue is the limited availability of cash. It is unlikely that in the near future WAJIB members will access credits to improve the forest productivity, for example by investing in planting. Also it is not likely that any investment is being considered to improve the recovery rate by adopting appropriate timber processing techniques, for example portable sawmills. However,

Wily and Dewees (2000) remarked that the relevance of community forestry is rooted in the challenge of securing forested areas, rather than in the need to undertake extensive and costly silvicultural operations. Therefore the possibilities for investment, in other words improvement of recovery rates and silvicultural interventions, should be conducted only if improvements can be sustained by WAJIB members.

A major setback within the WAJIB communities is the social division of labour and responsibility within traditional Ethiopian society. This leads to the fact that local decision-making and benefit-sharing clearly favour well-off, male community members. Conroy and co-workers (2002) found a similar situation in joint forest management projects in India: participation and acceptance of women were equally high as those of men, but their decision-making and benefit-sharing was much lower. In fact, the participation of women in community forestry is a key issue, as it has a positive influence on forest governance and resource protection (Agrawal et al, 2006)

In summary, conditions for sustainable forest management by the WAJIB members are favourable. However, technical advice on silvicultural management issues, access to credits to implement improved timber processing techniques and advice on the harvestable stand increment are crucial to achieve sustainable forest management. The following recommendations are proposed to improve the WAJIB approach:

- **Improving the productivity of the forest.** The actual productivity of the heavily disturbed natural forests is very low. With silvicultural interventions, for example stimulation of the natural regeneration and liberation of potential crop trees, the increment and therefore the harvestable volume can be improved substantially. The limited silvicultural knowledge of forest-dwellers shows the demand for silvicultural and technical guidelines, developed with and for the forest-dwellers to improve the management of the forest resources. Thus the focus should be on the needs and possibilities of WAJIB communities (FAO, 2004).
- **Improving the productivity of forest utilization techniques.** The recovery rate of the forest products, processed by the traditional techniques of axe and pit-sawing is, at 10–30 per cent, very low. An improvement through the introduction of new techniques, like a portable sawmill, already practised in other parts of Ethiopia (Demesa, 2002), could contribute to increase recovery rates.
- **Supporting vertically integrated local forest industry.** The economic value of forest products increases significantly along the value chain. Establishing a vertically integrated local forest industry, such as wood workshops for carpentry, could provide higher income for the WAJIB members and additionally create quality jobs for educated labour (de Blas et al, 2008).
- **Installation of a fair benefit-sharing system.** Fair benefit-sharing regulations based on additional by-laws within the forest-dweller associations are needed to secure equal rights for poor forest-dwellers, among which many are women. A different option to support females could be to develop alter-

native income generation possibilities, for example non-wood forest products, exclusively for women. Rojahn (2006) found income possibilities for the following products on local markets in the area: brown cardamom (*Kororima*); *Gesho*, a condiment for making a local drinks; *Desha*, used to clean the oven; *Ensosela*, used for decorating the skin; as well as mats and baskets. Furthermore, bamboo resources, growing mostly along streams in the Bale Mountains (Fichtmüller, 1997), are an unused resource with a high potential for poverty reduction (Ruiz Pérez et al, 1999).

## Exclusive forest user rights against the tragedy of the commons?

Community forestry has been implemented in many different forms in South America, Africa and Asia, with varying foci: to protect threatened forest areas, to empower local people or as part of a poverty-reduction strategy. However, many initiatives resulted in top-down approaches that facilitated environmental protection rather than poverty reduction of local communities (Charnley and Poe, 2007). The same problems have occurred in former attempts at community forestry in Ethiopia, often leading to failure (Mekonnen, 2000).

Nevertheless, the Ethiopian government supports community forest management in forest reserves of high conservation priority (Wily, 2002). Yet, due to shortages of funds, equipment, education and public awareness about community forestry, the policy still cannot be implemented efficiently nationwide (Regasa, 2000). Despite these shortcomings, there are few competing concepts to maintain the forest at large, even inside national parks (Watson, 2007). However, William (1997) claims that community forestry is a traditional and accepted form of land use in east Africa and effective in combating deforestation. The Integrated Forest Management Project achieved maintaining the forest cover by limiting access to the forest and enabling forest-dwellers to protect their resource (Amente et al, 2006).

Finally, it is questionable whether it is socially acceptable to provide exclusive user rights to a small number of people, while the majority is excluded from an essential, former public resource. However, Kessy (1998) defined the 'ability to exclude others' as a main requirement for community forestry. Hardin (1968) points out that open-access resources inevitably face over-utilization, as every user maximizes profits. Considering the interest of the government in maintaining the forest at limited costs, exclusive forest user rights may be a reasonable tradeoff between social and ecological needs in Ethiopia. Therefore, the future of the Bale Mountains forests highly depends on the commitment of the Forest Administration to share responsibilities for and benefits from the forest with local forest-dwellers and surrounding communities. Currently the conditions for that are favourable in the Bale Mountains of Ethiopia.

# REFERENCES

Agrawal, A., Yadama, G., Andrade, R. and Bhattacharya, A. (2006) 'Decentralization and environmental conservation: Gender effects from participation in joint forest management', CAPRi Working Paper No. 53, downloadable from www.capri.cgiar.org/pdf/capriwp53.pdf

Amente, G., Huss, J. and Tennigkeit, T. (2006) 'Forest regeneration without planting: The case of community managed forests in the Bale mountains of Ethiopia', *Journal of the Drylands*, vol 1, pp26–34

Atteslander, P. (2000) *Methoden der Empirischen Sozialforschung*, 9th edition, Gruyter, Berlin, Germany

Bekele, M. (2001) 'Country report – Ethiopia', Forestry Outlook Studies in Africa (FOSA), FAO, Rome, www.fao.org/DOCREP/004/AB582/AB582E02.htm, accessed 6 March 2003

Charnley, S. and Poe, M. R. (2007) 'Community forestry in theory and practice: Where are we now?', *Annual Review of Anthropology*, vol 36, pp301–336

Conroy, C., Mishra, A. and Rai, A. (2002) 'Learning from self-initiated community forest management in Orissa, India', *Forest Policy and Economics*, vol 4, pp227–237

de Blas, D. E., Pérez, M., Sayer, J. A., Lescuyer, G., Nasi, R. and Karsenty, A. (2008) 'External influences on and conditions for community logging management in Cameroon', *World Development*, vol 37, no 2, pp445–456 http://dx.doi.org/doi:10.1016/j.worlddev.2008.03.011

Demesa, G. (2002) 'Forest utilization and comparison between traditional lumber conversion techniques and a portable sawmill in Arero', MSc thesis, SLU, Swedish University of Agricultural Sciences, Uppsala

EFAP (1994) *Ethiopian Forestry Action Plan, Final Report: Volume I – Executive Summary*, Ministry of Natural Resource Development and Environmental Protection, Addis Ababa, pp14

FAO (2003) *State of the World's Forests 2003*, Food and Agriculture Organization of the United Nations, Rome

FAO (2004) 'Simpler forest management plans for participatory forestry', working paper, Forestry Policy and Institutions Service, Forestry Policy and Information Division, Forestry Department, Food and Agriculture Organization of the United Nations, Rome

Fichtmüller, S. (1997) 'Useful plants in the vicinity of Dodola, Ethiopia', unpublished IFMP document

Friis, I. (1992) *Forests and Forest Trees of Northeast Tropical Africa. Their Natural Habitats and Distribution Patterns in Ethiopia, Djibouti and Somalia*, Royal Botanical Gardens, Kew, UK

FSC (2008) Forest Stewardship Council website, www.fsc.org/vision_mission.html

Fufa, D. (2002) 'Analysis of benefit sharing mechanisms among Forest Dweller Association (WAJIB)', BSc thesis, Wondo Genet College of Forestry, Wondo Genet, Ethiopia

Hardin, G. (1968) 'The tragedy of the commons', *Science*, vol 162, pp1243–1248

Hohlweg, C. (1998) 'Impact of silvipastoralism on natural regeneration of main tree species in Afromontane forests in the Bale Mountains, Ethiopia', MSc thesis, Freiburg University, Freiburg, Germany

IFMP (2000) 'Methodology of tree cover assessment for blocks allocated to forest dweller associations', unpublished IFMP document, Dodola, Ethiopia

ITTO (2002) *International Tropical Timber Organization Guidelines for the Restoration, Management and Rehabilitation of Degraded and Secondary Tropical Forests*, ITTO Policy Series No. 13, Yokohama, Japan

Kessy, J. (1998) 'Conservation and utilization of natural resources in the East Usambara Forest Reserve: Conventional views and local perspectives', Dissertation, University of Wageningen, Netherlands

Kubsa, A. (2002) 'Granting exclusive user rights to the forest dwellers in the state owned forest: The WAJIB approach in Ethiopia', *PFM Newsletter*, Addis Ababa, pp2–7

Lamnek, S. (1993) *Qualitative Sozialforschung. Band 2: Methoden und Techniken*, 2nd edition, Weinheim, Germany

Mekonnen, A. (2000) 'Valuation of community forestry in Ethiopia: A contingent valuation study of rural households', *Environment and Development Economics*, vol 5, pp289–308

Regasa, T. (2000) 'Environmental issues', in C. Fellner (ed) *Ethiopia – An Introduction into Culture, Economics, Politics and Cooperation*, Brandes and Apsel, Frankfurt am Main, Germany, pp9–21

Ruiz Pérez, M., Zhong, M., Belcher, B., Xie, C., Fu, M. Y. and Xie, J. Z. (1999) 'The role of bamboo plantations in rural development: The case of Anji County, Zhejiang, China', *World Development*, vol 27, no 1, pp101–114

Rojahn, A. (2006) 'Incentive mechanisms for a sustainable use system of the montane rain forest in Ethiopia', Inaugural-Dissertation der Christian Albrechts Universität zu Kiel, Berlin

Schönhuth, M. and Kievelitz, U. (1993) 'Partizipative Erhebungs und Planungsmethoden in der Entwicklungszusammenarbeit', *Schriftenreihe der GTZ*, no 231, Wiesbaden, Germany

Trainer, J. (1996) 'Forest inventory of indigenous forest pilot area. Inventory results', IFMP document, Dodola, Ethiopia

Unkovsky, S. (1998) 'Model calculation of minimum forest area per household for forest-dwellers in the forest priority area of Adaba-Dodola', IFMP document, Dodola, Ethiopia

Varian, H. R. (2003) *Intermediate Microeconomics. A Modern Approach* (sixth edition), University of California at Berkeley, Berkeley, CA

Watson, C. (2007) 'Direct consumptive use valuation of the ecosystem goods and services in the Bale Mountains Eco-Region, Ethiopia', master's thesis, Faculty of Natural Sciences, Centre of Environmental Policy, Imperial College London, London

Wiersum, K. F. (1999) 'Social forestry: Changing perspectives in forestry science or practice?', PhD dissertation, Wageningen University, Wageningen, The Netherlands

William, G. S. (1997) 'Community forestry management: Experiences from East Africa', in M. Victor et al (eds) *Community Forestry at a Crossroads: Reflections and Further Directions in the Development of Community Forestry*, RECOFT Report No. 16, Regional Community Forestry Training Center for Asia and the Pacific, Bangkok, Thailand, pp71–76

Wily, L. (2002) 'The political economy of community forestry in Africa – Getting the power relations right', *Forest, Trees and People*, Newsletter No. 46, pp4–12

Wily, L. and Dewees, P. (2000) 'From users to custodians: Changing relations between people and the state in forest management in Tanzania', Policy Research Working Paper No. 2569, World Bank, Washington, DC

World Bank (2010) 'World Bank data: GNI per capita, Atlas method (current US$)', http://data.worldbank.org/indicator/NY.GNP.PCAP.CD

# Index

361